Systems Analysis and Design: A Comparison of Structured Methods

Dorothy J. Tudor & Ian J. Tudor

First published 1995

NCC Blackwell
108 Cowley Road
Oxford OX4 1JF
UK

Blackwell Publishers Inc.
238, Main Street, Cambridge, MA 02142
USA

ISBN 1-85554-351-6

Library of Congress data has been applied for.

Typeset in 10.5 on 12pt Times Roman by ColourTech, Bradford, West Yorkshire
Printed in Great Britain by Hartnolls Ltd, Bodmin, Cornwall

This book is printed on acid-free paper

Contents

Preface

Structured Methods are now very widely used in the development of computer software and many people are faced with the choice of one method against others. However, there seemed to be a shortage of books comparing and contrasting structured methods with each other. Furthermore, with such an array of methods to choose from, systems developers are faced with a sometimes confusing set of techniques and terminologies which have essentially the same purpose but *appear* to be different within the alternative methods. This book addresses this problem, which ultimately ought to be resolved through a joint standard terminology but, for commercial reasons, perhaps will not. Competitive commercial factors will necessarily bring about some convergence of methods but vendors will attempt to maintain their individuality behind their own 'brand' of jargon and toolset.

The authors have long been aware of the need for this book from consultancy questions which we were frequently asked and from our running of training courses in a variety of methods. In fact, we could not understand why a book of this nature had not already appeared on the market – until we tried to write it. We realised part-way through that the task was a wicked one. We needed to present enough of each method to allow an understanding sufficient for comparison, whilst keeping within a number of pages which would not require a free wheelbarrow to be supplied with the book. Our intention has been to distil down the essence of each of the methods covered, trying to lose nothing vital in the distillation. It has not been our intention to provide exhaustive coverage of the methods nor an academically rigorous comparison. Rather, we have tried to encapsulate the main features of each method and give a flavour of its usage.

We hope that the book will be useful on several levels. For the new student of structured methods, we hope that the early chapters will provide a good foundation in the basic structure, the underlying principles and the most common techniques. Then we hope that the description of each method will be sufficient as a starting point for the discovery of any one of them, upon which further research can be laid, if necessary. For the student or project manager needing to perform a comparison of methods, we hope that the chapters which introduce DESMET and Euromethod will give food for thought. They give information about a method for comparison and detail the assistance available from Euromethod in translating between the different terminology and concepts. The comparison chapter should provide a guide to the contents of the methods and their differences of approach and as such be a good foundation for further methods of comparison and evaluation work. For the hardened structured methodologist we hope you find a few new and useful diagrams and ways of looking at things, and have some fun agreeing or disagreeing with our view of the future in the final chapter. For the avid bookworm, we hope you find it a good read!

Acknowledgements

Judith Fynn (CCTA)

Jerry Humphries (The National Computing Centre)

Ken I'Anson (CCTA)

Jeet Khaira (BIS Information Systems)

Barbara Kitchenham (The National Computing Centre)

David Law (Seneca Services)

Steve Newton (NCC Blackwell) for patience beyond the call of duty.

Richard Seed (Barclays Bank)

Stella Simpson (NCC Blackwell).... for her painstaking search
for the truth

Glynis Ward

The Target Reader

Students of structured methods or computer science, corporate business users of structured methods and those with a responsibility for the setting of development standards and the selection of methods.

The Contents of the Book

Chapter 1 Who Needs a Structured Method Anyway?
This chapter establishes the history and driving forces behind the development of Structured Methods. It also introduces the Midlinks Motel scenario, which forms a common thread through the book upon which methods and techniques are hung.

Chapter 2 A Method for Selecting a Method – Dream or Reality?
Is there any way to objectively compare methods which raises the odds of selecting the most suitable method for adoption by a particular organisation? This chapter looks at the similarities and differences of approach of the chosen structured methods. It then discusses the factors which must be considered when selecting a structured method and looks at a methodology for evaluating methods and the assistance available from Euromethod in bridging between methods.

Chapter 3 The Common Techniques
The perspectives of data, function and event which underlie the structured methods are discussed. The techniques which appear in more than one method are covered.

Chapter 4 Soft Systems/Multiview
Soft Systems has developed through Checkland's work at the University of Lancaster and is concerned with the the first problem encountered by the analyst, namely problem specification. If the problem can not be identified where is the solution to come from? Soft systems is presented in conjunction with the Multiview method, developed by Antill, Wood-Harper and Avison.

Chapter 5 Structured Systems Analysis and Design (Version 4)
SSADM V4 is the method which has been developed as the standard method for use by UK government departments. It is also becoming more widely accepted in the private sector. Its large, and growing, customer base alone justifies its inclusion. It is one of the methods with an influence on the development of Euromethod. This chapter has been reviewed by the Technical Committee of the International SSADM Users' Group and conforms with the concepts of SSADM V4.

Chapter 6 Information Engineering

Information Engineering was the product of the combined thoughts of Martin and Finkelstein but from those early beginnings these co-developers have diverged into their own brands of Information Engineering, and other vendors' variations have also emerged. Its initial objective of being designed in parallel with a CASE tool to support the method and its impact on Euromethod make it worthy of consideration.

Chapter 7 Yourdon Structured Analysis and Design

Yourdon Structured Method is the modern day descendant of the early writings of De Marco and Yourdon in the late 1970s. The method has grown and matured in the intervening years. It contains many techniques in common with the other methods which are considered but also has additional techniques of its own. Its wide usage both in the UK and elsewhere have justified its inclusion.

Chapter 8 MERISE

MERISE is the method of preference in France and is included because its impact upon the development of Euromethod has brought its name, if not its content, to the ears of many. It is a well-developed method with an interestingly different angle on the presentation of a dynamic view of the system, and an all-embracing framework of three concurrent development cycles.

Chapter 9 The Comparison

The methods which have been considered individually are now compared directly against a set of features chosen to highlight differences in approach, framework and techniques of the methods.

Chapter 10 Focus on Methods and Tools –The Future

Here we attempt to examine the future of Structured Methods in general with reference to those hot topics which look destined to affect them. These include Object Orientation, CASE Tools, Euromethod, RAD. Reverse Engineering and Business Process Redesign. We end with a consideration of what is still wrong with our system developments, how to effectively implement structured methods, and the points which form the methodplan for success.

Appendix A Midlinks Motel Scenario

This section contains details of the hotel scenario in the form of narrative and interview notes, in order to aid the understanding of the techniques presented in the individual chapters.

Appendix B Bibliography

Conclusion

It has been quite a major effort to bring together all of the information for this book. We have to thank our many friends and colleagues who have painstakingly ploughed through our process hierarchies, delved diligently into our dialogue designs and improved the state of our state transition diagrams. It is thanks to their help and encouragement, in spotting those undesirable "design features" which we had unwittingly built in, that this book was ever completed. Yes, resisting all invitations from the spell-checker to change CATWOE to CATTLE and Bachmann to Batman, here it is! If we have succeeded in giving a feel for the similarities and differences between the structured methods of the moment and provided a basis, and perhaps even an incentive, for further, detailed study of any particular method, then we are happy.

About the authors

Dot Tudor BSc MBCS
Ian Tudor BSc MSc

Dot Tudor and Ian Tudor, founders of TCC, are methods consultants who have been in partnership since 1984.

Dot founded TCC in 1982, having gained programming, systems analysis and project management experience in a variety of commercial environments, from pharmaceutical manufacture and distribution, through the automotive industry to local government, both before and after the advent of structured methods. She has given consultancy and delivered training in a wide variety of methods both nationally and internationally. Her work has addressed the systems development life cycle from initial strategy to implementation and beyond. Her involvement in structured methods for systems analysis and design began at the turn of the 1980s, when MERISE was just a lad and SSADM was nothing more than a glint in its founders' eyes.

Ian came to the computing industry in 1984, having trained as a biologist, with postgraduate qualifications in education and computer science. He has had experience within the health service and in university research. He has a fresh and sometimes heretical approach to systems development (having not had the burden of programming in assembler) and keeps abreast of the innovations arriving on the market.

Together Dot and Ian bring to the book a wealth of training and project experience in a spectrum of structured methods wider than the range presented in this book.

1

Who Needs a Structured Method Anyway?

1.1 Overview

The very fact that you are reading this book suggests that you at least feel that you may need to know about structured methods. This chapter seeks to convert those feelings to conviction by laying down the evidence in support of the need for Structured Methods in general. The choice of a particular method is more complex and will depend upon such factors as your organisation's culture, its size and therefore the available budget and not least the level of experience held by the organisation's Systems Analysts. These factors and others related to your organisation are unknown to the authors. However by presenting an overview of a selection of the more widely used methods together with criteria which are useful for selecting a method, this book will offer some guidance to you in your quest. Your final choice of method may not be represented in this text but the issues which are raised here ought to contribute to the manner in which you approach the task.

1.2 Method or Methodology?

At the outset we are faced with a quandary: Structured Method or Methodology? Even between authors the **same** Structured 'something-or-other' is referred to by one as Method and another as Methodology! Is there a universally accepted definition of method and methodology? The short answer is 'no' but to illustrate the confusion surrounding the terms it is worth examining a number of definitions which have been offered.

The Concise Oxford Dictionary of Current English defines a Method as 'a special form of procedure especially in any branch of mental activity'. Methodology per se is not defined. However we may try and draw some inference from the definition attributed to other 'ologies'. Biology is the 'Science of physical life', Sociology is the 'Science of the development and nature and laws of society', Psychology is the 'Science of nature, functions and phenomena of human soul or mind'. From this it appears that 'ology' means 'Science of'. Science is defined in the same source as 'branch of knowledge, organised body of the knowledge that has been accumulated on a subject'. Therefore we arrive at the definition of {Structured} Methodology as

'Science of {Structured} Method' which translates to 'branch of knowledge, organised body of the knowledge that has been accumulated on {Structured} Methods'.

Checkland (1981) defines a methodology as 'a set of principles of method which in any particular situation have to be reduced to a method uniquely suitable to that particular situation'. He goes on to say that a methodology is 'intermediate in status between philosophy, using that word in a general rather than a professional sense, and a technique or method. ... A methodology will lack the precision of a technique but will be a firmer guide to action than a philosophy'.

Avison and Fitzgerald (1988) answer our dilemma thus: 'A methodology is a collection of procedures, techniques, tools and documentation aids...which consists of phases...but a methodology is more than merely a collection of these things. It is usually based on some philosophical view, otherwise it is merely a method, like a recipe'.

Yourdon (1992) defines a methodology as 'a step-by-step "battle plan", or cookbook for achieving some desired result' whilst a method 'is a step-by-step technical approach for performing one or more of the major activities identified in an overall methodology'.

From a marketing point of view, if a Methodology is more than a Method then commercial factors will encourage vendors to use the term Methodology for competitive advantage whatever the definition of the terms. In the absence of a universally accepted definition, Method and Methodology have a tendency to be used interchangeably.

Faced with such a choice, which ball should we pick up and run with? In order to eliminate the distraction introduced by the confusion of definitions for method and methodology we have opted to use the term Method exclusively in this text save in the discussion above and for the discussion of DESMET (Chapter 2).

- **A set of inter-connecting parts, together with the interconnections**

- **A set of related procedures, with an objective**

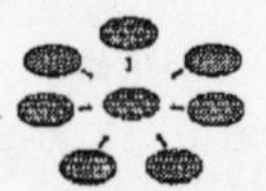

- **many systems may interact as one system**

Figure 1.1 What Is a System?

1.3 What Is a System?

Before turning attention to Methods for Systems Development it is important to understand the system environment within the organisation within which the analyst has to work. To do this we need to answer the question, 'What is a system?' Some definitions of a system are given in Figure 1.1.

A 'system' is an organised way of doing something specific. The definition above emphasises the sense of purpose within a system and the way in which systems frequently overlap with and interrelate to one another. The problem is often defining the boundaries of our system because of the overlap and inter-dependency of systems.

The systems which most commercial systems analysts will be involved with are Information Systems. These are often, although not always, well-suited to computerisation.

Information systems have a number of characteristics which are depicted in Figure 1.2.

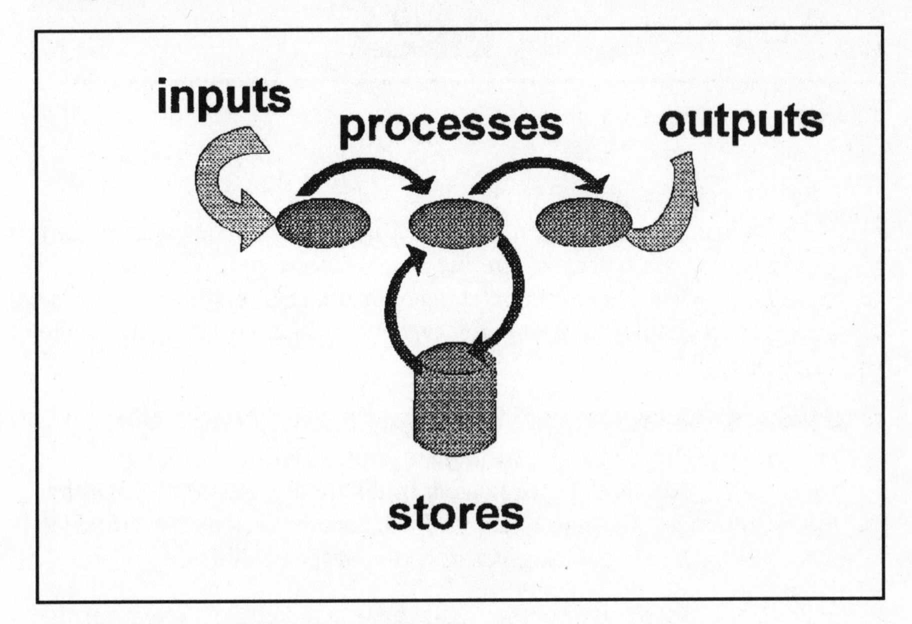

Figure 1.2 Characteristics of Systems

Such systems always have:

- inputs;
- outputs;
- processing;
- storage (files).

The systems analyst has to identify and define the above aspects of the system in detail.

Within any organisation there are a number of other types of system (Checkland 1990) which interact with each other and which the analyst should be aware of. These are shown in Figure 1.3 and are further described below:

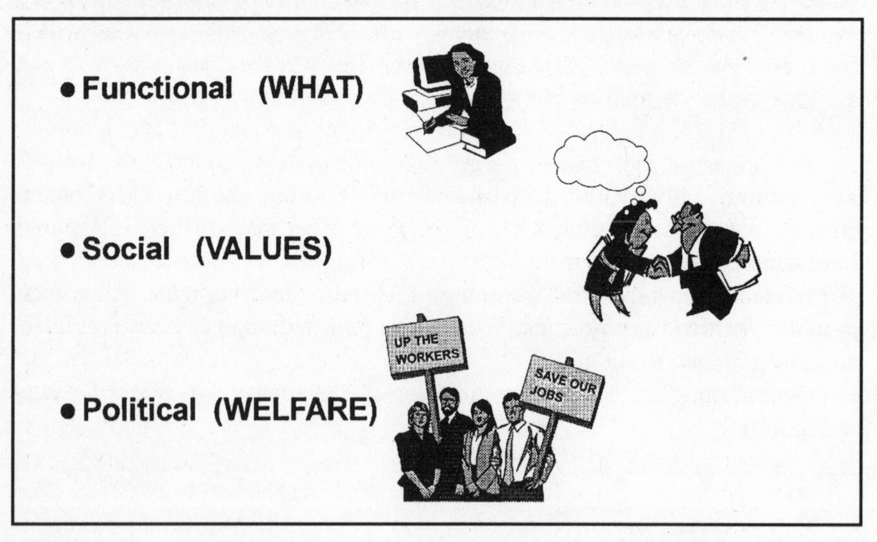

Figure 1.3 Types of System

- a **Functional System** is **what** people do within the organisation to further the activities of the organisation. This is the aspect which may incorporate a computer system, but this is only a part of what people actually do. Other supporting clerical/human interface systems must be provided to enable the computer system to function correctly (or to function at all!);
- a **Social System** is concerned with the **values** which people hold. What do people think is good, bad etc. in their relationships with other members of staff either on equivalent or different grade positions within the organisation. The introduction of a computer system can influence this considerably by affecting staff roles and responsibilities;
- a **Political System** is involved with **people's welfare.** What are the relationships of unions with each other and management within the organisation? The impact of the computer on personal interests such as job prospects and status within the organisation may be crucial in the acceptance and subsequent security of the system. It is extremely unlikely that a manager will be kindly disposed toward a new computer system which cut his staff by half in an organisation in which status is measured by the number of staff under a manager's control.

The structured methods which we shall be considering focus mainly on the computerised aspects of the functional system. However, the analyst must be keenly aware of the interaction with the other types of system in operation.

1.4 The Approach and Purpose of This Book

1.4.1 The Approach of This Book

The approach of this book is to justify the need for a Structured Method, to present a way to select a method and to examine the methods detailed below. Following consideration of the common techniques each method is reviewed and a feature comparison of the chosen methods, based upon the framework presented below, is presented in Chapter 9. Each method has been presented here only in enough detail to allow the reader a sufficient overview of the framework and techniques to appreciate the comparison. Examples in each method have been related to a consistent case study (see Appendix A) to facilitate comparison. The hotel case study was chosen since it represents a situation with which we hope our readership can identify. The chosen methods continue to evolve becoming refined in the light of experience of their use and hence we present here a snapshot of them at the time of writing.

1.4.2 The Purpose of This Book

This book describes the approach of each of these methods, highlighting their similarities as well as their differences. It is hoped that this will enable the reader to better understand the nature of the methods, their frameworks and techniques. It is our fervent wish that the reader may be inspired to investigate in further detail the methods included here and Structured Methods in general. It may even encourage you to consider using features from several methods to form a composite practical method for Systems Development.

1.5 The Objectives of Systems Development

In general terms, the objectives of systems development are to produce a system:

- which is a **working, reliable system,** within bounds specified by the business;
- which will do what the user **requires**. This means what is required to meet business objectives and those of the user in his/her business area;

- at a price which **can be justified**. The business benefits which are expected to accrue from the new system must be offset against the cost of developing and implementing the system.

1.6 The Problems of Systems Development

As depicted in Figure 1.4 Systems Development has been acknowledged to suffer a number of failings.

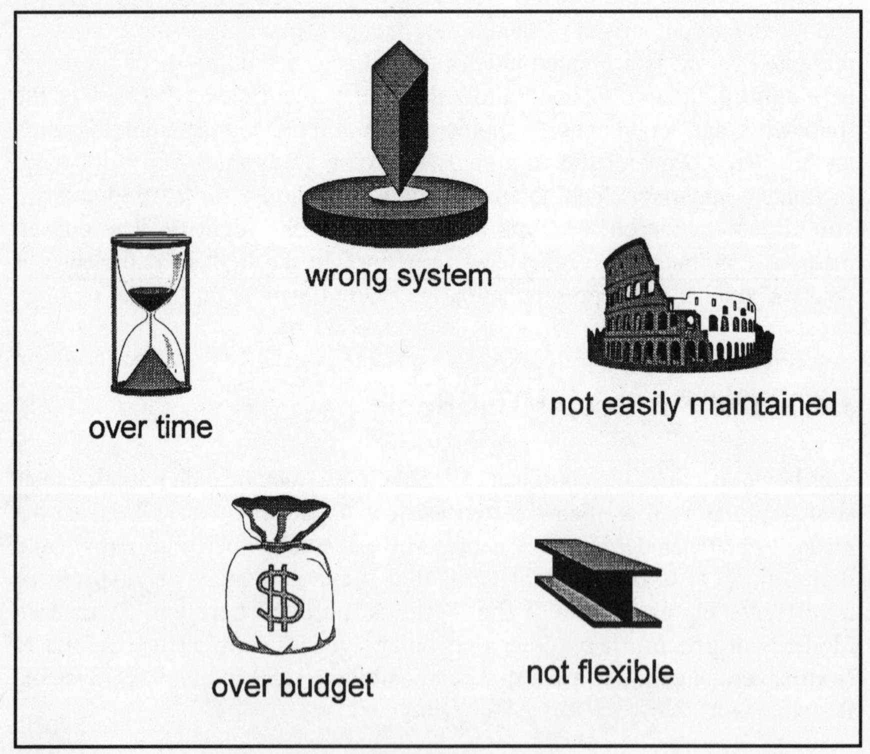

Figure 1.4 The Problems of Systems Development

The initial forays into systems analysis and design were largely borne out of a technical programming background and therein lay the source of the problems which have dogged systems development. Failed communications have often led to technically correct programs delivering the wrong system which did not do what the users wanted it to do. These errors, discovered only when the system was implemented, proved very costly. Hence workable systems were eventually delivered in an untimely way and over-budget. The recognition of these failings determined the need for better control over systems development. The development of structured methods in combination with the concept of a systems life-cycle were the weapons developed to attack the problems.

1.7 The Systems Life Cycle

1.7.1 Stages in the Life Cycle

Because the environment within which an organisation functions is subject to continual change, the systems within it, whether manual or computerised, also have to change. In order to understand and keep control of this change, it is useful to visualise the stages which a system goes through over time, from its initial conception to its eventual replacement. These stages form the System Life Cycle:

- **Initial Strategy**: identifying the initial problem or idea and assessing the justification for further action against the business objectives;
- **Feasibility Study**: a high-level overview analysis of the problem area to identify the boundary of the area for investigation and the outline requirements;
- **Requirements Analysis**: an in-depth analysis to establish the environment and the exact business requirements;
- **Systems Analysis**: a definition of what the business requirements mean in terms of a new system;
- **Specification**: the detailed description of the requirements in a form in which they can be interpreted by the technical designer;
- **Design**: the detailed technical plan for the new system;
- **Development**: the coding of the processes for the new system, if a computer system has been specified;
- **Testing**: the trial of the component parts and then the whole system to ensure that it works (that we have built the system right) and that it does what is required (that we have built the right system);
- **Implementation**: the conversion of procedures from current working practices to new ones, and of data from current forms of storage to new formats;
- **Maintenance**: minor modifications to the system to optimise its performance, improve its usability or accommodate small changes in the environment;
- **Review**: when, due to changes in the business environment, major modifications would be needed to retain business effectiveness, a full review is required, which may result in major re-working or even complete replacement of the system.

Any implication that these phases of the life cycle are totally discrete and sequential is not intended. We shall return to this point later.

1.7.2 Types of Life Cycle

There are many variations on the theme of a life cycle for systems development. Two of the more common ones are considered below:

- the Waterfall Model;
- the 'V' Model.

A more complete discussion can be found in Sommerville (1992).

1.7.2.1 The Waterfall Model

The Waterfall Model takes the main stages within systems development and represents them diagrammatically as a series of sequential steps with the flow of time and information from left to right. This is shown diagrammatically in Figure 1.5.

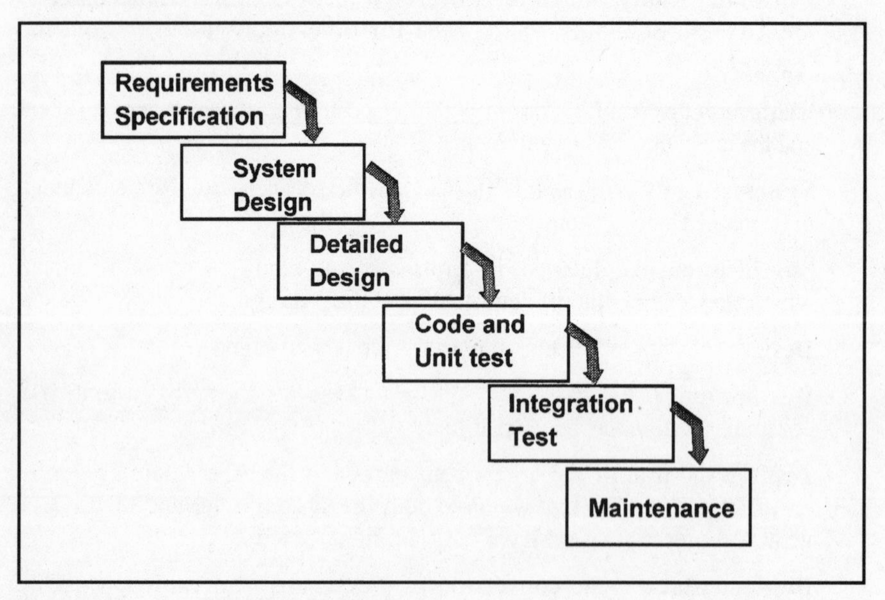

Figure 1.5 The Waterfall Model

1.7.2.2 The 'V' Model

The 'V' Model is, again, a diagrammatic representation of the life cycle and is shown in Figure 1.6. However, its additional strengths over the Waterfall Model are that:

- the products, which result from each stage, are passed to the following stages;
- the products, against which various levels of testing of the system can be performed, are identified.

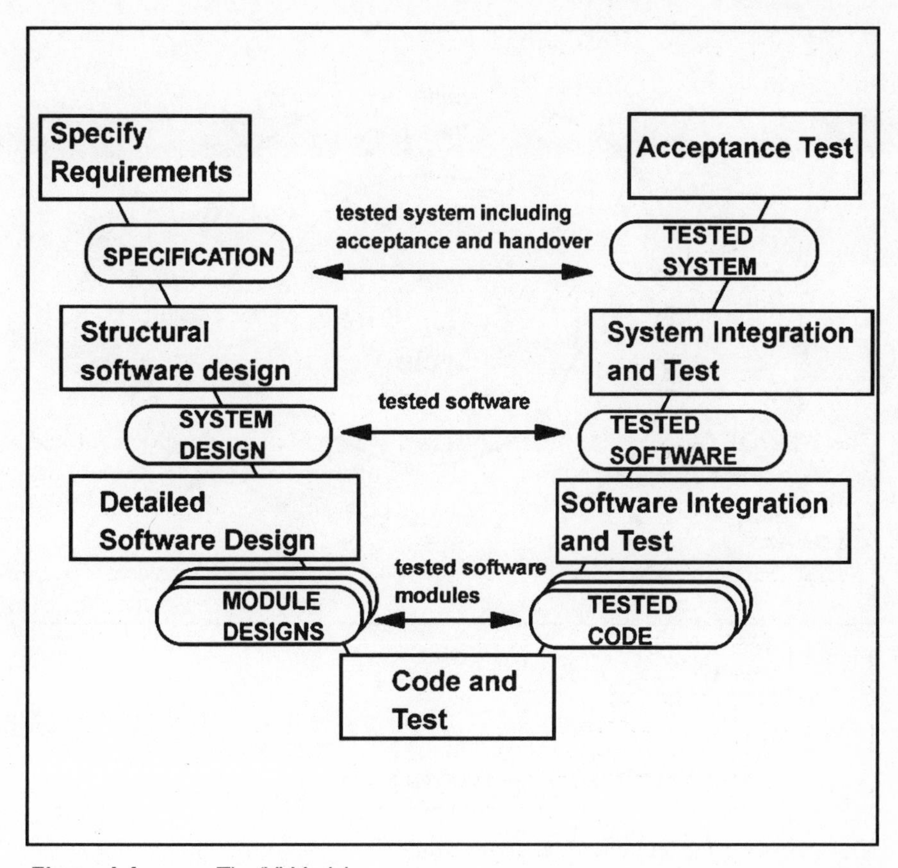

Figure 1.6 The 'V' Model

In the diagram, the products are shown in the 'lozenges' and the stages as rectangles. The flow of time is down the left-most leg of the 'V' and up the right-most.

1.7.3 Iteration Within the Systems Life Cycle

The linear view of the life cycle of a system is rather an over-simplification. For example, a system will have certain requirements which are known from the outset. Additional requirements will have been established during the requirements analysis and systems analysis stages. However, during specification or even later stages of the life-cycle, further requirements emerge due to a better understanding of the business area, or due to the changing environment of the system. This may necessitate further feasibility study to decide whether to incorporate the new requirements, further analysis and specification, and possibly further design, coding and testing. This iteration does not invalidate the life cycle. Rather, the life cycle provides a mechanism through which it can be recognised and controlled. Figure 1.7 illustrates in a general way this iterative nature of the Systems Life Cycle.

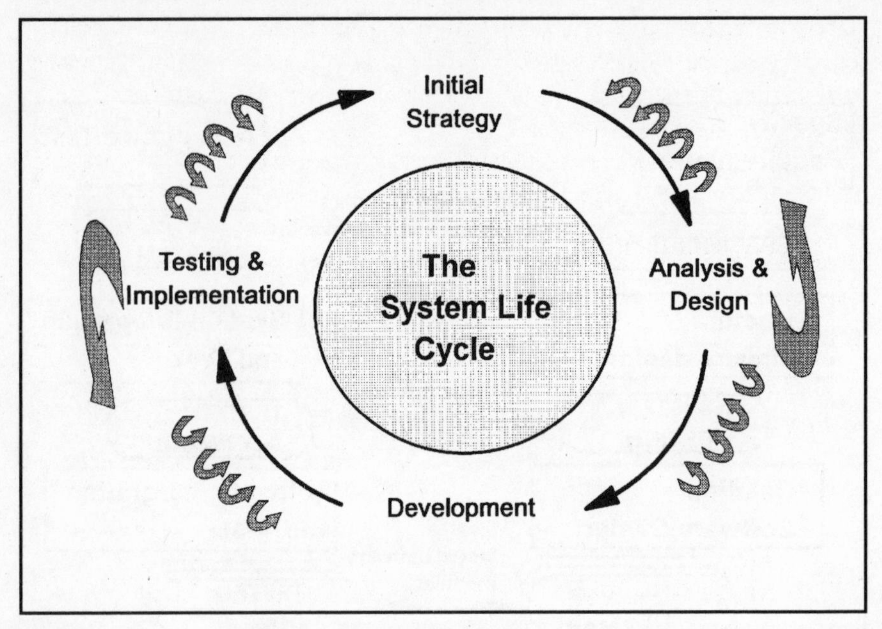

Figure 1.7 Iteration in the System Life Cycle

1.8 The Business Life Cycle

The life cycle for an individual project must be seen within the wider context of the life cycle for the business (Figure 1.9). The business life cycle has the following characteristics:

- the **Business Life Cycle** extends from the overall business strategy plan through to implementation of the many systems which support the strategy;
- the individual **System Life Cycles** within the business life cycle could be based on one of a number of possible models. (Waterfall, 'V' Model). One business strategy and information systems strategy would typically result in the identification of many projects, each with its own system life cycle. These projects may be developed in parallel, or with some degree of concurrency or in series.

1.9 The Role of Structured Methods

In the discussion above the problems of systems development were identified and it was noted that Structured Methods have evolved to help address these problems. The essence distilled from those identified problems is **increased cost**. How great this increased cost will be in absolute terms depends upon a number of factors such as size of the system and how deviant from the user requirement the system is. Barry Boehm (1981) in his book 'Software Engineering Economics' illustrates the relative costs of fixing an error in relation to the phase of the software life-cycle in which the error is discovered. Dependent upon the size and complexity of the project, an error uncovered at the implementation phase may cost between 10 and 100 times more to fix than if that error were uncovered in the Analysis of Requirements phase. Figure 1.8 shows the cost to correct an error versus the stage in the life cycle in which the error is discovered.

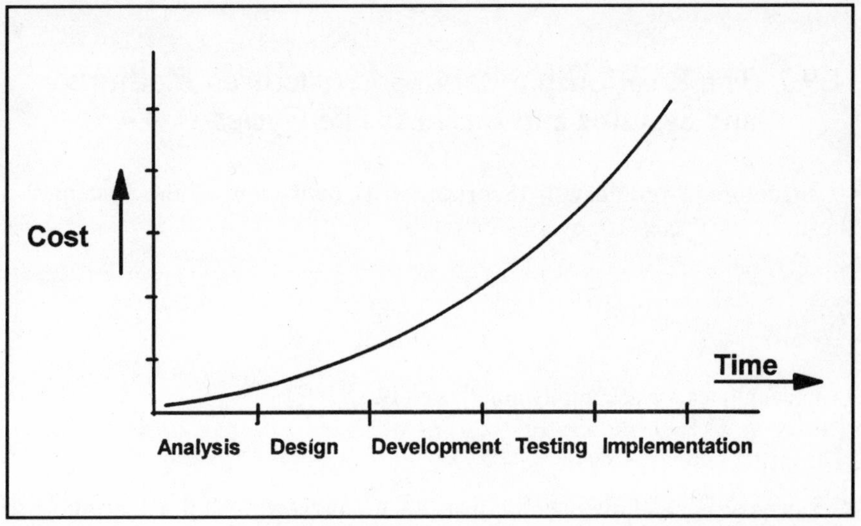

Figure 1.8 Costs to Fix an Error Versus Stage in Which the Error Is Discovered

1.9.1 What Is the Purpose of a Structured Method?

The purpose of a Structured Method is to provide a framework within which the systems developer can produce an effective solution to a business problem which requires the use of a computer system and a set of techniques with which to do this. To do this the method must do the following:

- **aid communication** between all those involved with systems development: management, users, analysts, programmers etc.;
- provide a **set of techniques** so that tasks can be performed in a standard way, using proven ways of working;
- provide for **effective review** to identify errors earlier in the system life cycle rather than later;
- permit **flexible** business and technology changes by, for example, separating the Design phase from the Analysis phase of the systems development;

- enshrine a **development strategy** to eliminate 'ad-hoc' problem-solving;
- indicate when and to what extent **user involvement** is required, encouraging and enabling it to occur;
- ensure that sufficient effort is put into the **analysis of the business** to ensure that the system delivered addresses business and user requirements appropriately.

The relative emphasis placed on these objectives will depend on the particular structured method.

1.9.2 The Relationship Between Structured Methods and Systems and Business Life Cycles

A Structured Method will incorporate its own view of the system and business life cycles and overlay onto this:

- stages;
- tasks;
- techniques for accomplishing those tasks.

The stages and tasks also:

- provide a framework for project management and assessment of progress;
- give a basis for estimation of the scale of a project.

The coverage of these life cycles varies from method to method and is addressed in relation to the methods within this book in Chapter 9.

1.10 The Methods Examined in This Book

This book examines five of the foremost Structured Methods which are currently available mostly in commercial environments and compares how they address particular problems associated with the phases of software development within an *organisation*. (Throughout the book organisation will be taken to mean the whole business or a particular area or department within which the system is to be developed.) The five methods are Soft Systems Method (SSM)/Multiview (MV), Structured Systems Analysis and Design Method Version 4 (SSADM(V4)), Information Engineering (IE), Yourdon Structured Method (YSM) and MERISE. They were chosen for inclusion in this text for the following reasons:

- **Soft Systems Method/Multiview** places great emphasis on the human aspects which have to be considered for successful Systems Development and Implementation;
- **SSADM Version 4 (SSADM (V4))** is of importance as it was the method developed for the UK government as the standard method for system development;
- **Information Engineering (IE)** has been adopted by many large international companies, its importance being that there are powerful Software Tools for code-generation which support the method. This current availability of tools means that no organisation can afford to ignore the method;
- **Yourdon Structured Method** is included because of the significant impact which it has had on the development of structured methods and because of its extensive usage internationally;
- **MERISE** is the French equivalent of the UK SSADM (V4) and is important for its role in the development of Euromethod.

Of these methods SSADM, IE and MERISE are included in the set of methods which impact upon the development of EUROMETHOD (see Chapters 2 and 10).

1.11 A Framework for Comparison

Each method is described independently of the others, within its own chapter. The comparison of the above methods is the subject of Chapter 9.

For our purposes the framework which will be used to compare the methods is the business life cycle together with elements from the framework for comparison defined by Avison and Fitzgerald (1988). Within this the business life cycle shall be defined to have the following characteristics:

- the **business life cycle** will extend from the overall business strategy plan through to implementation of the systems to support the strategy;
- the individual **systems life cycles** within the business life cycle could be based on any of a number of possible models. Rather than adopt any of these models (Waterfall, Putnam Waterfall, Prototyping Circles) this study will distil the elements of various models to form a generalised system life cycle. Within this general model each stage in the life-cycle forms a good foundation for the next stage but still allows iterations back in the cycle. The business life-cycle framework is summarised in Figure 1.9. The business life cycle is simplified in form and therefore only three individual systems life cycles are shown. The authors also recognise that certain terms may not be used within certain development environments, for example program coding may be described as code generation where a software tool accomplishes this task within the life cycle. It should also

be noted that although the figure appears to suggest that the individual systems are developed in parallel this is not necessarily the case. Additionally any implication that the phases of the life cycle are totally discrete and sequential is not intended.

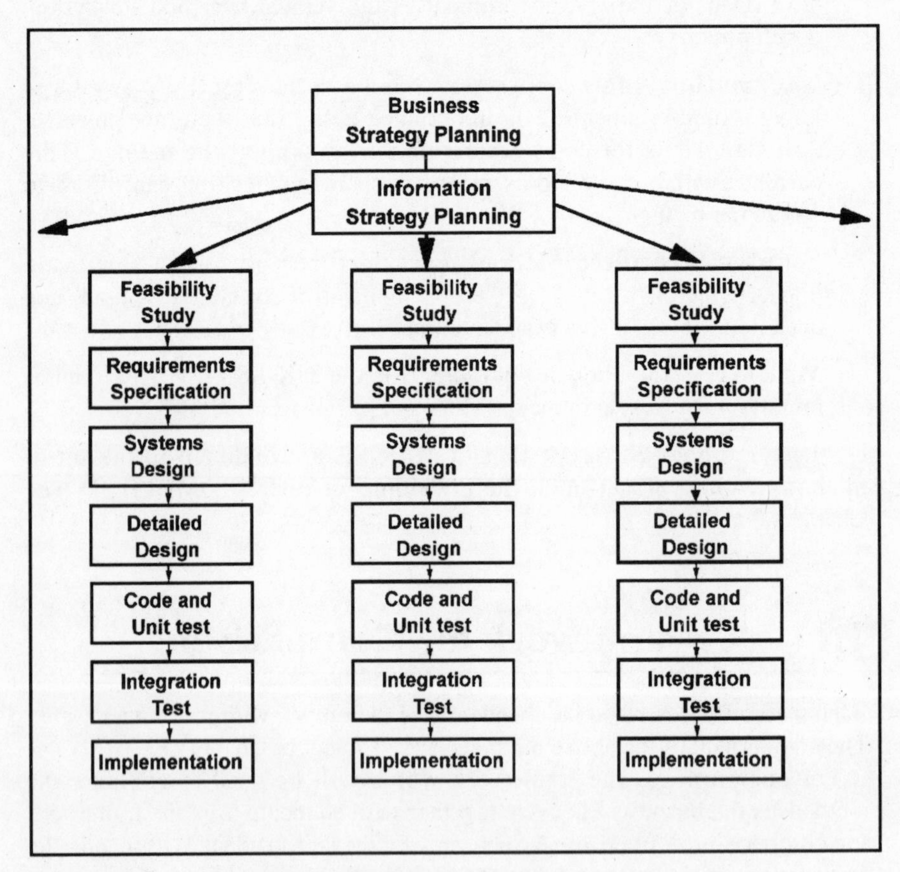

Figure 1.9 Framework –The Business Life Cycle

Within this framework this text will:

- determine the extent of life cycle covered by the methods and the philosophies which underlie them;
- determine how highly structured the methods are and what size of system they are directed toward;
- examine the techniques as used by each method and identify any differences in their implementation;
- investigate the degree of user involvement within the methods;
- look at the level of CASE tool support for the methods;
- address the confusion surrounding the terminology as used by the chosen methods.

1.11.1 The Outcome of the Comparison

The comparison presented in Chapter 9 will present a feature comparison as described below:

- a 'business life cycle' with the methods overlaid on it to show where the methods operate;
- figures showing diagrammatic comparisons of any variations in implementation of the individual techniques;
- a consideration of the size of system to which the individual methods are particularly suited;
- comparative comment on the features outlined above.

The comparison will not be value-laden since the level of importance placed upon any feature can only be determined by its importance to a particular organisation.

1.12 Summary

In this chapter we have looked at:

- what a system is;
- the approach and purpose of this book;
- the objective of systems development;
- the problems of systems development;
- life cycles for systems development;
- the role of structured methods;
- the methods examined in this book;
- a framework for comparison.

We have seen various life cycles, used to bring a measure of control to the systems development process by defining stages through which systems development progresses. We have seen that structured methods bring tasks, and techniques for doing the tasks, into the stages of the systems development life cycle. The effect of these approaches to systems development has been to bring the creation of computer software for business purposes from the realms of an imperfect art to approach the precision of an engineering discipline.

2

A Method for Selecting a Method – Dream or Reality?

2.1 Overview

The use of methods and software tools to assist in the development of Information Systems has become widespread in recent years. The selection of an appropriate method can be a complex, confusing and expensive issue. In this chapter, we look at what structured methods have in common, from the perspective of their approach, framework and general technique. We identify their respective 'blue canaries', i.e. the aspects claimed to make them different. We then consider, in overview, the DESMET evaluation methodology (method for evaluating methods!) which gives structure, procedures and guidelines for performing a commercially appropriate evaluation of specific methods or software tools. Finally, we look at the impact of Euromethod on selection and comparison of methods and the assistance which its products should offer.

2.2 What Do Methods Have in Common?

In this section, we shall look at the strong similarities which exist between methods. An English scholar once observed that 80 per-cent of the English language is derived from Latin, and another 50 per-cent from Ancient Greek. This confusing statement can only be true if there is an interrelationship between Greek and Latin: a common base. Similarly, the structured methods we consider here started from the same seeds. They have a common base and exert a continuing influence upon each other. Thus there is a common core of ideas which can be identified in each of the methods. These relate to the following concepts:

- abstraction;
- diagrammatic modelling techniques;
- user involvement.

2.2.1 Abstraction

Abstraction is simplifying an area of study by concentrating on specific aspects whilst temporarily disregarding others. Abstraction is commonly used in engineering to isolate details at a specific level. For instance, when designing an integrated circuit, electronic engineers first consider the logical level in order to verify the functionality of the circuit, without considering the cooling system. Similarly, in structured methods, it is usual to consider at least:

- a physical level of description of the system;
- a logical (conceptual) level of description.

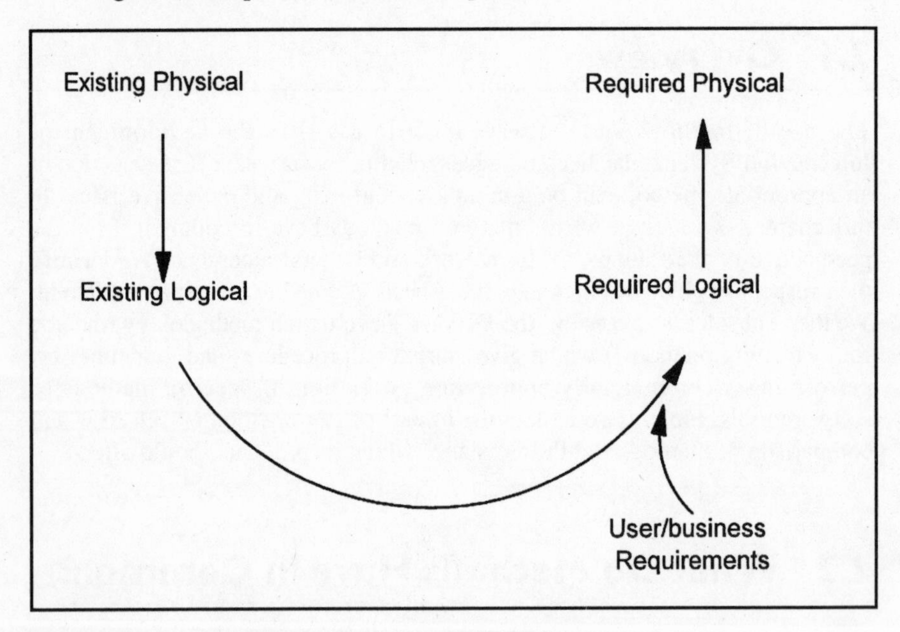

Figure 2.1 Physical and Logical Levels of Abstraction

The **physical** level incorporates **how** things are, or will be, achieved and by **whom, where, when**. The **logical** level (also called **conceptual** or **essential**) describes **what** the business requires without reference to where, by whom, when and how. The operational level resides part-way between the physical and logical, incorporating aspects of **what**, by **whom** and **where** (but not how) the system will operate.

The process of systems development is traditionally seen as a progression from modelling the existing physical system to the existing logical system, to the required (future) logical system and finally arriving at a model of the required (future) physical system. This is illustrated in Figures 2.1 and 2.2.

One method, MERISE, considers separately an additional level of abstraction, the organisational level. This extra level of abstraction is justified on the grounds that data can have different meaning depending upon the organisational area and context in which it is found. It is argued that the detail of this level is too rich to allow its easy incorporation into the physical, where other structured methods put such considerations. In theory, more levels could be introduced, if their complexity justified this (see Figure 2.2).

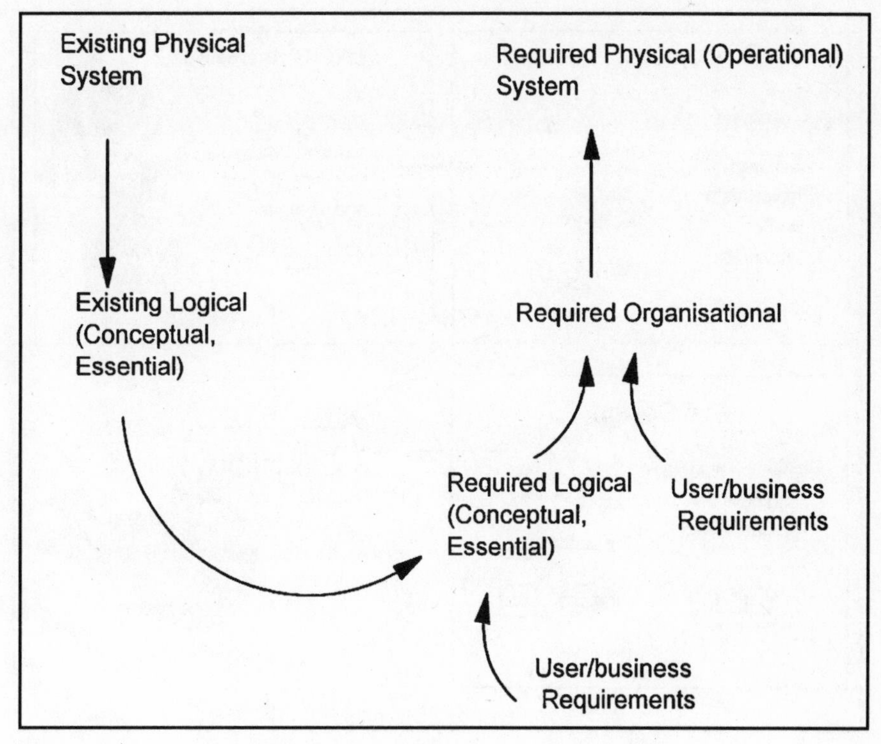

Figure 2.2 Physical, Logical and Organisational Levels of Abstraction

On occasions where there is no existing system, it may be necessary to identify a 'surrogate' existing system by looking at systems of the same type in other organisations, or it may be necessary to start from the required logical model.

2.2.1.1 A Common Architecture for the Design of Information Systems

Another view of the systems development process is described in the ANSI-SPARC report (Tsichritzis and Klug 1975). This proposed three levels of architecture: conceptual, internal and external. This is sometimes referred to as the '3-schema architecture' and is a rationale consistent with SSADM (V4), and other methods. This sees aspects of systems development as contributing to three areas of definition of the *required* system (Figure 2.3):

- the **Conceptual Model:** this encompasses the logical view (and models) of data and processes for the required system;
- the **External Design:** this is concerned with the required physical system and involves the design of user functions and the human computer interface (dialogues, menus etc.);
- the **Internal Design:** this contains the technical design of the physical database and programs.

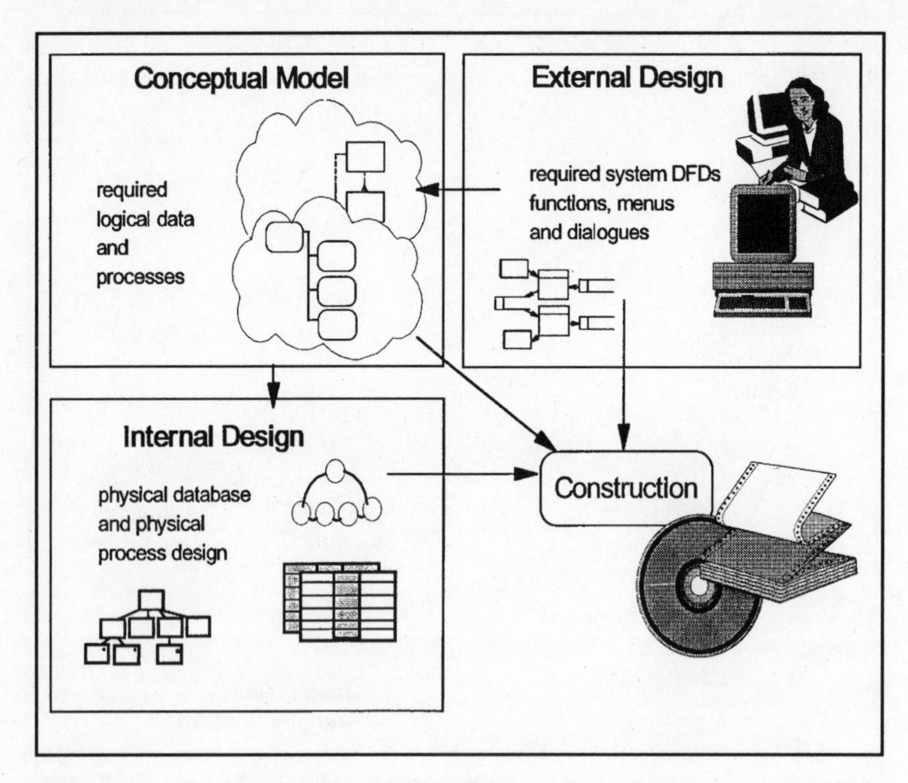

Figure 2.3 The 3-Schema Architecture

This view minimises the emphasis on the existing system, whilst acknowledging that it must be understood within the system of business activities, with which the 3-schema architecture interacts. This system of business activities may be derived from the current system or may be modelled on a standard set of processes for the type of business in question.

2.2.2 Diagrammatic Modelling Techniques

Embedded in the concept of abstraction and the three-schema architecture is the diagrammatic modelling of the system. Models (diagrammatic representations) of the developing system are drawn at each level of abstraction. Often there are many different, but interacting, models for each abstraction level. The major modelling effort is directed at the required system logical (conceptual) level in order to define WHAT the developing system should do.

Early structured methods (De Marco, Yourdon) concentrated on modelling the system in terms of processes and the data flowing between them. It soon became apparent that, in all but small, real-time systems, the structure of data was important. Techniques to handle data structure were added (entity modelling being the principal one). It then emerged that it was not just data and processes which were important, but the timing and effects of the interaction between them. Thus, now that structured methods have matured, they have adopted a three-sided approach to specification of the system:

- **data**;
- **processes (functions)**;
- **events** (and the effect of these on processing and data).

This is illustrated in Figure 2.4.

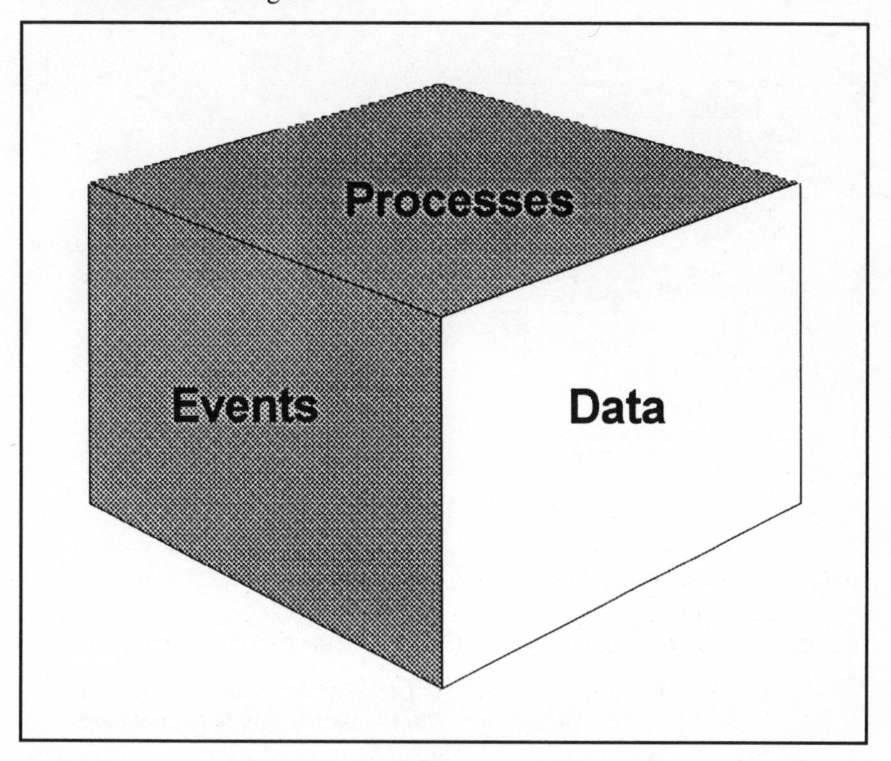

Figure 2.4 The Three-View Approach to Modelling a System

Rochfeld and Tardieu (1983) identify three ways in which interactions between processes and data can be viewed:

- **a state-oriented approach**, which specifies which *after* states may be reached from a specific *before* state;
- **a command-oriented approach**, which places emphasis on the *procedural* description of commands for changing data from a before state to an after state;
- **an interaction-oriented approach**, which focuses on the specification of rules and constraints controlling the interaction between *events*, processing and data (i.e. event driven).

These are all just slightly different views or angles on the same picture. The concepts which underlie all of them are:

- event (trigger, synchronisation);
- process;
- data;
- effects on data;
- before and after states of data.

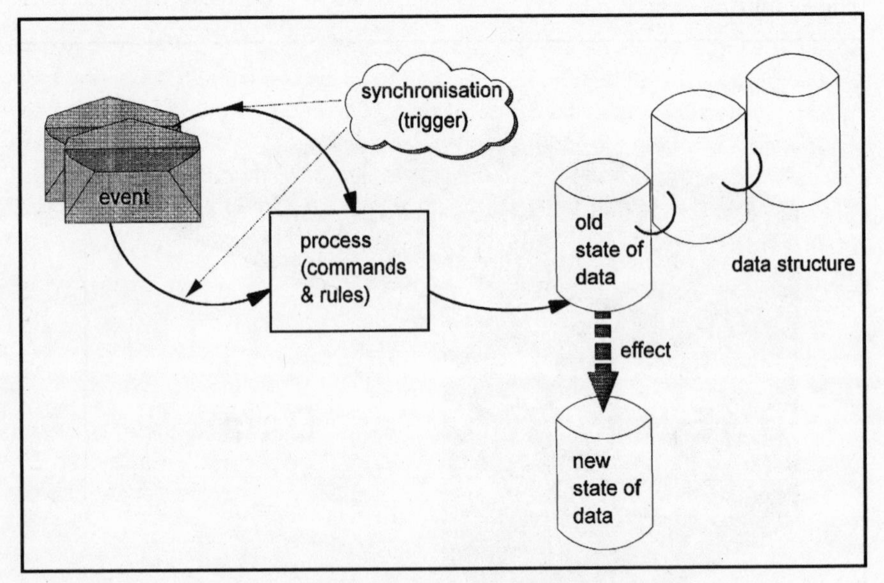

Figure 2.5 The Factors Modelled by Structured Techniques

What the modelling techniques need to model is the following situation:

> an ***event*** in the outside world will trigger a ***process***. This will cause an ***effect*** on ***data*** in a given ***state*** and (if this is an updating event) will transform data into another recognised ***state***.

Different diagramming techniques will bring into sharp focus certain aspects of this picture, whilst embodying or implying the others in a less central way. Some concentrate on the state-changes of data, others concentrate on processing, others focus on the data structure.

We may model the processing and data, with effects implied within the process descriptions and inputs; we may concentrate on states of the data and analyse events and processes from this perspective. Each of these views has areas of overlap with the others. Thus, the use of two or more perspectives, as in most methods, will allow a valuable degree of cross-checking.

2.2.3 User Involvement

The principle of user involvement is fundamental to all structured methods. The users know the needs of their business area, and what is, or is not, likely to work within it. The methods differ in their degree of formal definition of the way users are involved, but all agree that they are there in an information-providing, specification-checking and decision-making capacity.

2.3 How Do Methods Differ?

When everyone else's canary is yellow, the way to stand out is to have a blue one. The key thing to note is that they are all canaries! All structured methods have the same basic approach. Even object-oriented methods (see Chapter 10 for an overview) have fundamentally the same elements, albeit packaged slightly differently. However, they do have their differences. When identifying differences from a technical perspective of techniques, approach and framework, their particular 'blue canaries' are in the areas of:

- **The methods and the business life cycle:** we look at life-cycle coverage, from strategy to implementation, and the support given for project tasks outside of the analysis and design activities, such as quality assurance and project management;
- **The underlying philosophy of the methods:** here we consider whether the method embraces the science paradigm (divide and conquer) or the systems paradigm (the whole is greater than the sum of its parts). These are further explained in Chapters 4 and 9;
- **The user role in the method:** here we address the level to which the methods formally define user roles and user participation;
- **The 'structuredness' of the methods;** 'structuredness' is the term we have used (following Avison (1988)) to describe the formality of the available manuals and documentation and the prescriptiveness of approach;
- **What size of system each method is aimed at:** here we assign a notional size to the idea of small, medium and large systems and assess the level or levels to which each method is appropriate;
- **The techniques within the methods:** we look at the particular diagramming techniques chosen to address the function, data and events views of the system;
- **CASE tools and the methods:** we examine the maturity and scope of CASE tool support available.

The above areas represent, from our chosen technical perspective, the main differences between the methods studied in this book. In Chapter 9, they reappear as the set of features upon which we have chosen to compare the methods. However, we emphasise that anyone intending to evaluate the methods in order to make a cost-effective choice of one specific method for their organisation's use would also need to consider additional factors such as:

- relevance: applicability to the size and type of projects undertaken by the organisation;
- ease of use: relative simplicity or complexity;
- ability to highlight errors;
- availability and cost of training;
- availability and cost of manuals, user-group support, consultancy expertise;
- efficiency: direct savings to be expected, as part of a cost–benefit analysis;
- applicability to current staff expertise;
- social acceptability, i.e. do the staff who will be affected want it (!).

The relative importance of evaluation features will depend upon the specific nature and circumstances of the organisation for which the evaluation was being performed, and must be assessed in relation to this.

2.4 DESMET: A Methodology for Evaluating Methods and Tools

In the title of this chapter, we asked whether a method for choosing methods was a dream or reality. At the end of June 1994, it would seem that the dream became reality with the arrival of the DESMET (**D**etermining an **E**valuation methodology for **S**oftware **ME**thods and **T**ools) evaluation methodology. The DESMET methodology aims to enable independent and consistent evaluations to be made of any software development methods or tools. It also aims to enable evaluators to choose the evaluation method most appropriate to their purpose and environment. The DESMET project started in Autumn 1990 and its products were made publicly available in June 1994, having undergone trials in both commercial and academic environments.

DESMET was sponsored by of the United Kingdom Department of Trade and Industry (DTI) and the Science and Engineering Research Council (SERC). The partners in its development were the National Computing Centre (lead partner and supplier of the methodology), The University of North London, GEC–Marconi Software Systems and BNR Europe.

2.4.1 What DESMET Provides

DESMET provides a structure, procedures and guidelines for the objective evaluation of all types of methods and tools used in the development and/or maintenance of software-based systems. It covers:

- setting up an evaluation project;
- quantitative methods of evaluation;
- qualitative methods for evaluation;
- a measurement system.

It provides assistance in the assessment of the appropriate evaluation method(s), and raises the question of whether an evaluation will be cost-justified in the particular circumstances. It gives guidance upon the initiation and management of the evaluation project, within a defined project life-cycle, and provides templates for the statistical analysis of results obtained.

2.4.2 DESMET Evaluation Methods

DESMET proposes eight types of evaluation, of which one or more may be appropriate in any particular evaluation project:

- formal experiment (quantitative);
- case study (quantitative);
- survey (quantitative);
- formal experiment (qualitative);
- case study (qualitative);
- survey (qualitative);
- qualitative effects analysis (hybrid qualitative/quantitative);
- benchmarking (hybrid qualitative/quantitative)

These are shown in Figure 2.6 and further described below. DESMET makes a fundamental distinction between the way an evaluation is organised (e.g. case study, experiment etc.) and the way the evaluation is assessed. For example, the evaluation may be organised as a case study and be assessed by feature analysis.

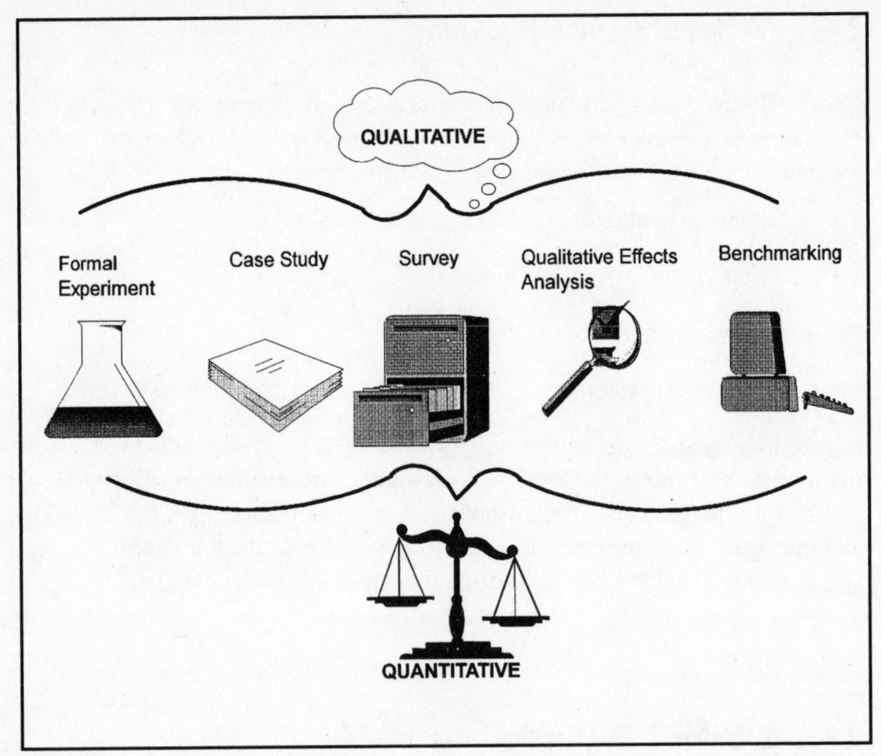

Figure 2.6 Qualitative and Quantitative Evaluation Methods within DESMET

2.4.2.1 Formal Experiment

Experiments allow control and manipulation of the evaluation project in order to test specifically defined behaviour. DESMET recommends factors well-established in the performance of scientific experiments, such as the setting up of a control, repeatability of the experiment, control of variables. It mandates the definition of a null hypothesis, and then, very sensibly, allows the setting up of an alternative hypothesis, generally more acceptable in terms of gaining authorisation to perform the evaluation! (See Figure 2.7) DESMET also gives advice on experiment design and analysis of the resulting data.

2.4.2.2 Case Studies

Case studies are projects where the evaluation of a method or tool is performed during a real systems or software development project. As such, there will be limited potential for replication. It is not usually possible to manipulate such projects to isolate the required behaviour. Rather, the evaluator must sample from the behaviour exhibited.

Figure 2.7 The Null Hypothesis and the Alternative Hypothesis

2.4.2.3 Surveys

Surveys involve the collection of data from completed projects. They are suitable for mature methods or tools, where much information is already available in the marketplace. With a sufficiently large population to survey, it should be possible to generalise results and perform statistical analyses on the data collected, provided that the data has been collected in a way that makes it comparable.

2.4.2.4 Qualitative Effects Analysis

This involves the collection of (usually subjective) assessments of the effects of the use of a method or tool on attributes of the developed system. It may also be possible to quantitatively measure benefits of these effects.

2.4.2.5 Benchmarks

These are objective measurements of selected features. The selection of which tests to perform is a subjective element, but it is possible to get a very precise comparison of the selected features. Benchmarks are appropriate when the features to be compared can be expressed as a machine manipulation, i.e. there is no human interaction.

2.4.3 The Choice of an Evaluation Method

The choice of the appropriate evaluation method(s) for a particular circumstance will depend upon:

- the goals of the particular evaluation;
- the expected impact of the evaluation itself;
- the exact nature of the method or tool to be evaluated;
- the impact of selection or rejection of the method or tool;
- the maturity of the method or tool under evaluation;
- the learning curve associated with the method or tool;
- the evaluation capability of the organisation performing the evaluation.

Evaluation will be of limited use in an organisation where every project necessarily uses different methods, as the results will not be able to be generalised. If the evaluators have no experience of collecting metrics, quantitative evaluation may prove difficult.

2.4.4 A Summary of the DESMET Modules

DESMET consists of the following modules, each implemented as a handbook. Where marked (*) the modules were not available at the time of writing, but, subject to commercial sponsorship, are planned for the future.

Modules for Setting Up the Evaluation Project

- Evaluation Method Selection (EMS);
- Investment Analysis and Justification (INVA) (*);
- DESMET Maturity Assessment (DMA);
- Managerial and Sociological Issues (MANSOC).

Modules for Quantitative Evaluation

- Experimental Design and Analysis (EXPDA);
- Case Study Design and Analysis (CSDA);
- Survey Design and Analysis (SURVDA) (*).

Modules for Qualitative Evaluation

- Feature Analysis (FEA);
- Qualitative Effects Analysis (QEA) (*).

Modules for the Measurement System

- Data Collection and Metrication (DCM);
- Data Collection and Storage System (DCSS).

These are explained in more detail below.

2.4.4.1 Modules for Setting Up the Evaluation Project

Evaluation Method Selection (EMS)
This is the user-interface module to enable the evaluator to assess the need for an evaluation and identify the best method for that evaluation. In order to match the evaluation project to the appropriate evaluation method(s), DESMET identifies factors to be considered.

Investment Analysis and Justification (INVA) (*)
This would provide guidance on the assessment and measurement of costs of the evaluation exercise itself and expected costs and benefits of the implemented method or tool.

DESMET Maturity Assessment (DMA)
This allows the organisation considering performing an evaluation to assess its ability to do an evaluation and to effectively use a method or tool after evaluation. DESMET presents a maturity model against which the organisation can assess itself. This is not the well-known SEI maturity model (Please see Yourdon's *Decline and Fall of the American Programmer* for a quick reference, if this is not well-known to you!) but is a more specific set of questions geared to evaluating specific aspects of organisational maturity in relation to performing evaluations.

Managerial and Sociological Issues (MANSOC)
DESMET asserts that an evaluation methodology is incomplete if it fails to acknowledge the human factors which will inevitably affect an evaluation. It discusses factors such as the 'Hawthorne' effect: the phenomenon that the very process of observing a situation has an effect upon the situation – a fact well-known to the time-and-motion studier. It also considers the effect of the champion of a method or tool on the evaluation team, and the effects of team conflict.

This module identifies the factors which may have a distorting effect on the evaluation process and gives guidelines for the planning and performance of the evaluation designed to minimise, or at least highlight, these effects.

2.4.4.2 Modules for Quantitative Evaluation

Experimental Design and Analysis (EXPDA)
This booklet provides guidelines for determining whether formal experiment is appropriate and, if so, for designing an experiment which will provide meaningful results related to appropriate factors.

Case Study Design and Analysis (CSDA)
This is seen as one of the key modules, because case study (evaluation during a real project) is often the most feasible approach to evaluation. DESMET provides checklists to assist the evaluator in setting the hypothesis, defining the metrics to be captured, identifying the interactions between variables and assessing the results.

Survey Design and Analysis (SURVDA) (*)
This will give guidance on the capture of data for evaluation from different organisations and the factors to be considered in the comparison and collation of this data.

2.4.4.3 Modules for Qualitative Evaluation

Feature Analysis (FEA)
This is likely to be the most often-used method for evaluating methods and is a recommended technique for that purpose.

Feature Analysis is the comparison of the method or tool being evaluated against a set of features which have been identified as important to the organisation. DESMET provides a framework and scoring method for this. It also provides a sample list of features for guidance, although it is stressed that it is the evaluator's responsibility to define the features of interest to the particular organisation, and project's circumstances. DESMET recommends a scoring system on a scale of minus one to five or a simple yes/no answer for the evaluation of these features, according to the nature of the feature. Figure 2.8 shows, in outline, the type of evaluation table used in Feature Analysis.

QEA – Qualitative Effects Analysis (*)
This module would assist in bridging between objective and subjective evaluation. It would provide a mechanism for the direct estimation of the effects of the method or tool on defined quality attributes of the project.

2.4.4.4 Modules for the Measurement System

Data Collection and Metrication (DCM)
This module consists of three handbooks to assist the organisation in setting up a measurement programme: the first handbook addresses specifying

Method under evaluation:
Your Structured Analysis and Design Method (YSADM)

Feature	Importance	Description	Scale of marking	Feature's Score
Support for risk identification	High	Full support	2	*0*
		Some support	1	
		No support	0	
Support for project management	Low	Full support	2	*0*
		Some support	1	
		No support	0	
Support for quality assurance	Medium	Full support	2	*2*
		Some support	1	
		No support	0	

Figure 2.8 Feature Analysis of a Structured Method

measurement goals and the selection of metrics, the second is concerned with the establishment and maintenance of a data collection system and the third gives guidance on the analysis of software measurement data.

Data Collection and Storage System (DCSS)
This is a PC-based software package originally developed as part of the ESPRIT MERMAID project. It gives support in areas of: data model definition, data collection, validation and storage metrics calculation and storage. It is able to support any software development life cycle.

2.4.5 DESMET and Structured Methods Comparison

DESMET is aimed at the evaluation of software development methods and software tools. Structured methods would usually be best assessed by *quantitative* rather than *qualitative* means. However, the problem with this is the difficulty and expense of setting up a fair, practical, quantitative experiment, case study or survey. The fact that structured methods potentially impact upon the full length and breadth of the project life-cycle will render an experiment or case-study a lengthy undertaking. There are seldom two or more projects or sub-projects sufficiently alike to allow equitable comparison. In a commercial situation, the managerial and sociological issues cannot be easily controlled and should not be underestimated. Surveys may be possible for mature methods, but the collection of sufficient data from which to draw statistically significant conclusions would be time-consuming. The comparability of data from different sources could also prove difficult to establish.

Feature Analysis is a practical qualitative alternative for the comparison of structured methods. In Chapter 9, we address our comparison by identifying features and contrasting the chosen methods in relation to those features. We do not attempt to assign scores, however, as the relative importance of the features we have selected would vary from one organisation to another. Indeed, individual organisations would be advised to produce their own list of features for comparison.

2.5 Euromethod and Its Impact on Comparison of Methods

Euromethod is an IS methods framework, aimed at improving the relationships between suppliers and customers, particularly in IS procurement projects. Its purpose and origins include the 'harmonisation of methods'. Euromethod and its contents are more fully explained in Chapter 10. However, in outline, the Euromethod documents which will assist in the comparison of methods are:

- a METHODS BRIDGING GUIDE, which provides guidance to the project manager in populating a project plan with the correct products from a chosen development method;
- a CONCEPTS DICTIONARY, which defines the concepts and terminology used in Euromethod, in alphabetical order. It is intended that each methods' vendor will supply a bridging guide for their method, mapping their concepts and terminology to the standard definitions in the Euromethod dictionary;
- a DELIVERABLE MODEL, which gives a framework for defining the products (deliverables) of the methods. This will be the main mechanism for comparison of the products of different methods.

These are illustrated in Figure 2.9.

However, it should be noted that Euromethod will not provide a detailed translation aid between different methods. It is important to note that whilst Euromethod will provide a means of mapping deliverables and activities between methods, it will not provide a means of translating diagrams or techniques from one method to another. Such translation falls within the scope of standards such as the CDIF (CASE Data Interchange Format) project, which is currently working towards a standard for interchange of data between software tools which support systems development methods. Euromethod concentrates on high-level deliverables.

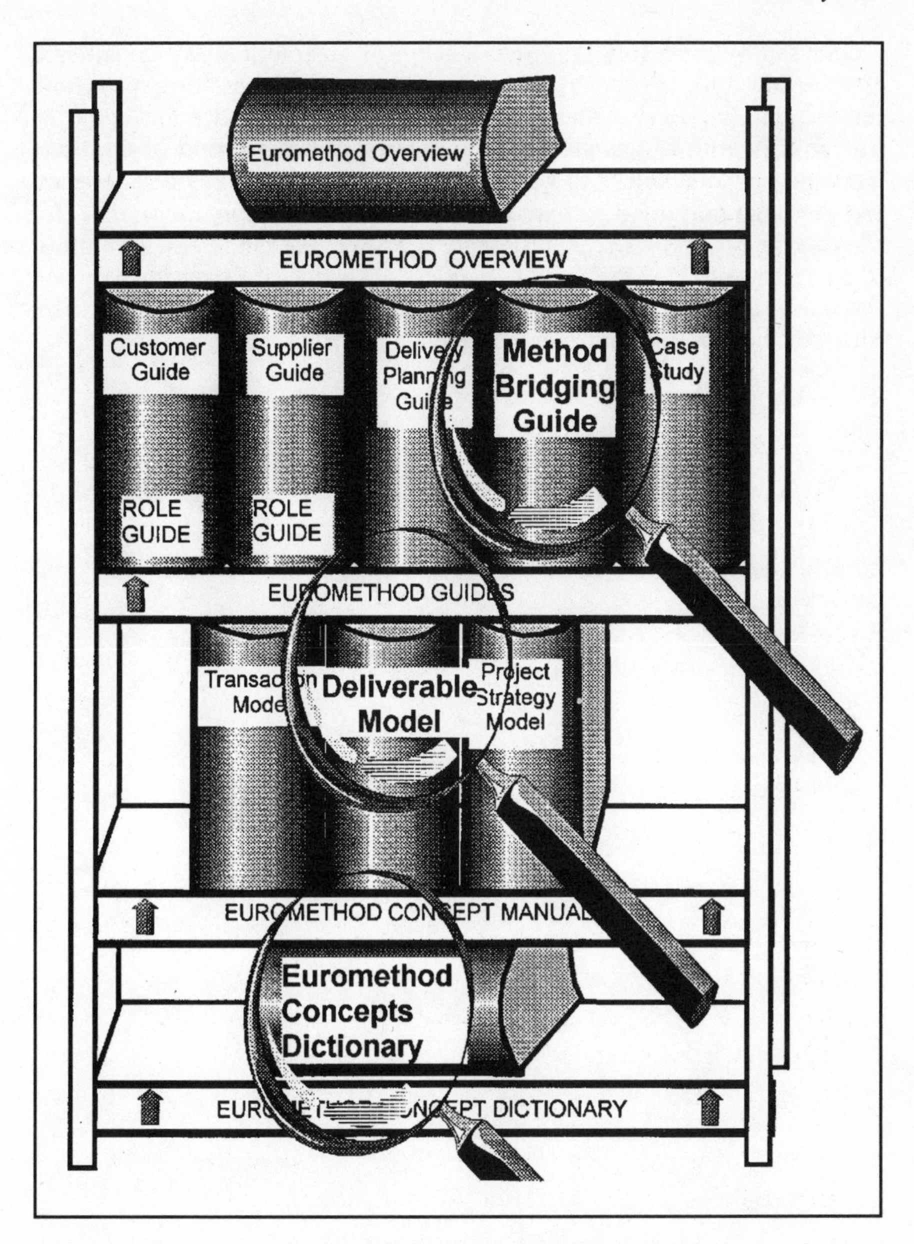

Figure 2.9 The Euromethod Documents, Highlighting Those Which Assist in the Comparison of Methods

2.6 Summary

In this chapter, we have looked at the areas of common ground between structured methods and identified some relatively small differences of approach. It was apparent that there is a common core approach to the views addressed by modelling and the concept of abstraction. We considered DESMET as an approach for performing a commercially appropriate evaluation of specific methods or software tools. Finally, we looked at the documents within Euromethod which present a definition of common concepts and terminology to which method purveyors can provide interfaces. At the beginning of the chapter, we asked the question 'A Method for Selecting a Method – Dream or Reality?'. With the recent release of most of the elements of DESMET, and with an initial version of Euromethod having been delivered in June 1994, with refinement work on-going, it would seem that a method for selecting a method **is** reality – virtually!

3

The Common Techniques

3.1 Overview

This chapter describes the major techniques which are found in more than one of the methods in this book. It explains these in a generic way, in order to provide a standard description against which a particular method's specific use of a technique can be compared. Other techniques are detailed within the chapters describing the methods which use them. The chapters dealing with specific methods identify the position of the techniques within the framework of the method.

Each technique is designed to model a system from one or more of the fundamental views of structured methods: function, event and data. They may be used at different levels of abstraction or detail and for either the current system or the required system. The concepts of function, data, event and abstraction were addressed in Chapter 2.

The techniques covered in turn in this chapter are:

- Data Flow Diagramming, including Logicalisation;
- Entity Modelling;
- Relational Data Analysis;
- Function/Entity Matrix.

In each case, we shall see:

- what the technique is and its aims;
- symbols and conventions used;
- guidelines for performing the technique.

3.2 Data Flow Diagramming

3.2.1 What Is a Data Flow Diagram?

Data flow diagrams (DFDs) are a set of diagrams to graphically depict how data travels within a system. They show the processes carried out, the flows of data (and possibly materials) between those processes and to and from data stored within the system area. They also show flows of data to and from the people, departments and organisations external to the area of study.

They can be used to show an existing system and **how** it works physically; an abstraction (logical or conceptual view) of the system, illustrating **what** is going on without reference to how and **where** it happens or **who** does what. They can also be used to describe a new system, if required, in logical, organisational (showing who does what and where, but still not showing how) or physical terms (showing how the implementation is to be done.) An example of a data flow diagram is shown in Figure 3.1.

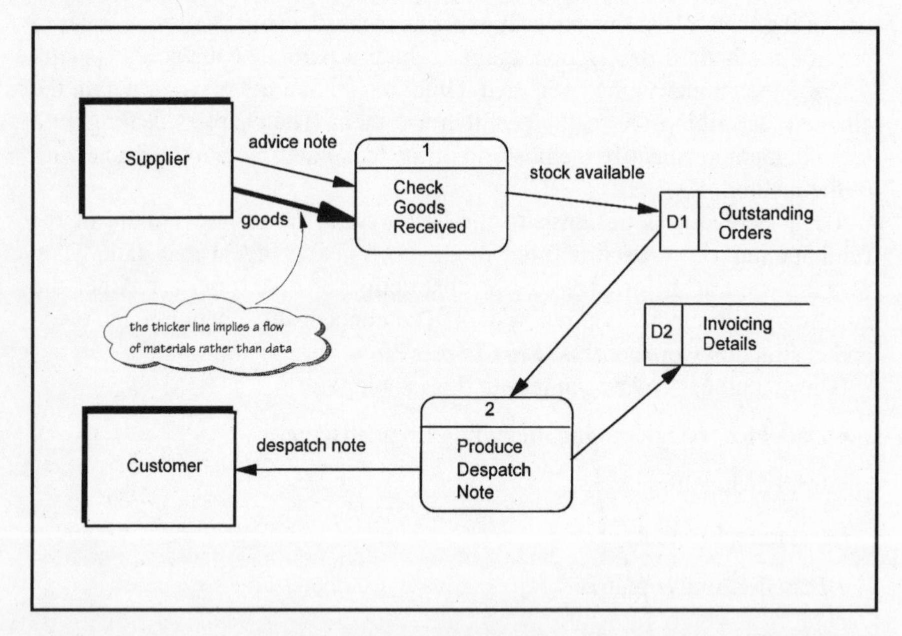

Figure 3.1 An Example of a Data Flow Diagram

3.2.2 Symbols and Conventions for Data Flow Diagramming

Data flow diagrams have only a few symbols and straightforward conventions, which make them readily understandable. For this reason, they can prove very useful in discussions with users.

The four main symbols are shown in Figure 3.2. These are:

- process;
- store (this can be data or materials);
- flow (this can be data or materials);
- external entity.

The symbol shapes vary from method to method. However, the principles remain the same.

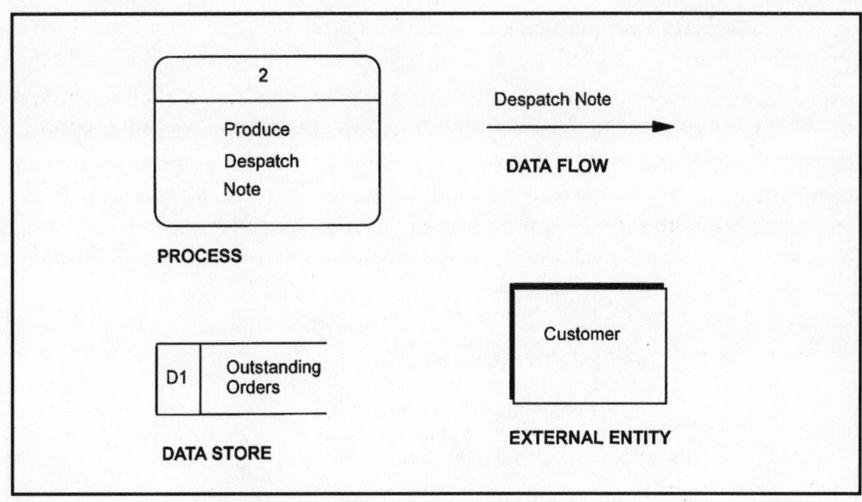

Figure 3.2 Symbols Used in Data Flow Diagrams

3.2.2.1 The Process Symbol

The process (Figure 3.3) is represented here by a rounded rectangle ('soft box'). The process described must perform a transformation of the input into output. The description of the process should be a verb, followed by an object clause. This is a useful convention since it deters the analyst from giving processes general names like 'Accounts Department Jobs' and promotes more detailed thinking about actually what happens in the accounts department. It is so easy to slip into general naming without really understanding the processes. For this reason, naming such as '*Perform* Accounts Department Jobs' should also be avoided!

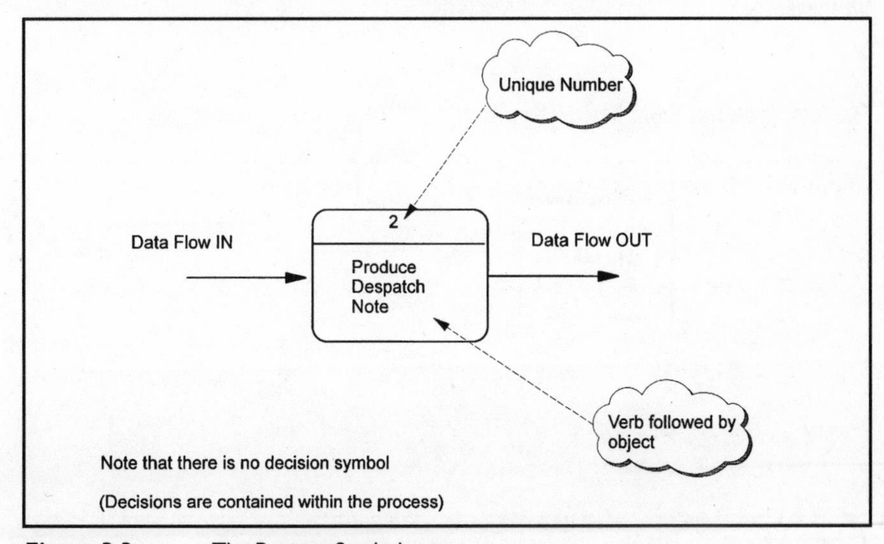

Figure 3.3 The Process Symbol

3.2.2.2 The Data Flow Symbol

The data flow (Figure 3.4) is depicted by an arrow, showing direction of flow. Flows of materials can also be shown in the same way, using a slightly thicker, or dotted, arrow to distinguish them. However, it is usually the data, rather than materials, which the analyst is trying to track. Material flows should be used only if they are helpful in discussions with users.

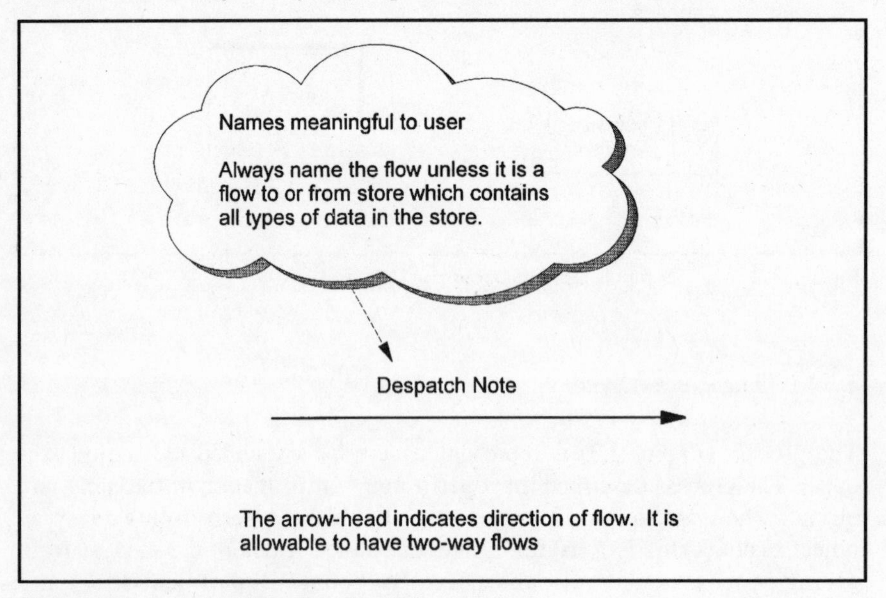

Figure 3.4 The Data Flow Symbol

The naming of the flows should be as explicit as possible, and should not be duplicated unless they are genuinely the same flow at the same stage of processing.

3.2.2.3 The External Entity Symbol

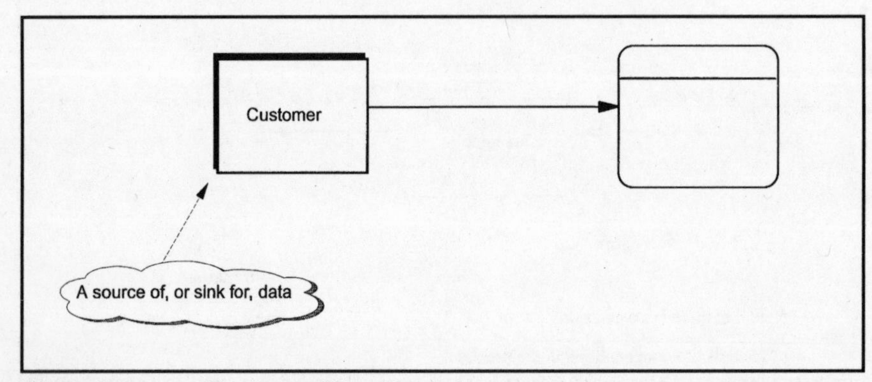

Figure 3.5 The External Entity Symbol

The external entity (Figure 3.5) is depicted here by a shadowed rectangle. External entities are those people, departments, organisations which either provide information to the area of study (sources) or which receive information from the area of study (sinks). An external entity can be both a source and a sink. For example, a customer who sends orders into the system and receives goods and invoices from it is both source and sink.

The external entity is, by definition, outside the boundary of the current area of investigation.

3.2.2.4 The Data Store Symbol

The data store symbol (Figure 3.6) is an open-ended rectangle. The same symbol can be used to show stores of material, if required, or in some methods the open end is closed for this purpose. The naming of the store should be clear and, if describing an existing physical system, should reflect the name the user is familiar with. Information which comes to rest within the system area, or is held for reference, is shown as a data store.

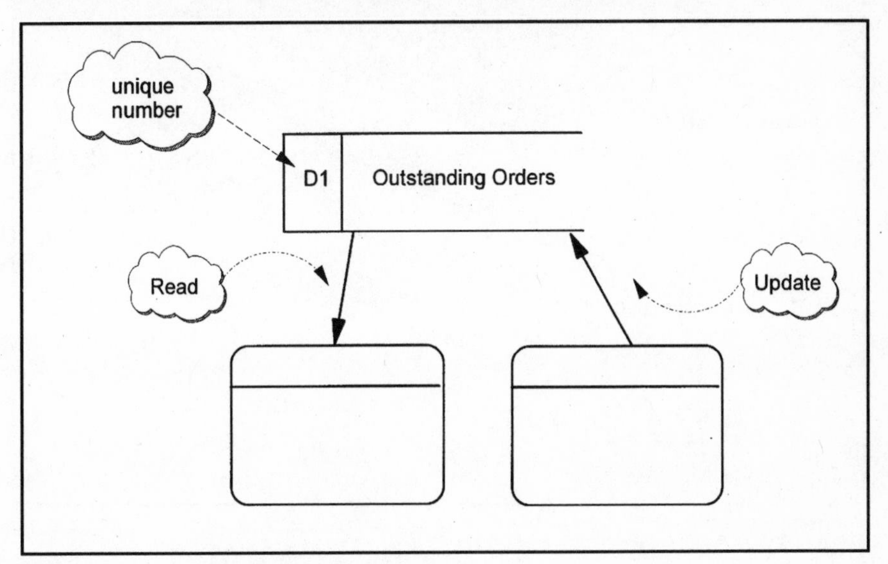

Figure 3.6 The Data Store Symbol

When reading from a store, the flow should be shown as coming from the store; when updating, it should be shown going to the store.

3.2.3 A Set of DFDs

Putting together the symbols seen above produces a diagram similar to Figure 3.1 above. Of course, within an area of study, there would normally be more than one or two processes! In fact, there are usually so many processes that they would not intelligibly fit onto one sheet of paper, unless it was the size of a house wall. Therefore, the recommendation is that a **hierarchical set** of diagrams is produced, with no more than seven (plus or minus two) processes at any one level (based on Miller's magical number seven, plus or minus two) (see Figure 3.7).

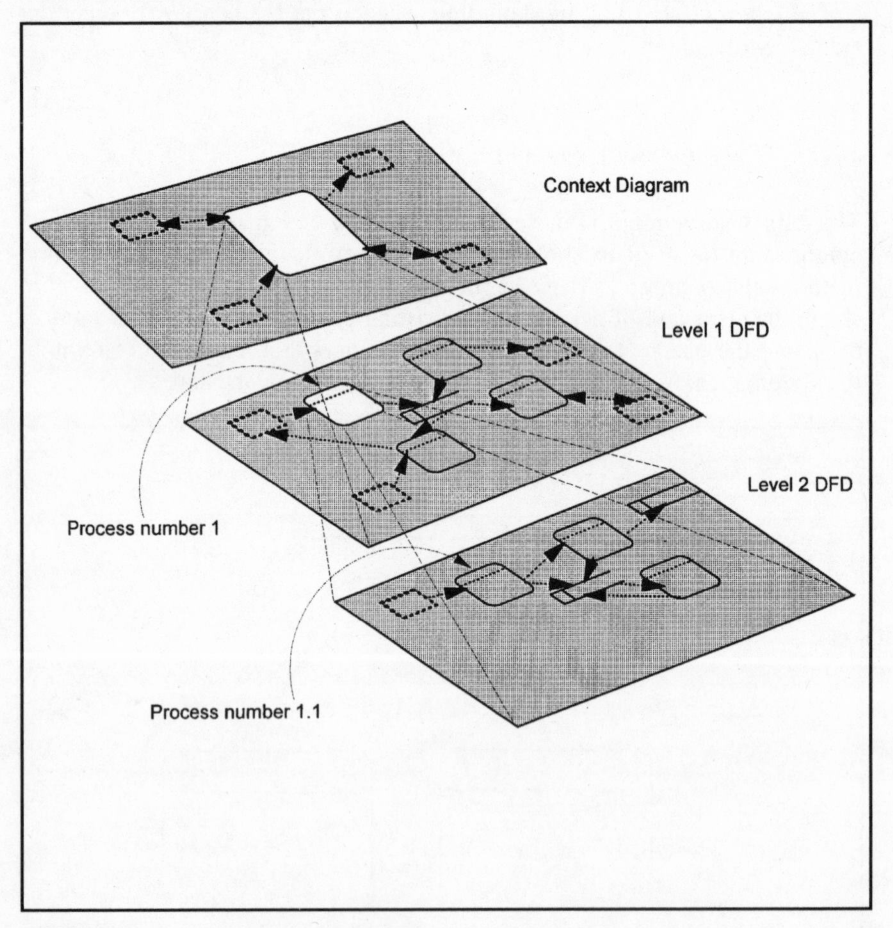

Figure 3.7 A Levelled Set of DFDs

Thus, at the first level, there may be, let us say, five processes, one of which is 'Despatch Goods'. Then on a further diagram, 'Despatch Goods' may be further broken down into further processes:

- Produce Despatch Note;
- Run Overnight Invoice Production;
- Produce Weekly Despatch Statistics.

The very top level diagram in a set of DFDs would appear with one overall process and just the externals and data flows communicating with it. This level is referred to as the **Context Diagram** (Figure 3.8).

Method under evaluation:
Your Structured Analysis and Design Method (YSADM)

Feature	Importance	Description	Scale of marking	Feature's Score
Support for risk identification	High	Full support	2	*0*
		Some support	1	
		No support	0	
Support for project management	Low	Full support	2	*0*
		Some support	1	
		No support	0	
Support for quality assurance	Medium	Full support	2	2
		Some support	1	
		No support	0	

Figure 3.8 The Context Diagram

3.2.4 Guidelines for Developing a Level 1 DFD

The following points provide a good checklist when developing DFDs:

- **identify key data flows:** this information can be gleaned from the user by asking what are the major documents coming into and leaving the area of study;

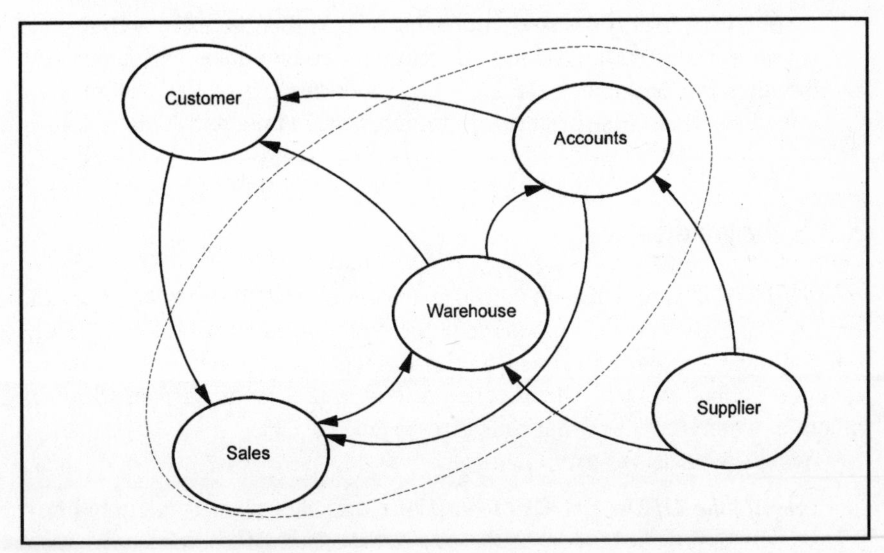

Figure 3.9 An Initial DFD

- **creating an initial DFD** (see Figure 3.9): not to be confused with a context diagram, an initial DFD is not really a full DFD, in that it shows no stores. It is merely a **document flow diagram** between departments or functional areas of the business. However, it is very useful for getting an overall picture of the area of study and is readily accepted by users because they can relate to their own department and documents.

- **agree the system boundary:** the initial diagram is a useful diagram on which to draw the system boundary and unambiguously agree it with users, and the steering committee for the project;
- **identify processes and data stores:** so far we have identified only departments and documents (data flows). This step takes us to a further level of detail. By following the input documents, in turn, through the system, the processes through which they pass and the data stores they use become apparent. Finally, the analyst must check any output documents which were not a result of those input documents, to discover the processes which generated them and the stores involved;
- **check for completeness:** omissions can be readily spotted in the DFD by looking for:
  - stores with inputs but no outputs, or outputs but no inputs;
  - processes which seem not to use the data entered, or seem to produce outputs for which they do not have input information.
- **establish level of detail:** it is hard to know, when first drawing the DFD, how much to break a process down. Consequently, the first attempt at a data-flow diagram may have many more than the recommended 'seven plus-or-minus two' processes. The analyst should re-group processes and create lower-level DFDs to expand the detail;
- **review with users:** the strength of the DFD is that the user, with a small amount of explanation of the diagramming conventions, can understand the diagrams and check the analyst's understanding of the system more effectively than if the system description were mainly narrative.

3.2.5 Logicalisation

Logicalisation distils the essential processing and data flows from the physical DFDs. Logicalisation is the process of abstraction from the DFDs of what is done, for current logical DFDs (or what is required to be done for required logical DFDs). All reference to how a process is performed, or by whom or where or when it happens is removed.

Logicalisation serves to:

- *simplify the DFDs:* the act of removing the who, where, when and how aspects significantly reduces the size of the set of DFDs. Many processes will be duplicated because of geographical location: these can be combined in the logical view. Many data stores are duplicated in the physical view: again, this duplication is removed. The remaining model is a representation of what the business does, irrespective of how, and will remain the same over a range of physical implementations. Thus it is a sound foundation on which to build a new system;

- *allow a 'blue sky' view:* The movement away from exactly how and where things currently happen and who does them makes the logical DFD a better vehicle for a completely fresh look at (a 'blue sky' view of) what needs to happen for the future system and allows the locations for processing and the manner in which it is done, to be re-thought. The logical DFD gives a good starting point for Business Process Redesign, described in Chapter 10.
- *highlight and bring together common processing:* the removal of specific locations and people from association with specific processing allows the bringing together of similar or identical processing. For example, sales order processing, carried out at ten locations nation-wide could appear as ten subtly different physical processes on the physical DFDs. On the logical DFDs the realisation that what was being done was the same would allow the bringing together of these processes into one.

3.2.6 Guidelines for Logicalisation

Logicalisation should follow the guidelines given below, in order that all areas are addressed and that nothing is missed from the physical DFDs. There is a growing trend amongst systems analysts (and some methods advocate this) to move straight to a logical DFD without ever developing a physical DFD. This can be dangerous as there is then no starting point with which users can identify. They can recognise their own jobs in a physical data flow diagram. There is also no cross-check that nothing from the physical system has been missed. Thus, this type of intuitive logicalisation is discouraged. However, the diagramming of the physical system should be kept in proportion as a means to arrive at the logical view. Projects have been known to spend 80 per-cent of their available development and implementation time analysing and documenting the current physical system, which is certainly unlikely to be cost-effective!

The following guidelines recommend a systematic way of moving from a physical set of DFDs to a logical set. The analyst should:

- rationalise the data stores;
- regroup processes;
- rename processes and flows;
- cross-check to ensure nothing has been missed.

The guidelines operate on the lowest-level diagrams within the physical DFD set and affect the elementary processes.

3.2.6.1 Rationalise the Data Stores

This has two aspects: the removal of transient stores and the replacement of physical stores with entities. Transient stores exist purely as a holding area for information moving from one person or location to another. In a logical view they are not necessary, since physical locations and separation of people's jobs are not retained. Entities will have been identified during entity modelling. Replacement of physical stores with the appropriate entities give a separation of the data which removes duplication often found between physical stores. The entities should be created as early as possible. In the physical system, data is not always stored as soon as it is available, but travels around on documents. It enters processes which do not need it simply because it accompanies other data which is needed by those processes. Such 'travelling data' should be treated as being stored on its appropriate entity as soon as it is available and should be accessed specifically when required by a process.

3.2.6.2 Regroup Processes

Processes which are joined by flows in the physical DFD can be combined if all data to begin the second process is available from the first. The effect of this and earlier guidelines make it unlikely that processes will be directly connected by data flows in the logical DFD. Where the same basic process is found in different locations, these can be combined. If a process merely re-organises data (e.g. a sorting, merging, filing) or is only present due to the physical method of working (e.g. reconciliation between sources of the same data) it can be omitted.

3.2.6.3 Rename Processes and Flows

Processes and flows should be given non-physical names. For example, rather than calling a flow *sales report SA151*, something like *sales variance details* would be better. This also does not necessarily imply a hard copy or a particular layout to the reader. References to physical documents and ways of performing jobs should be removed.

3.2.6.4 Cross-Check for Completeness

The analyst should finally check that, in the enthusiasm to simplify the physical DFDs, nothing essential has been lost. The only processes lost from the physical DFDs should be those legitimately accounted for by the guidelines. The logical DFDs must contain all of the functionality and data of the physical DFDs. This can be checked by reading each of the logical processes, using the entity model for data access. The logicalisation operates on the lowest level processes. These will need to be re-combined into higher level processes. This should be done on the basis of ease of understanding by users and similarity of use of data from the entities.

3.2.6.5 Physical to Logical Transformation – An Example

The diagram in Figure 3.10 is part of a physical DFD for the processing of sales orders. The validation of stock and the assigning of stock are carried out by different departments (perhaps for historical reasons). The order travels to the warehouse with no copy being kept by either of the departments through which it has passed.

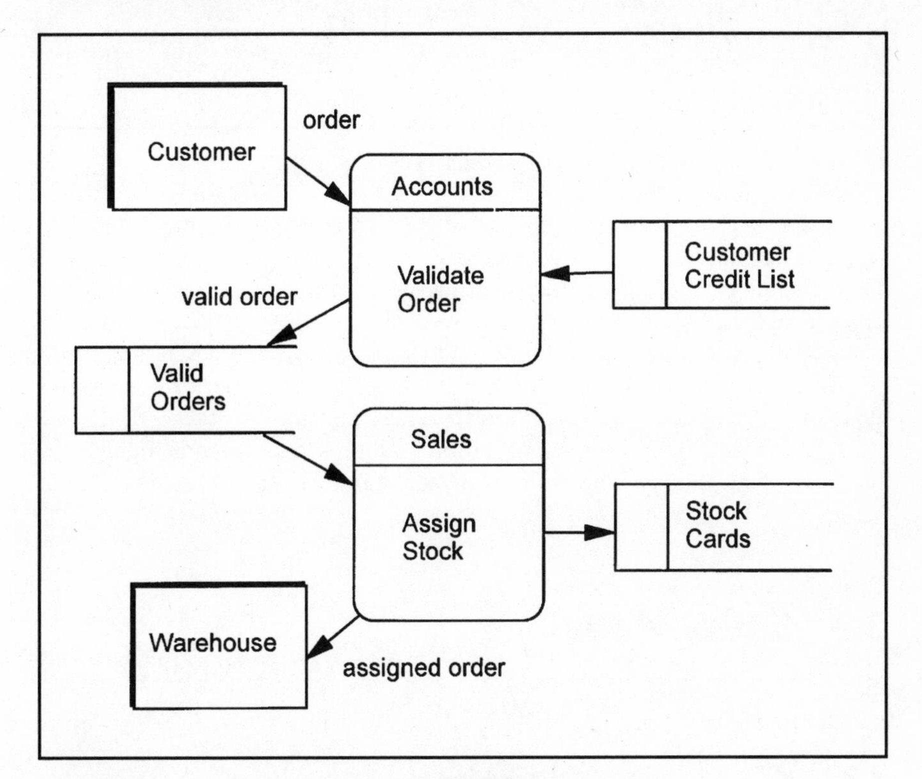

Figure 3.10 An Extract from a Physical DFD

The effect of application of the guidelines for logicalisation is to remove the transient store 'Valid Orders', to combine the two processes, which would now be joined by a data flow, and to replace the reference stores with entities. The 'assigned order' data flow has been replaced with the logical description 'picking details' which would be described in the data dictionary as containing the specific attributes needed by the warehouse to pick goods required for the order. The order details are stored immediately.

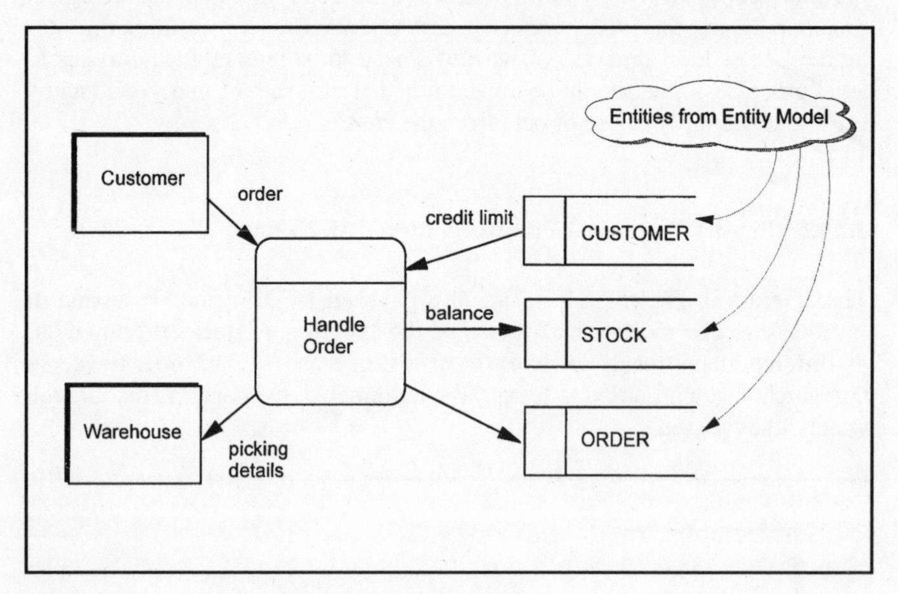

Figure 3.11 An Extract from a Logical DFD

The Logical DFD derived from Figure 3.10 is shown in Figure 3.11. It is simpler, having only one process instead of the original two. It also allows the order information to be stored earlier, perhaps to be available for earlier analysis in the required system.

Logicalisation simplifies the current physical system, removing duplication of processing and data and provides a sound basis for building required functionality and data as a progression to a new system.

Data-flow diagrams are a powerful tool, but only one of the techniques in the analyst's armoury. The next major technique which we shall cover is Entity Modelling

3.3 Entity Modelling

3.3.1 What Is Entity Modelling?

Entity Modelling is a 'top-down' technique for analysing and understanding the data within the area of study. It begins from the global view of 'what groups of things do we need to keep information about within the project area or within the organisation as a whole?' and works its way down to the detail of entities and their interrelationships.

Entity Modelling is not as precise a technique as Normalisation (Relational Data Analysis), which is described later, but is much easier to use to get the main groupings of data. It is surprisingly easy to discuss with senior management, provided that you de-jargonise it first. If you substitute the word 'thing' for entity, and ask 'What things do you keep information about?', you will get answers like: customers, orders, invoices, products which are, in fact, the main entities of the project area.

Entity Modelling:

- provides a basis for physical data file or database design. Together with normalisation, it provides an unambiguous model of the data items, groupings and their inter-relationships;
- provides flexible, maintainable data structures: Data structures are generally less likely to change fundamentally than the processing which uses them. A business may change the way in which it processes orders, but it will seldom change (willingly) the fact that it receives orders or that it needs to keep specific details about them. The fundamental ways in which a business views its data is relatively stable. Entity modelling seeks to collect and diagram the detailed structure of the data. By modelling the data and allowing file design to be driven by the structure of the data rather than by the processes, we build a database which is aligned to the business needs and flexible to change because it avoids building dependencies between data groupings which are contrary to the business view of the data;
- avoids data redundancy: the entity model expresses the data as logical (conceptual) groupings of data items, with their inter-relationships. It removes redundancy of data by ensuring that each data item (attribute) is stored with the entity to which it truly belongs;
- promotes better understanding of the business area: the performance of entity modelling prompts the analyst to ask searching questions about the nature of the data and thus understand the business area better. The Entity Model can also be a useful communication tool during discussions with users.

3.3.2 Entity Modelling Symbols and Terminology

ERM ERD

Entity Modelling, also called Entity Relationship Modelling or Entity Relationship Diagramming is a graphical technique for data analysis. The symbols used in the Entity Model, or Entity Relationship Diagram (ERD) vary from one structured method to another. Here we have opted for the widely used hard-cornered boxes, with lines and 'crow's-feet' to illustrate relationships and their degree. Where a different symbol set is used by a particular method, this is explained in the chapter concerned.

Entity Modelling is defined in terms of:

- entity type;
- entity occurrence;
- attribute;
- value;
- relationship.

3.3.2.1 Entity Type

An **entity type** (usually shortened to just **entity**) is a thing of importance to the business, about which it holds, or anticipates the need in future to hold, information.

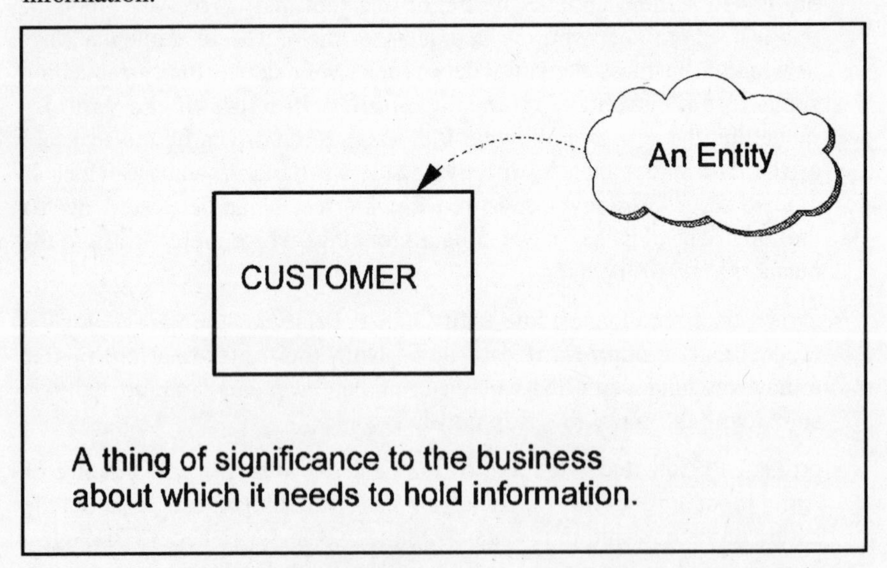

Figure 3.12 An Entity Type (Entity)

The entity types which can be identified depend on the area of study, but may be such things as:

- CUSTOMER;
- ORDER;
- INVOICE;
- JOB TYPE;
- GRADE.

An entity is shown in a box, as in Figure 3.12. By convention the name is usually shown in capital letters and always in the singular.

3.3.2.2 Entity Occurrence

An **entity occurrence** is one instance of an entity type. If the entity type is thought of as a file of information (e.g. customer file), then an entity occurrence is one customer, and all the information related to that customer (name, number, address etc.)

3.3.2.3 Attribute

An **attribute** is a data item holding information about an entity. In the entity CUSTOMER, customer number may be an attribute. Other examples are customer name, customer address. An attribute is any detail which serves to:

- identify;
- describe;
- qualify;
- classify;
- quantify;

an entity.

An attribute will have a specific, not necessarily unique, **value** for an individual occurrence of an entity. For example, the attribute *customer last name* may have a value such as Sartre or Tolstoy or Smith or Brown. The attribute customer number may have the value 112233 for a particular customer, which may be unique and thus sufficient to act as an identifier for that particular entity occurrence.

3.3.2.4 Relationship

Entities are logically (conceptually) separate groupings of data. However, they may have **relationships** with each other. The entity CUSTOMER holds information specifically about customers: the entity ORDER holds information about the sales orders received. However, the relationship of specific customers to specific orders has to be investigated and recorded. For example:

- A customer may place many orders;
- An order is specific to one customer.

Relationships may be:

- one to one;
- one to many;
- many to many.

These degrees of relationship are represented as illustrated in Figure 3.13. The degree of a relationship is often referred to as its *cardinality*.

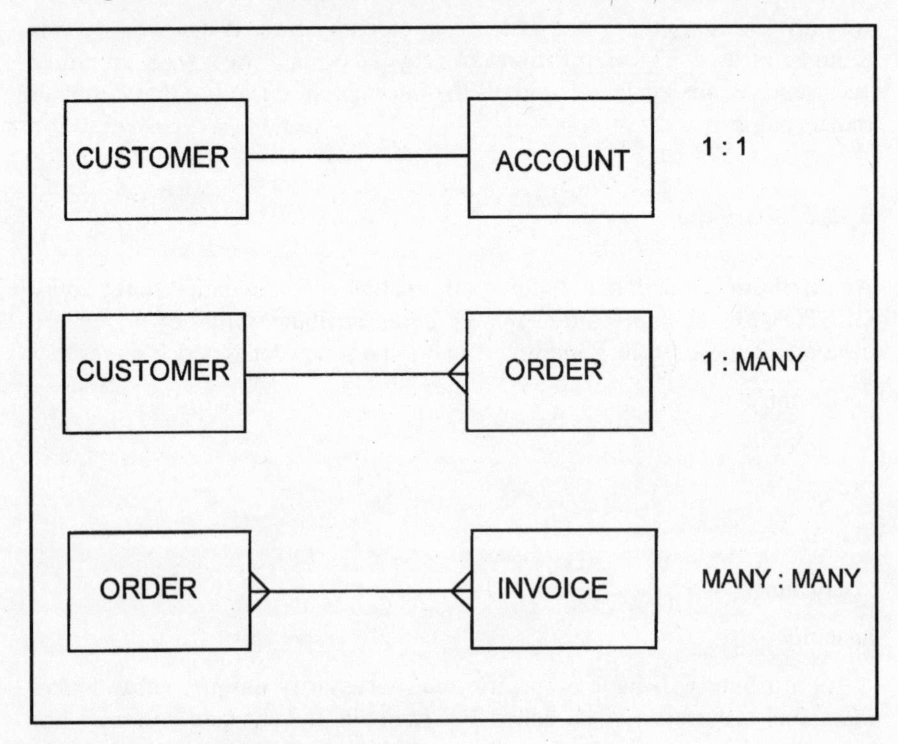

Figure 3.13 Degrees of Relationship

3.3.3 Guidelines for Entity Modelling

Entity Modelling is done by performing the following steps. Throughout the process, the users' involvement is sought in determining the entities and relationships. The major steps in the process are:

- identify initial entities;
- identify direct relationships;
- construct a diagram;
- determine degrees of relationship;
- add refinements;
- rationalise the model;
- validate the entity model against DFDs.

3.3.3.1 Identify Initial Entities

Initially, the user should be asked to state the major things about which data is held in the project area. The analyst should not seek to find all entities at this stage. The process of entity modelling itself will elicit others as it progresses. An initial five or six is usually sufficient to start. If the initial number exceeds about fifteen, the process becomes more difficult to manage and the analyst should partition these into areas and work on a small number at a time. Guidelines for selecting entities are given below:

- entities should be expressed as a singular noun;
- there must be more than one occurrence of the entity, otherwise it is not an entity in the context of the study;
- existing files are a good guide to entities, since most of the data needed is probably already being stored. However, an existing file may not be one entity but a collection of entities, or one entity may be split between several existing files;
- a good test of an entity is to try and identify a key for it, i.e. an attribute (or group of attributes) which uniquely identifies a single entity occurrence.

3.3.3.2 Identify Direct Relationships

A **direct** relationship exists between two entities when it is possible to describe their relationship to each other in business terms *without* having to involve any other entities to make the description complete.

3.3.3.3 Construct a Diagram

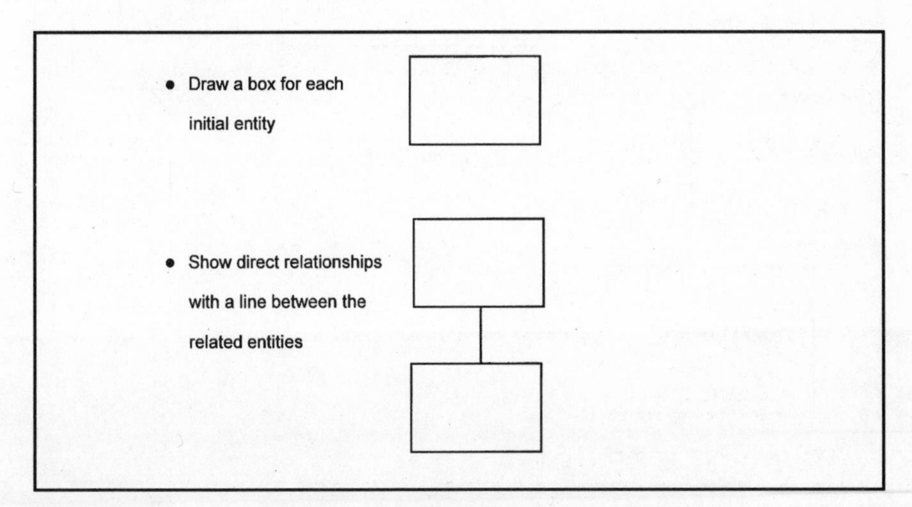

Figure 3.14 An Initial Diagram (degrees of relationship not shown)

The initial entity diagram just links entities where a direct relationship has been identified (Figure 3.14) The degree of the relationship is considered later. If there is doubt about the directness of some of the relationships, this will not have a disastrous effect provided that the entity model is checked and validated at a later stage, as suggested in the steps.

3.3.3.4 Determine Degrees of Relationship

For each direct relationship marked on the initial diagram, the question must be asked from **both ends** of the relationship:

> 'For *one* occurrence of this entity, *how many* potential occurrences are there of the related entity?'

The resulting relationship may be one to one, one to many or many to many:

- a one-to-one relationship usually implies that the same entity has been discovered twice, perhaps at different stages in its life (e.g. ORDER and DESPATCH may be one to one if every order results in one and only one despatch);
- a many-to-many relationship usually implies the omission of an intersection entity.

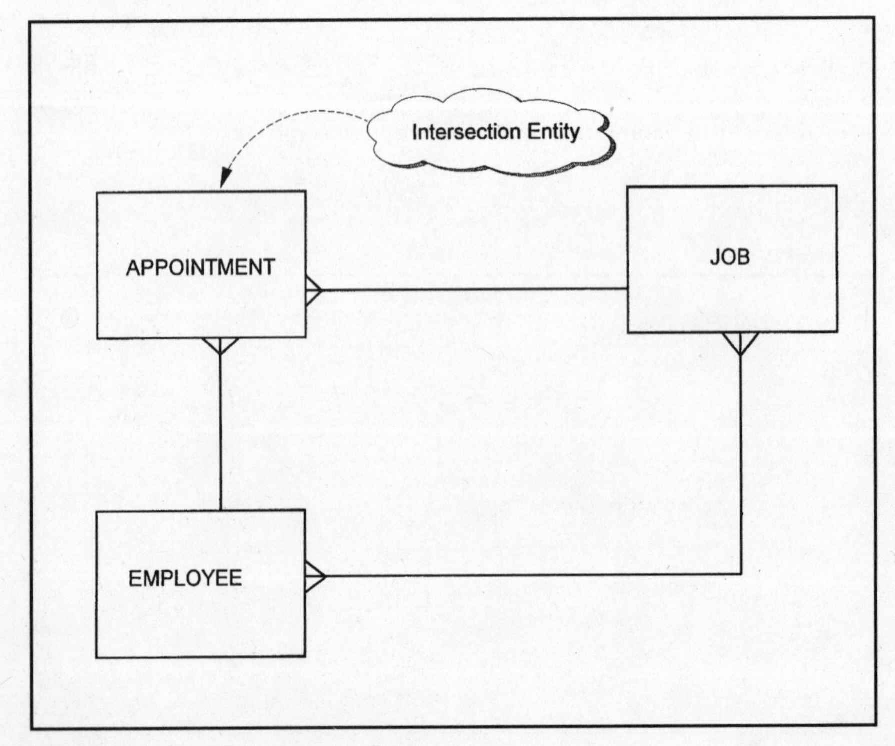

Figure 3.15 An Intersection Entity

In Figure 3.15 the entities EMPLOYEE and JOB are found to have a many-to-many relationship. An EMPLOYEE may be recorded as having many JOBS. Also, a particular JOB may be recorded as having been held by many employees (the business requires to hold this historical information). Thus there are two one-to-many links contained in the many-to-many link, and a new entity which links a particular employee with a particular job must be found. This extra entity is APPOINTMENT, which contains the dates of starting and ending a particular job for a particular employee. It will be found that wherever a many-to-many relationship occurs, an intersection entity can be used to resolve it. This intersection entity will always have a one-to-many relationship with the original entities, with the crow's feet attached to the intersection entity.

Whenever an intersection entity is inserted, the relationship of this new entity to all other entities in the model must be considered. This process may uncover the existence of yet more entities.

Only one-to-many relationships usually remain in the finished Entity Model.

3.3.3.5 Add Refinements

Optional Relationships

Relationships which have been identified between two entities may not necessarily always be present between the two entities. A CUSTOMER may send our company many sales orders, but it is possible to imagine a time (perhaps when the customer is new) when there would be no recorded orders for the customer. This situation can be represented showing the relationship as optional. The convention for this is a dotted line, as shown in Figure 3.16.

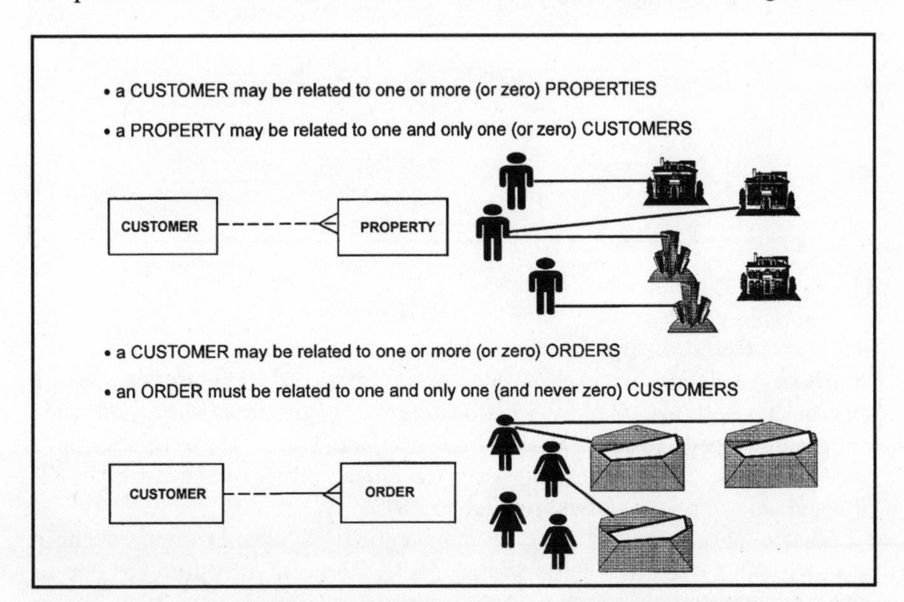

Figure 3.16 Optional Relationships

The relationship may be fully optional. In Figure 3.16 a customer who places sales orders with our company may also be recorded as renting a property from us (but does not necessarily have to be). On the other hand, we have properties recorded that are not rented out to customers.

Otherwise, the occurrence of one entity may be dependent on the existence of an occurrence of the other: a sales order would not be accepted if there were not a customer to attach it to.

Exclusive Relationships

Exclusive relationships occur when one occurrence of entity 'A' can be linked to one or more occurrences of entity 'B' *or* one or more occurrences of entity 'C', but not to both entity 'B' and entity 'C' simultaneously. In Figure 3.17, exclusivity is shown on the Entity Model with an arc across the affected relationships.

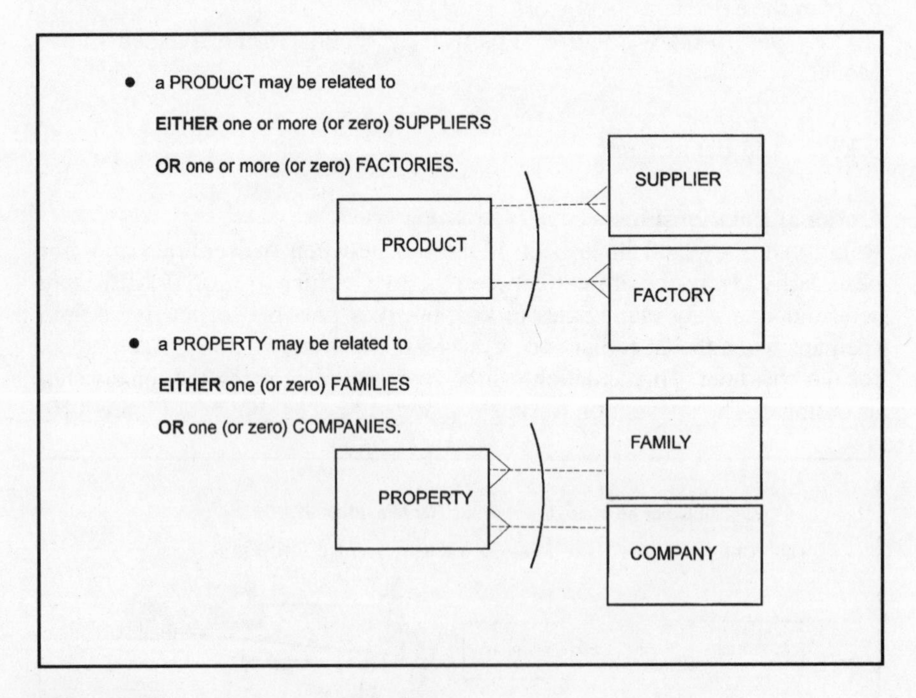

Figure 3.17 Exclusive Relationships

Recursive Relationships

A recursive relationship exists when one occurrence of an entity can have a relationship with one or more other occurrences of the *same entity* and when information needs to be kept about these relationships. These relationships are represented by drawing a pig's ear (technical term!) on the corner of the affected entity box (as shown in Figure 3.18).

These relationships can occur for many entities. Typical examples occur in the entity PRODUCT, where one product may be a substitute for one or many other products, or where one product is a component part of one or many other products.

Substitution may occur when, if the customer asks for a box of 100 envelopes with product number 22222, we could supply two of product number 11111 which represents the same envelopes in packs of 50. A product may have many potential substitutes, and may in turn be a substitute for many other products. We need to represent these relationships in the entity model so that in the eventual physical system we build the appropriate links between such products.

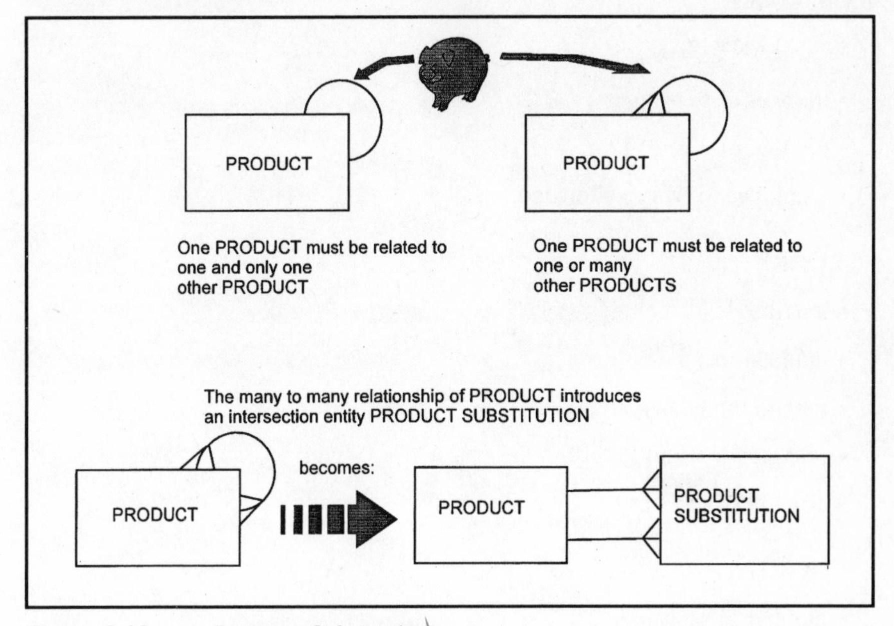

Figure 3.18 Recursive Relationships

Another common example of the recursive relationship is what is commonly known as a 'bill of materials' where one product is made up of many other component parts. Thus a car engine may be identified by one part number, but the valves, piston rings, etc. etc. have separate part numbers, as well as being included in the engine. For example, the piston ring (part number 12345) is part of an engine (part number 56789). The representation of this is still achieved by using the 'pig's ear' symbols.

Entity Subtypes and Supertypes

The easiest way to explain the concept of entity sub-types and super-types is to take an example. The entity VEHICLE has been identified in our project area. It has attributes:

- chassis number (identifier);
- year of registration;
- date of purchase;
- site at which based;
- date of last service;

- insurance group;
- manufacturer;
- model number.

We have three types of vehicle: coaches, lorries and company cars. We need to keep the above set of basic attributes about these, but in addition we need the following:

COMPANY CAR

- number of passengers;
- number of wheels (3 or 4 (!));
- employee to whom allocated.

LORRY

- length;
- unladen weight;
- maximum laden weight;
- refrigeration unit (Y/N);
- articulated (Y/N).

COACH

- number of seats;
- type identifier (5-star luxury coach to 1-star basic).

An entity has sub-types when there are characteristics of an entity which refer to the whole group, but other attributes which refer to subsets of the group.

Naming of Relationships

As an aid to understanding the data it is helpful to name the relationships between entities (Figure 3.19). This is occasionally and quite wrongly treated as a trivial task which is delegated to the most junior analyst. However, if done properly, the act of naming the relationships can uncover the existence of many possible relationships between two entities (Figure 3.20).

The relationship of DRIVER to CAR could quite legitimately be any of those shown in Figure 3.20, depending upon what the business is trying to keep track of. A driver may drive many cars over a period of time (although in the diagram the fact that the driver is recorded as being related to only one car confirms that the users are only interested in keeping a record of the car he/she is currently driving). Additionally, the business may be more interested in the number of crashes a driver has than the number of journeys made.

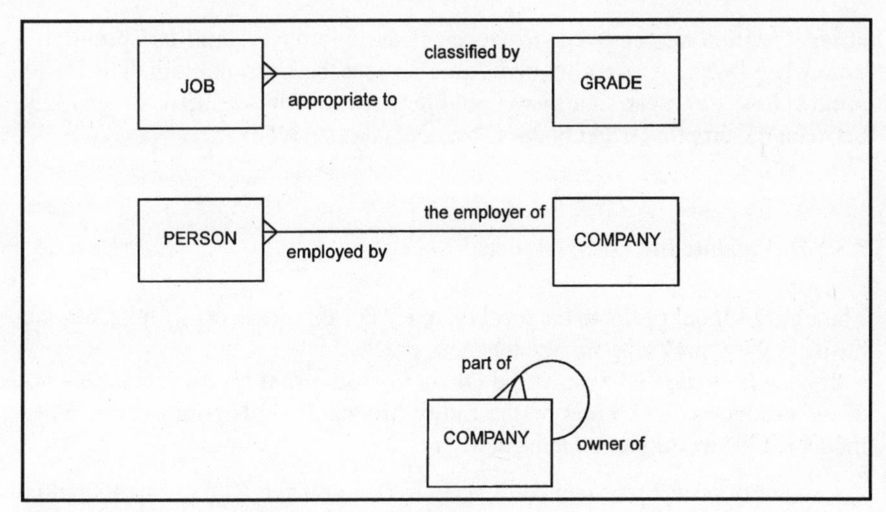

Figure 3.19 The Naming of Relationships

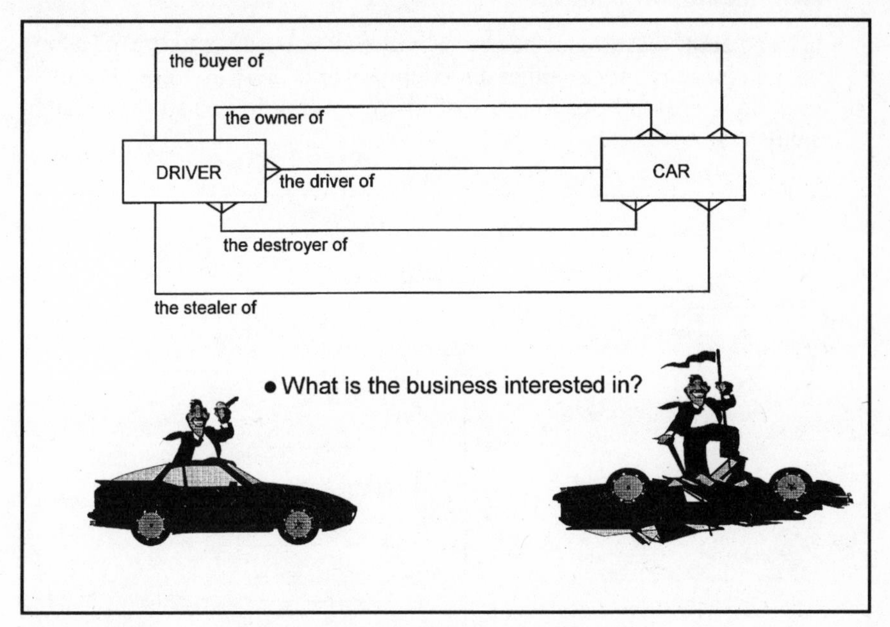

Figure 3.20 Multiple Relationships Between Two Entities

3.3.3.6 Rationalise the Model

Having followed the steps above and determined entities and relationships and added intersection entities, there may be relationships which are not needed. These **redundant relationships** should be removed once it has been established that they are adding no necessary information. In fact, to leave them in could actually give wrong information. In the example, if an absence is only relevant in association with a project, the link between employee and absence should be removed. However, if an employee could be absent at a time when he/she was not involved in a project the link should be left in. To remove it would mean that there could be situations where there was no link between the employee and his/her absence! (Figure 3.21)

3.3.3.7 Validate the Entity Model

The Entity Model needs to be checked to ensure that it is correct, matches the business view, and supports the business processing.

It must be validated against the DFDs to ensure that all data requirements of the processes can be met by the Entity Model. For this purpose, the entity model can be treated as a database where:

- an entity occurrence can be directly retrieved if the value of its identifier is known (for example, we can directly access a specific customer if we know the customer number);
- once an entity occurrence has been located, it is possible to 'travel' along the relationship lines in either direction (towards or away from the crow's foot) to access all occurrences of other entities linked to the original entity;

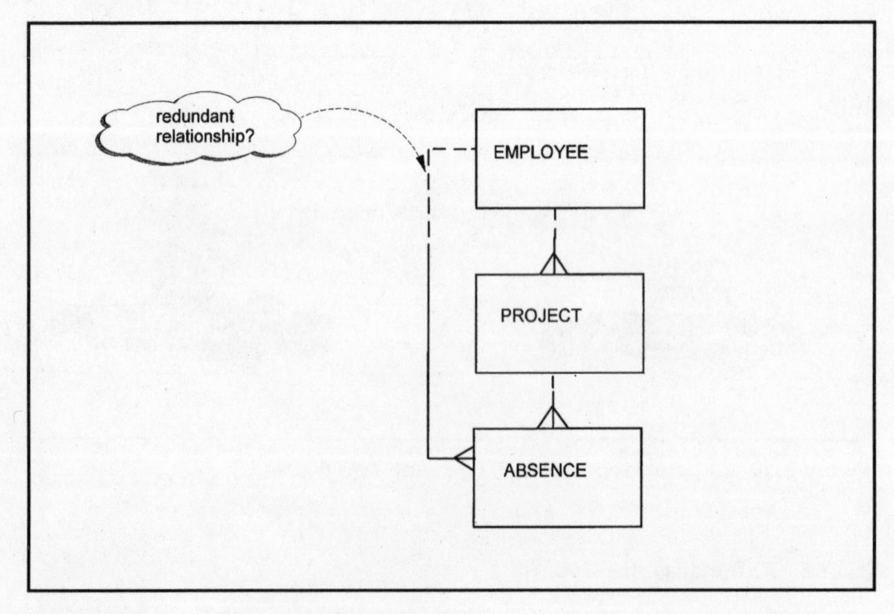

Figure 3.21 Redundant Relationship

- with the users: although they will have been involved throughout the development of the Entity Model, the final model should be checked with them. Entity Models are not an easy concept for users to understand since they do not mirror the physical way in which data is held. However, the analyst can 'read' the relationships and entity contents to check these.

The performance of entity modelling, together with the complementary technique of Relational Data Analysis (Normalisation) gives the analyst a deeper, more accurate understanding of the data and, during validation, of the processes which use it. It can thus highlight omissions and improve the quality of the analysis product. We shall consider the normalisation technique next.

3.4 Relational Data Analysis (Normalisation, Third Normal Form Analysis)

3.4.1 What Is Relational Data Analysis?

Relational Data Analysis (RDA) is also alternatively called Normalisation or Third (Fourth and Fifth) Normal Form Analysis. It has been used by data analysts for many years as a means to identifying the meaning and inter-relationships of data items, as a precursor to building file and database structures. RDA has its origins in mathematical set theory, through the work of Dr Edgar Codd in the early 1970s. Its step-by step approach makes it easy to perform. It raises questions about the data items within the area of study which improve the analyst's understanding of both the data and the business processes which use it. It allows the analyst to:

- gain detailed knowledge of the data. RDA prompts questions about the data, and thus helps to capture the user's detailed understanding of the data;
- validate the 'top-down' Entity Model. Entity modelling will give a good overview of the data within the system, but entity analysis has no mechanism for bringing out the finer detail such as data items within each entity. RDA provides this detail. Some entities which are not easy to determine from the top-down approach become clear from RDA;
- identify data interdependencies. It is important for ease of maintenance of data structures that interrelationships between entities and data items are fully understood. RDA highlights these interdependencies;
- remove unnecessary/redundant data. Redundant data creates problems of maintenance. The logical structure produced by RDA holds data items once only unless they are keys (identifiers);
- develop a basis for physical data structures. The data model which results from RDA provides a relational structure of tables of data, which is a sound basis not only for the specification of a relational database, but also for other databases (hierarchical, network) or for conventional file structures.

3.4.2 RDA Terminology

Certain terminology which RDA uses relates back to its origins in mathematical set theory, for example in the use of words like 'relation', 'tuple', 'domain'. These, and other terminology used in RDA, are explained below.

3.4.2.1 Relation

A relation is a table of data, with the following properties:

- no two rows are the same;
- the order of rows and columns is not significant;
- each column is uniquely named, as an attribute;
- all attributes are atomic (not further divisible and do not contain other attributes).

Figure 3.22 shows a relation.

A *normalised relation* is derived by performing RDA and is, in practical terms, the same thing as an entity (as derived from entity modelling);

The nearest *physical* equivalent of a relation is a *file* of many records all of one type (see Figure 3.23).

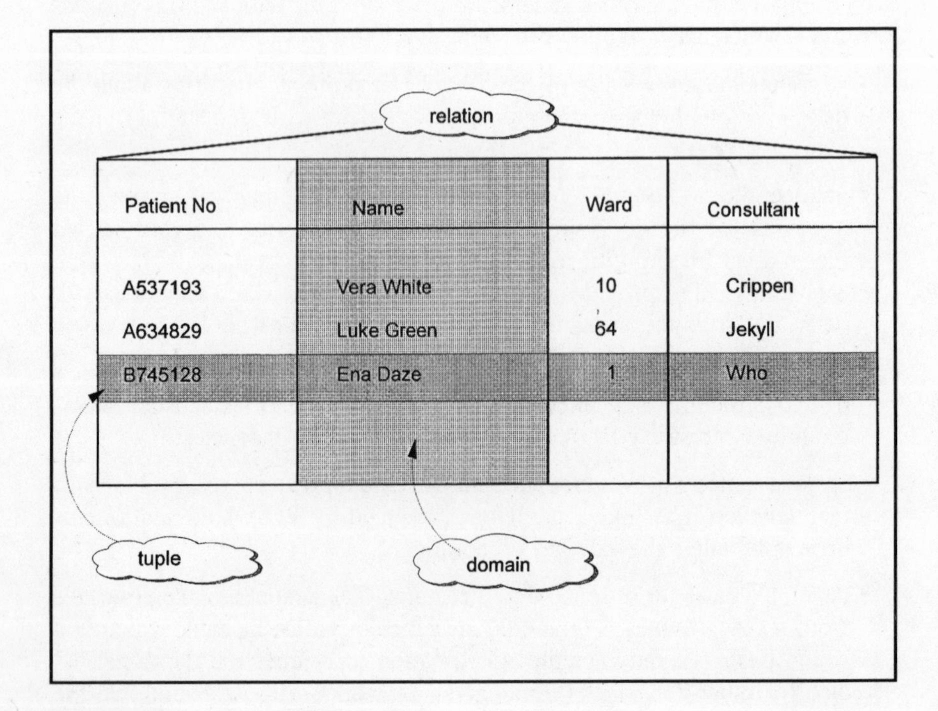

Figure 3.22 A Relation

3.4.2.2 Tuple and Domain

A **tuple** is a *row* within a table, and is similar to a record within a physical file.

A **domain** equates to a *column* within the table. The physical equivalent is a field which occurs within each record of a file. The term *domain* implies the whole population of allowable values within the column.

RELATION

Description	Physical Storage Equivalent	Relational Term
Table	File	Relation
Row	Record	Tuple
Column	Field	Domain

Figure 3.23 RDA Terminology

3.4.2.3 Types of Keys

One feature of a relation is that it has a key *(identifier)*. A key is a single attribute, or a set of attributes, which uniquely identifies a particular row within the relation. The key chosen to identify individual rows is called the **primary** (or prime) key.

A **simple key** is a key consisting of only one attribute:

e.g. <u>Customer Number</u>;

A **compound key** is a key consisting of two or more attributes:

e.g. <u>Order Number</u>
<u>Part Number</u>

A **composite key** is a key consisting of two or more attributes, but where one of the attributes is meaningful only in relationship to the other:

e.g. <u>Project Number</u>
<u>Task Number</u>

where Task Number has values of 1, 2, 3, etc. within a particular project.

The key of a relation is usually identified by underlining the attribute, or attributes. This convention will be used in the example which comes later.

A further type of key which will be introduced during the process of RDA is called a **foreign key**. A foreign key occurs when the identifier of one relation also appears as an ordinary (non-key) attribute of another relation, in order to show a link between the two relations. This will be identified by an asterisk (*).

3.4.3 Guidelines for Relational Data Analysis

RDA is done by performing the following steps on a data source. That source may be, for example:

- the data items within a data flow;
- an input document;
- a screen of data;
- an output report.

The steps for normalisation are:

- list the data items and choose an identifier;
- remove repeating groups to form a separate relation;
- remove part-key dependencies to a separate relation;
- remove inter-item and inter-key dependencies;
- rationalise and check the results.

These steps are best explained with reference to an example. The example takes as its data source a history card which is kept for hospital patients' blood samples (Figure 3.24).

Haematology: Blood test history

Patient Name	Ena Daze	Hospital Number	NW127
Date of Birth	12/12/00	Hospital Name	Orchard Cottage Hospital
Ward	1	Consultant Name & No	Dr. Who: BBC001
Clinical Details	Phobla methodologica	Patient Number	B745128

Test Code	Test Date	Test Description	Date Reported
A	11/2/99	Full Count	12/2/99
A	12/2/99	Full Count	13/2/99
X	13/2/99	Cross Match	13/2/99

Figure 3.24 Source Document for RDA

3.4.3.1 List Data Items and Choose a Key

All data items on the source document are listed, and carefully named, particularly to distinguish between different numbers, and different dates. Any groups of attributes which repeat within the source are marked (here with a solid line, although a bracket may be used) (see Figure 3.25). The list of attributes is now in **unnormalised form**.

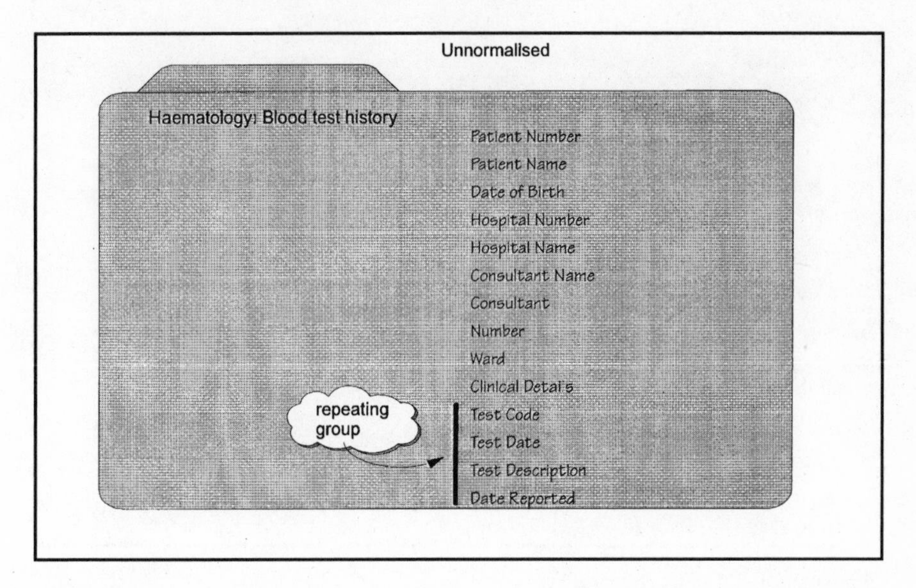

Figure 3.25: Unnormalised Data Items.

A key is chosen for the relation. Patient number is the attribute which identifies a single occurrence of the Haematology Blood Test History card, since the users confirm that there is one such card for each patient. Thus, this is a good key to start with. It should not theoretically matter which field is chosen as a key at this point, but it makes the process easier to look for a key to identify one occurrence of the source document.

3.4.3.2 Remove Repeating Groups

The removal of repeating groups to a separate relation puts the list of attributes into **First Normal Form** (see Figure 3.26).

The attributes:

Test Code;

Test Date;

Test Description;

Date Reported.

are a repeating group.

A key is chosen for this new relation. Test Code and Test Date provide a unique key for each occurrence of the repeating group, since a particular type of test is only performed once in a day for a particular patient.

In order to maintain the relationship of the attributes in the new relation to the original relation, Patient Number is also included as part of the key of the new relation.

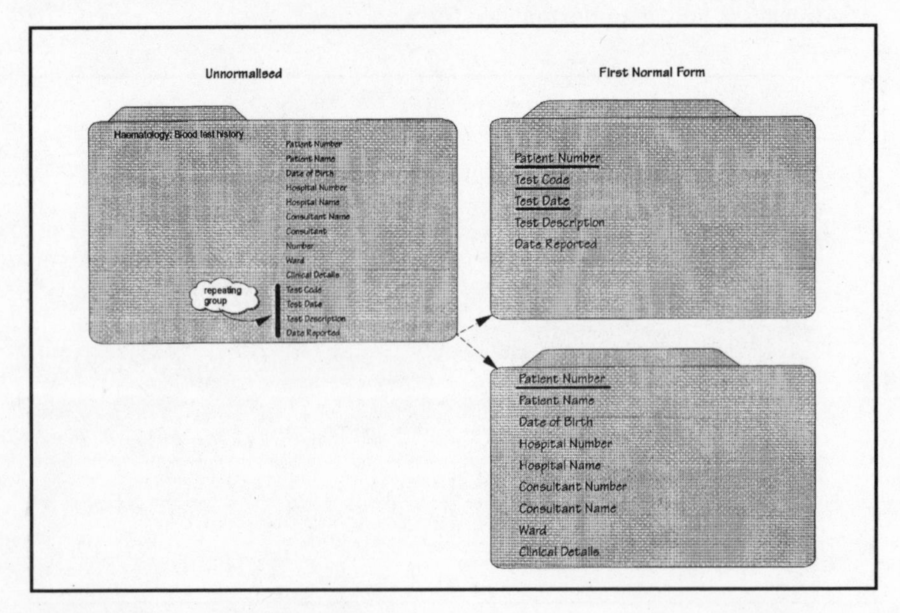

Figure 3.26 First Normal Form

3.4.3.3 Remove Part-Key Dependencies

In order to put the data into **Second Normal Form**, we must examine any relations with a compound key and check whether each attribute within the relation is dependent on part of the key or the whole key. In the example, Test Description is dependent on Test Code, i.e. Test Description remains the same for a particular value of Test Code, regardless of the value of Patient Number and Test Date. Test Code and Test Description form a new relation with Test Code as the key. Figure 3.27 shows Second Normal Form.

3.4.3.4 Remove Inter-Key and Inter-Item Dependencies

This step will convert the relations to **Third Normal Form**. It involves inspecting the attributes of all relations now in Second Normal Form, to establish whether any one attribute is dependent on any other within the relation, irrespective of the key (the inter-key dependency only applies to relations with three or more attributes in the key and identifies dependencies between subsets of the key).

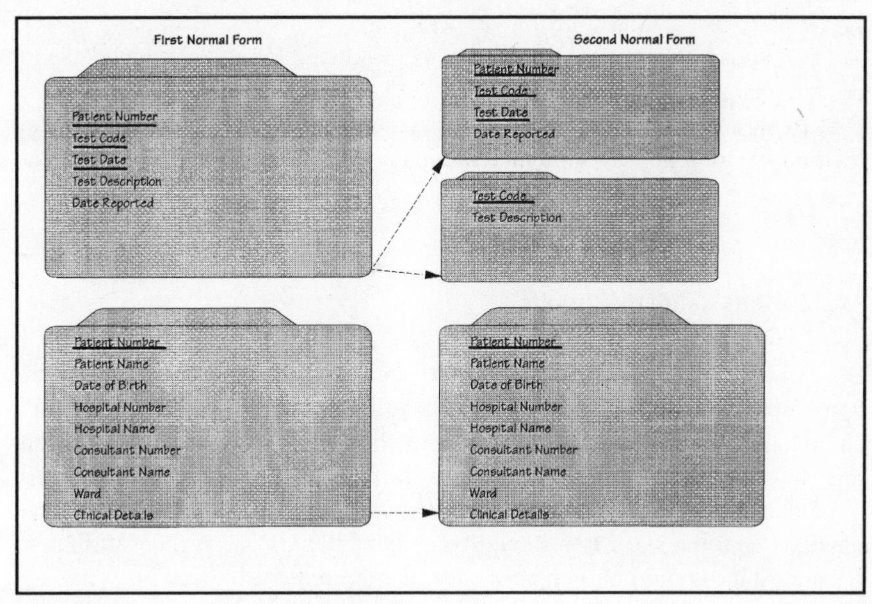

Figure 3.27 Second Normal Form

Inter-attribute dependency in the example occurs between:

Hospital Number *and* Hospital Name:

i.e. given the Hospital Number, we can identify the Hospital Name without reference to Patient Number;

and:

Consultant Number *and* Consultant Name.

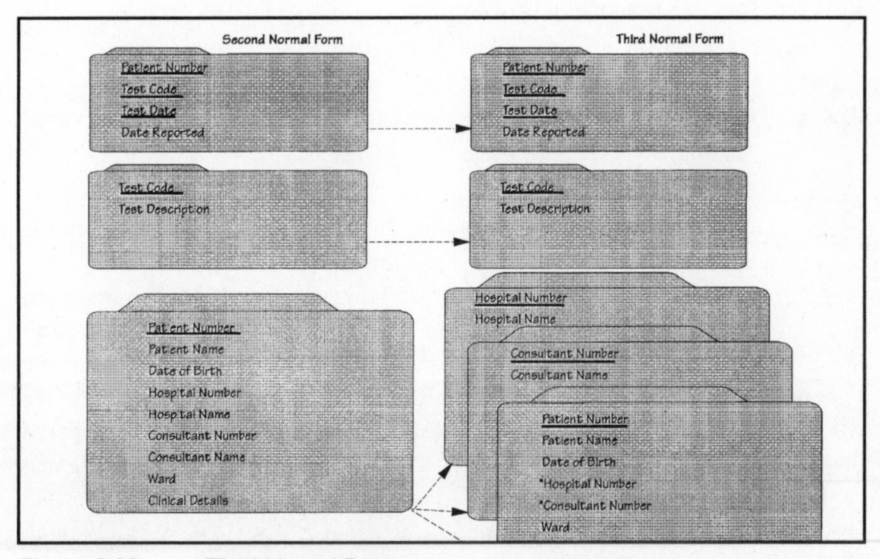

Figure 3.28 Third Normal Form

These pairs of attributes are moved to separate relations. Hospital Number and Consultant Number are also left in the original relation as foreign keys in order to maintain the link.

A foreign key is an attribute which is an ordinary non-key attribute in one relation but is a key in another relation. Figure 3.28 shows Third Normal Form.

3.4.3.5 Rationalise the Results

Data from as many sources as practical within the area of study should be normalised as described above, to give as full a picture as possible of the data and relationships. Individual structured methods have their own guidance on this. In doing this, the same relations (entities) are likely to be discovered more than once, but perhaps with different attributes. Where this is the case, relations with the same key should be combined, along with their attributes. Each attribute should only appear once, and be present in only one relation. The only exceptions to this are attributes which are keys or foreign keys.

3.4.3.6 Other Normal Forms

Other normal forms, specifically Fourth, Fifth and Boyce Codd normal forms, have been derived and for further explanation of these the reader is directed towards Chris Date's book (Date 1990). However, Third Normal Form, plus a re-check after application of the rules that each data item (attribute) will occur once and only once for a given value of the key, is sufficient for most practical purposes.

3.4.4 Drawing the Data Model

The final set of relations can be represented diagrammatically in an Entity Model, which can then be compared with the Entity Model derived from the top-down approach.

A simplified version of the rules for deriving the Entity Model are given below. Further refinements to these rules can be made for cases where there are composite keys, or to remove redundant relationships where there are keys comprising three, four or more attributes. However, the rules below are sufficient for most cases to obtain an Entity Model which can be compared with that derived from Entity Analysis. The SSADM (V3) Reference Manual is the best source for a more detailed set of rules for this transformation. (Rigorous application of this technique was a part of SSADM (V3), although not Version 4.)

The rules are applied as follows, and produce the diagram in Figure 3.29:

- Draw a box for each relation and give it an appropriate name. This gives entities PATIENT, HOSPITAL, CONSULTANT, PATIENT TEST and TEST TYPE. Figure 3.29 illustrates these, with keys shown in the bottom stripe of the entity boxes.
- Draw another box for any part of a compound key which does not exist as a simple key. In this case the compound key is Patient Number: Test Code: Test Date. Since Patient Number and Test Code already have their own entity boxes, these do not generate additional boxes. Test Date is not meaningful as a key in its own right (composite part of the key).
- Draw a one-to-many relationship between any box with a compound key and the boxes which have each of the attributes of the compound key as a simple key. The crow's foot should be at the end with the compound key. This gives relationships between PATIENT and PATIENT TEST; TEST TYPE and PATIENT TEST.
- Draw a relationship between any box which has a foreign key and the box for which the foreign key attribute is the key. The crow's foot should be at the foreign-key end. This gives relationships between HOSPITAL and PATIENT; CONSULTANT and PATIENT.

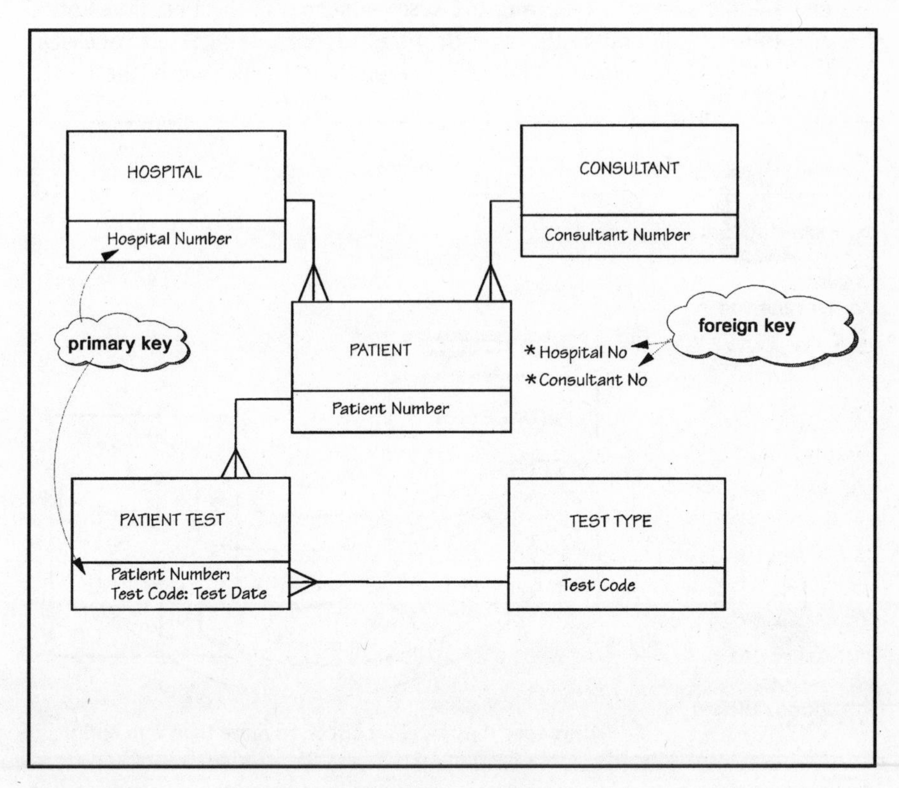

Figure 3.29 The Data Model, Drawn from RDA

3.4.5 Comparison of Entity Models

Where differences occur between the Entity Model derived from top-down Entity Analysis and that derived from RDA, the analyst must check the areas of difference with the user to ensure that the final Entity Model, which may be a compromise between the two approaches, is the one which best reflects the business needs.

RDA is an essential stage in analysis, prompting a complete understanding of the data in the area of study, which in turn leads to more flexible and maintainable data structures, modelled on the results of RDA.

3.5 The Function/Entity Matrix

The Function/Entity Matrix is included here because, although it is not, of itself, a major technique in many methods, it underlies the dynamic investigation and definition of data in all. Many different diagram-based techniques, such as Entity Life Cycles, Entity Life Histories, Effect Correspondence Diagrams, Enquiry Access Paths, Process and Procedure Logic Diagrams, are based on information contained in a Function/Entity Matrix or its equivalent. The Event/Entity Matrix and Process Entity Matrix are just another way of expressing the same concepts as the Function/Entity Matrix. Figure 3.30 shows the way in which **events, triggers, processes, data** and **effects** are related. An *event* is something happening in the world

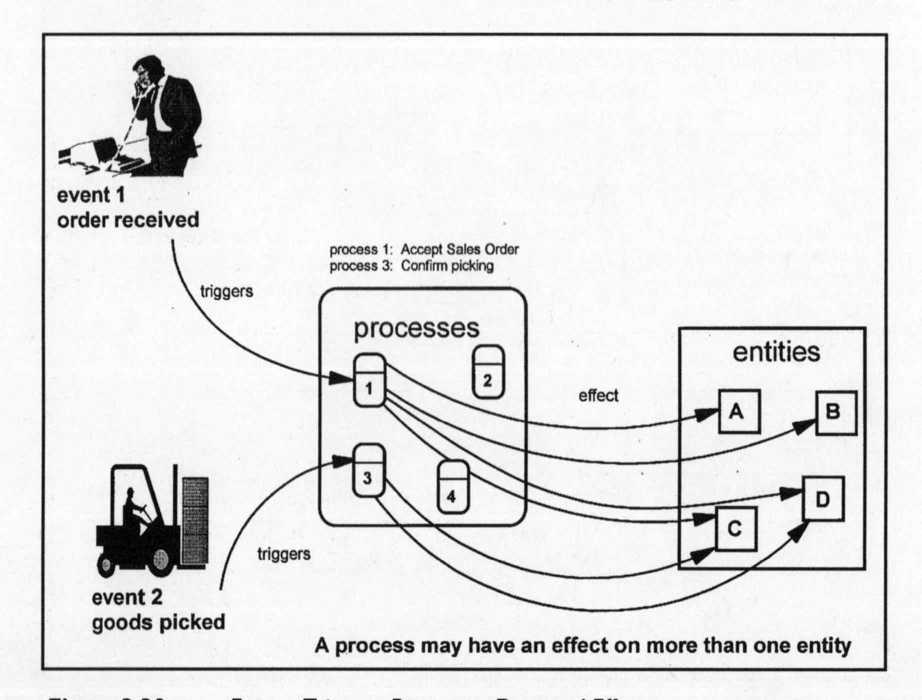

Figure 3.30 Events, Triggers, Processes, Data, and Effects

outside the system which prompts some kind of reaction from the system. An *event triggers* a *process* which may update, read, create or delete one or more entities. If the event causes the creation, updating or deletion of data, it is said to have an *effect* on the entity.

The Function/Entity Matrix is used to investigate the interaction between processing and data. It generally considers the logical processes and the logical data (entities) for the required system. The matrix is deceptively powerful. It contains, when complete, the full thinking behind the processing for the required system.

The Function/Entity Matrix is illustrated in Figure 3.31. The rows show the processes which affect an entity. This is the Entity Life History or Entity Life Cycle view (although Entity Life Histories would look at the event causing the process to be triggered rather than the process itself). The columns show the entities affected by a particular process. This is the Access Path of the process, or the view taken by the Effect Correspondence Diagram (SSADM), the Enquiry Access Path (SSADM) or the Process or Procedure Action Diagram (Information Engineering).

	FUNCTION					
ENTITY	Accept Sales Order	Confirm Picking	Confirm Despatch	Produce Invoice	Receive Goods In	Receive payment
Customer	C/R					
Order	C					
Order Item	C	U	U	U		U
Product	U					
•						

C - Create
R - Read
U - Update
D - Delete

Figure 3.31 Function/Entity Matrix (part-complete)

3.6 Summary

In this chapter, we have explained the major techniques which are the common core of the methods described in this text. Not all of the techniques are in every method. The individual chapters on particular methods give details of the techniques used in that method and the position of each technique within the framework of the method. They also detail other techniques specific to the individual methods.

4

Soft Systems Method/Multiview

4.1 Overview

Soft Systems Method (SSM) is derived from the work of Checkland (1981, 1990). The Soft Systems approach concentrates on the human aspects of a problem area, and the appreciation of the problem as a whole, rather than disintegrating it into small fragments. It seeks to improve human activity systems, which may or may not be associated with automated systems. Soft Systems is directed toward a better understanding of the problem area and because it is a general method for analysing complex systems it does not assume that a computer will form the solution to the perceived problems.

Multiview originated from the work of Wood-Harper, Antill and Avison (1985) and is further developed in Avison and Wood-Harper (1990). It is a method which includes the Soft Systems approach but is primarily concerned with solutions that involve computer-based information systems, i.e. a computer will form part of the solution to the problem area. In fact, Multiview has been used mostly for small-scale microcomputer applications involving software packages.

The inclusion of Soft Systems Method (SSM) as an integral part of the Multiview method (MV) affords the authors the opportunity to include some discussion of both within this chapter. To bring SSM, in isolation, into the context of the other methods in this text would clearly be unacceptable since it was never intended to encompass many of the techniques for Information System development which other methods employ. Thus, this chapter will consider the Multiview method and its interpretation of the Soft Systems approach.

Multiview is divided into five stages:

- Stage 1. Analysis of human activity (Soft Systems Method);
- Stage 2. Information analysis;
- Stage 3. Analysis and design of socio-technical aspects;
- Stage 4. Design of the human computer interface;
- Stage 5. Design of the technical aspects.

These stages encompass five views which together will lead to the development of a system which is complete in both human and technical terms. These stages are detailed below and are also shown in Figure 4.1. Any list of stages appears to imply some temporal sequence but this is not intended to be the case here. It is neither necessary to use them in the order given nor to complete one stage before another is started, hence permitting flexibility as appropriate.

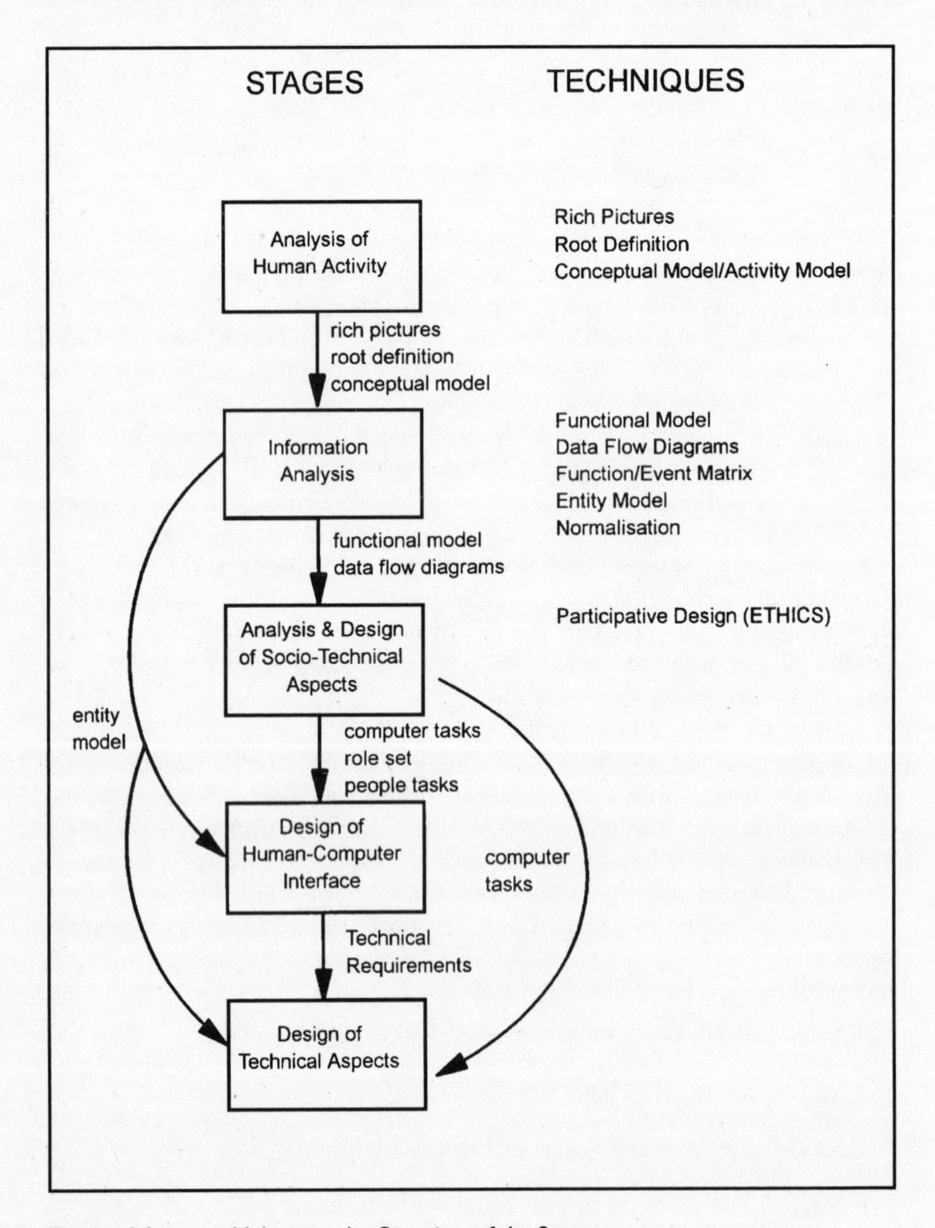

Figure 4.1 Multiview - An Overview of the Stages

Initially Multiview uses the Soft Systems approach to focus on the organisation and the problems within it, identifying what the system will be and what it will do (Requirements Analysis). This is followed by examination of the 'hard' aspects of the system, analysing the entities and functions of the new system.

The third stage is the participative role of the future users of the new system in the belief that if these users are involved with the system from its inception then the chances of successful implementation, operation and acceptance will be greater. Consideration then moves to the technical requirements of the user interface. This is determined by the background and experience of the users and the practical use to which the system is to be put in order to make their job easier. Finally in the last stage, the specific technical requirements of the new system are dealt with, taking account of such things as what type of computer is needed and whether a database to be used.

Rather than an organisation having a number of methods available to it upon which it can draw, depending on the situation surrounding the proposed new system, Multiview provides such a contingency approach within it. It is this flexible contingency within the method which sets MV apart from other methods such as SSADM and IE which operate in a tightly defined and more prescriptive manner. The tools and techniques present in Multiview are selected as needed as thought appropriate by the analyst.

As previously stated Multiview is divided into five stages, comprising five different views of the problem area into which the new system is to be integrated. Each stage has outputs which feed into other stages or which influence the final design directly.

4.2 Stage One – Analysis of Human Activity (Soft Systems)

The first stage of Multiview draws heavily upon the work of Checkland (1981, 1990) and his group's work upon SSM at Lancaster University in the UK.

Soft Systems is directed toward a better understanding of the problem area and because it is a general method for analysing complex systems it does not assume that a computer will form the solution to the perceived problems. Multiview, however, whilst giving attention to human activity systems, is directed toward computer-based systems as part of the solution to the problem area.

By direct contrast to other methods described in this text and for that matter subsequent stages within Multiview, Checkland suggests that Soft Systems does not suffer the failings which are fundamentally present in 'Hard' Methods. Hard methods analyse data and processes whilst minimising other components of the real world system such as the people who perform the actions within the system. The underlying rationale of these hard methods is based upon a scientific approach to problem-solving. They employ reductionist tactics ('divide and conquer') to break the problem area down into smaller manageable components. Now this is all very well but it does assume that in following the scientific method the combined properties of the sub-parts of the problem area are equal to the properties of the whole problem area. But how valid is this assumption? Soft Systems claims that this assumption forms a major barrier to truly understanding the complex problem situations in which human activity is involved. This barrier exists because some characteristics are lost from the analysis by virtue of the approach adopted by such hard methods.

Checkland developed the term **Human Activity Systems** from its somewhat casual usage in industrial engineering circles and defined it in terms of what activities are going on which together form a 'purposeful whole'. This purposeful whole or **holon** may be only one of a number of holons which when combined form a larger holon. As holons are combined to form larger holons, emergent properties appear which were not part of the sub-holons but which are nevertheless directed to the 'survival' of the organisation in which the holon resides (i.e the whole is greater than the sum of its parts). As an example of **emergent properties** consider the following:

> Our own muscular system is capable of contraction and heat generation whilst our skeletal system is able to impose shape and protect vital organs. When combined the emergent property of bodily movement appears, which is part of the characteristics of our Musculo-Skeletal System but not a part of either of the component systems.

To be considered a purposeful holon, the holon must consist of activities and infrastructure which, via communication and control (our nervous system in the above example), allows it to adapt and survive within its changing environment. This environment comprises the 'sub-holons' which combine to form the holon we are looking at, the sibling holons which interface and combine with it to form a 'supra-holon' and of course the supra-holon itself. To obtain a description of the purposeful holon, 'Human Activity System', it would be necessary for someone involved in, or affected by the Human Activity System to describe it. However, any one individual can only describe the system as they perceive it and therefore SSM consists of gathering a number of these perceptions or 'world views'.

The purpose of this first stage of Multiview, then, is to extract a composite soft systems 'world view' (sometimes known as Weltanschauung) of the organisation's problem area. This world view describes the requirements for the new system and influences subsequent stages of Multiview. For example, the Accounts System within a business may be required to maintain statutory company records, enabling the collection of revenue from the organisation's debtors as efficiently and quickly as possible. Another aspect of the world view may be the requirement to delay payments to its creditors for as long as possible.

The stage is divided into four phases:

- understanding the problem situation;
- building the systems models;
- comparison of the models to perceived reality;
- resolution of the comparison and implementation of the agreed solution to the problems.

4.2.1 Understanding the Problem Situation

A major output of this phase is the production of a **rich picture** of the problem area (see Figure 4.2). This picture is detailed in diagrammatic and pictorial form and is based upon objective observation by the problem-solver (analyst) together with subjective input from other interested parties, e.g. users, management, who will interface to the system. The rich picture will therefore hold both subjective and objective information. The development of the rich picture is intended as an aid to the analyst in understanding the problem area. In discussion with the interested parties, it may help to uncover and resolve misconceptions held by the analyst. The objective is to develop an overview of the problem area within the context of the environment in which the new system will operate. The information gathering necessary for its production is subject to all kinds of problems and its success depends largely on the interpersonal skills of the analyst, the experience of the analyst, the attitudes of the users, managers, etc. This problem is of course not unique to Soft Systems/Multiview!

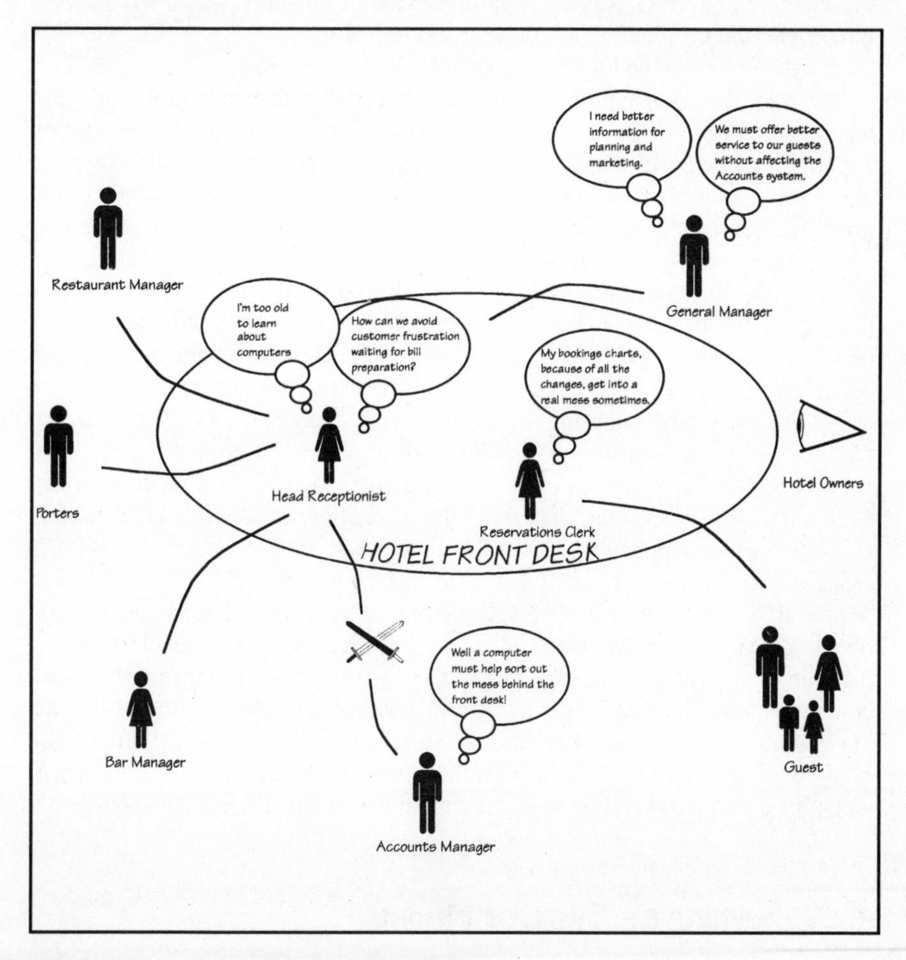

Figure 4.2 Rich Picture Midlinks Motel

In order to construct a rich picture the analyst looks initially for physical and logical structures within the organisation such as the number of departments and their physical location in relation to each other and perhaps the type of service or product which each provides. Overlaid upon this are the procedures which are being performed by the people. Immediate problems may simply arise through incompatible physical and procedural features of the 'problem situation'. Re-structuring the physical organisation may solve the problem. In order to visualise the problem area a rich picture (or perhaps several) is drawn to capture both **hard** and **soft** facts as perceived by the people involved (referred to as **actors**) and analyst. Hard facts are the practical aspects of the problem situation whilst soft facts are the emotional aspects displayed by the people involved in the problem situation, such things as 'will my job go if this new system is put in?' or 'my boss just wants the system to build his empire'. To develop the rich picture the analyst samples the views of the users of the present system to focus attention on key issues within the situation, the **primary tasks** that the organisation was created to fulfil. The Rich Picture presented in Figure 4.2 is for the Midlinks Motel case study, which will be used throughout the chapters which present individual methods. Details of the case study scenario are to be found in Appendix A.

The rich picture of Figure 4.2 shows one style for representing the problem area. It could equally have been drawn without graphical symbols, if thought by the analyst to be more effective that way. The form of representation, however, must remain simple and helpful in communication between the analyst and interested parties and must hold both *subjective* and *objective* information. It is of absolutely no use if the analyst's obscure and wonderful symbols are totally unintelligible to other analysts and to those users with whom communication is needed. A number of features exist in a rich picture as exemplified. Areas of conflict are represented here by crossed swords whilst 'thought bubbles' are used to represent the concerns of the actors within the problem area.

As an aid to discussion, the rich picture may be helpful in certain circumstances but care should be exercised. Some people may feel insulted if presented with such summary caricatures of their organisation or department. Indeed, careless inclusion of *soft* facts about some of the actors in the situation, facts very often supplied by one actor about another, may make the analyst's work difficult. One means of avoiding such problems may be to develop several rich pictures for use with specific actors such that if these were overlaid on top of each other the **full** rich picture would appear. The development of graphics applications together with the introduction of laptop/notebook computers will encourage their use in the development of rich pictures on site. However, use of clip art caricatures should be avoided, in favour of the computer-equivalent of 'stick-people', so that the actors in the problem situation are not offended by their personal representation in the rich picture.

4.2.2 Building the Systems Models

Using the rich picture the analyst focuses on problem themes, i.e. the specific causes of problems in the problem area. Having identified specific problem details, the analyst now proposes alternative relevant systems options which will have some effect upon the problem and from these selects, in discussion with the interested parties, the system which is likely to be most effective. Relevant Systems are systems which impact upon the problem area and the Selected Relevant System is the one believed to offer the best path to successful resolution of the problem situation.

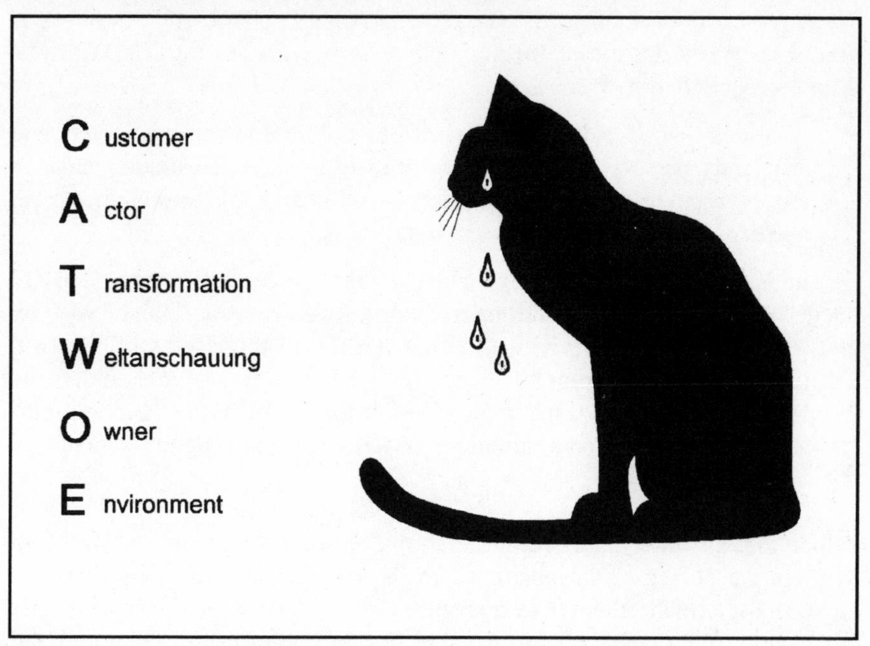

Figure 4.3 The Criteria for a Root Definition

From the alternative systems, the chosen 'view is used to develop a **root definition** which is a tightly defined and concise description of the Human Activity System. To ensure the development of a complete root definition a checklist of criteria, the **CATWOE** criteria, are used (see Figure 4.3).

Customer:	people affected by the system;
Actor:	people who perform the main activities of the system;
Transformation:	the processing of inputs to defined outputs;
Weltanschauung:	the 'world view' which leads to the root definition;
Owner:	the controlling body which authorises the system;
Environment:	the climate within which the system must operate.

The construction of a root definition focuses attention upon the problem situation in a specific way, such that a clear statement of the new system's objectives can be made. Discussion of the root definition reveals peoples' different perceptions of the current situation, which may not at first be obvious since they may, for example, use the same word to describe some aspect of the system but have quite different views, e.g. an order may mean a purchase order, a sales order, a piece of paper, a telephone call or maybe an instruction to do something! Indeed several possible root definitions may and probably will evolve through discussion each generated by different viewpoints of the problem situation. The analyst's problem is therefore to tease out of the alternatives the most suitable root definition, which may be one of, or a combination of, the alternatives. This selection process is iterative and is described further below. A root definition for Midlinks Motel's required system could be:

> A system owned and operated by Midlinks Motel to cost-effectively and efficiently perform front desk activities to make reservations, check in guests, prepare lists of requirements, add chargeable items, check out guests and reconcile cash for accounts.

The Root definition is used to construct the **conceptual model** (activity model) which will detail what the system will do (see Checkland 1984) by noting the required activities to fulfil the stated objectives of the root definition. The development of this model will show what ought to be happening to achieve the objectives stated in the root definition and will pinpoint where the real-world situation deviates from the logical conceptual model (see Figure 4.4).

A number of conceptual models are developed from the root definitions which encapsulate the different systems options which are available to address the objectives encapsulated in the root definition. These different options are termed 'alternative relevant systems'. These then prompt further discussion which culminates in the selection of a refined root definition and conceptual model which represents the chosen relevant system. The iterative impact of these discussions leads to a more fully formed root definition and conceptual model of the relevant system which is agreed to have the greatest impact upon the problem situation.

The root definition is examined and the *minimum list of verbs* which describe the main activities of the new system are identified and classified into groups of similar activities. This minimum list of verbs, derived from the root definition for Midlinks Motel, would be:

- make reservation;
- check in guests;
- prepare lists;
- add chargeable items;
- check out guests;
- reconcile cash;

Arrows are then used to join logically related activities, leading to the identification of sub-systems, and also the interdependencies between these activities.

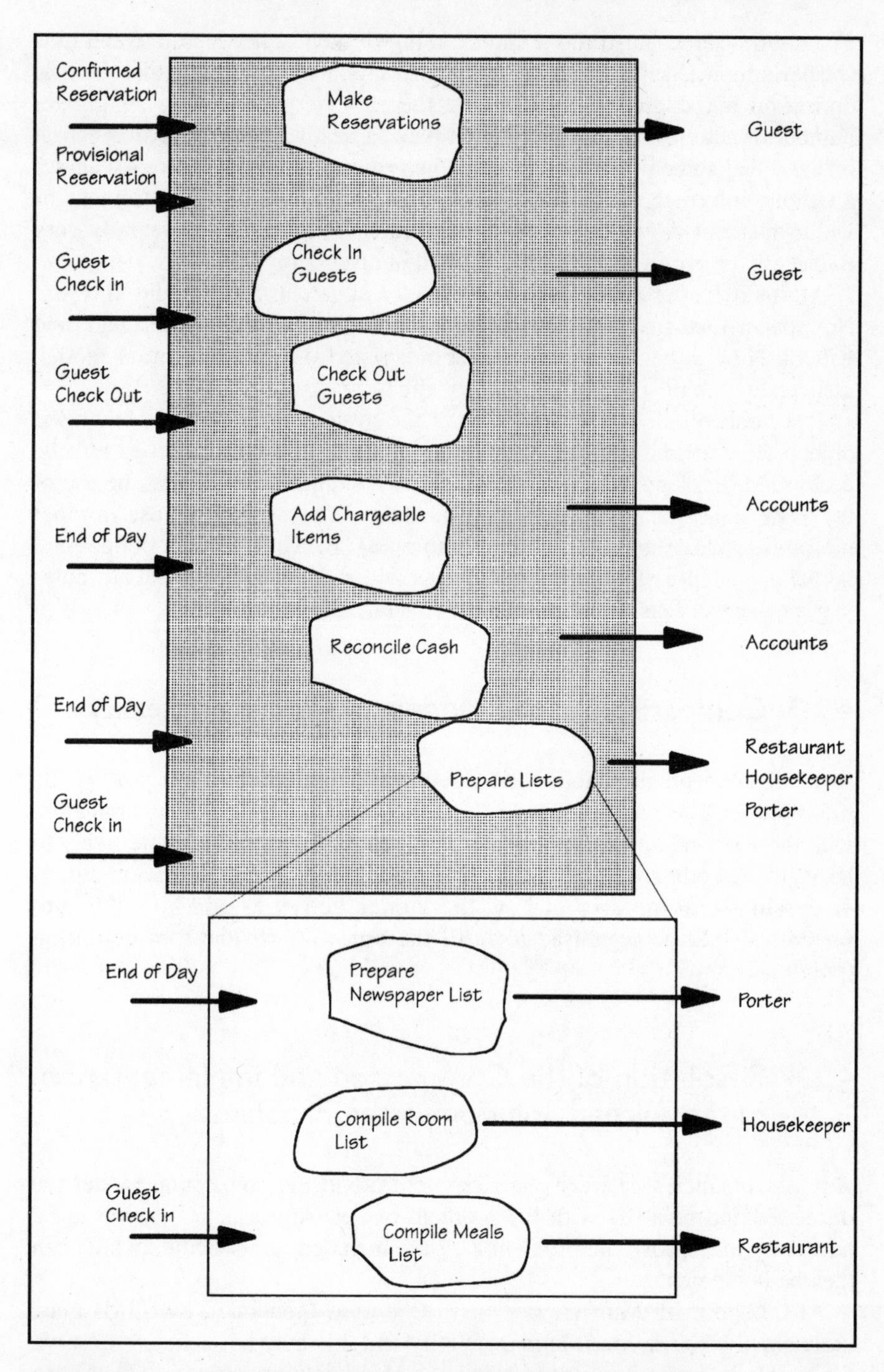

Figure 4.4 Two Levels of Conceptual Model for Midlinks Motel

The sub-systems of these main systems are detailed in expanded diagrams of the conceptual model and also via textual description. The conceptual model is thus composed of the activities of the human-activity system, the number of activities represented in the model being limited to between five and nine (Miller's limit)(Miller 1956). If fewer than this, then it is likely that the root definition is too general in form and therefore requires review. A greater number of activities would be too complex to handle at this point and here it is likely that some of the activities can be grouped together. Figure 4.4 shows a simple, informal, conceptual model of the activities identified from verbs within the root definition for Midlinks Motel. The diagram is initially very informally prepared and would be refined in discussion with users.

At this point certain sub-activities may be apparent, for example, 'Prepare Newspaper List' may have sub-activities including 'Contact Newsagent' and 'Check Newspaper Delivery'. These are listed on the conceptual model, associated with the primary activity to which they belong.

The conceptual model details what the system must **do** to achieve the objectives embodied in the rich picture and not how the activities will be performed (implementation independence). It should be noted that the use of the term 'conceptual' in Multiview is subtly different from its use in other methods, particularly MERISE (Chapter 8). In Multiview, a conceptual model is a picture of what is required, but will retain some physical attributes (e.g. newspaper lists) because it uses terms familiar to the users.

4.2.3 Comparison of the Models to Perceived Reality

Now the conceptual model is compared to the 'real world' described by the rich picture. The features within the conceptual model aid communication with the user and any agreement which is reached is therefore more likely to be valid. A comparison of this logical model with the real situation allows discussion with the user of how the human activities and how the sub-systems should be organised to fulfil the objectives of the root definition (previously agreed!).

4.2.4 Resolution of the Comparison and Implementation of the Agreed Solution to the Problems

Any discrepancies between the rich picture and the conceptual model are discussed and resolved with the problem-owner. Any changes agreed to be necessary to improve the problem area are included in a schedule which can then be implemented.

At this point in Multiview a possible output (generated by SSM) may, quite simply, be improved human activity and this may be sufficient to solve the problem, rendering further stages of Multiview unnecessary. That is to say, a computer will **not** form part of the recommended solution. If further work is required and a computer system **will** form part of the solution, then the root definition and the conceptual model are the outputs to stage 2, the Analysis of Information.

4.3 Stage Two – Analysis of Information (Entities and Functions)

The purpose of this stage is to analyse functions and data of the system described in the rich picture and the conceptual model, independently of **how** the system will develop. There are two phases within the second stage of Multiview which culminate in the development of two models:

- Functional Model;
- Entity Model.

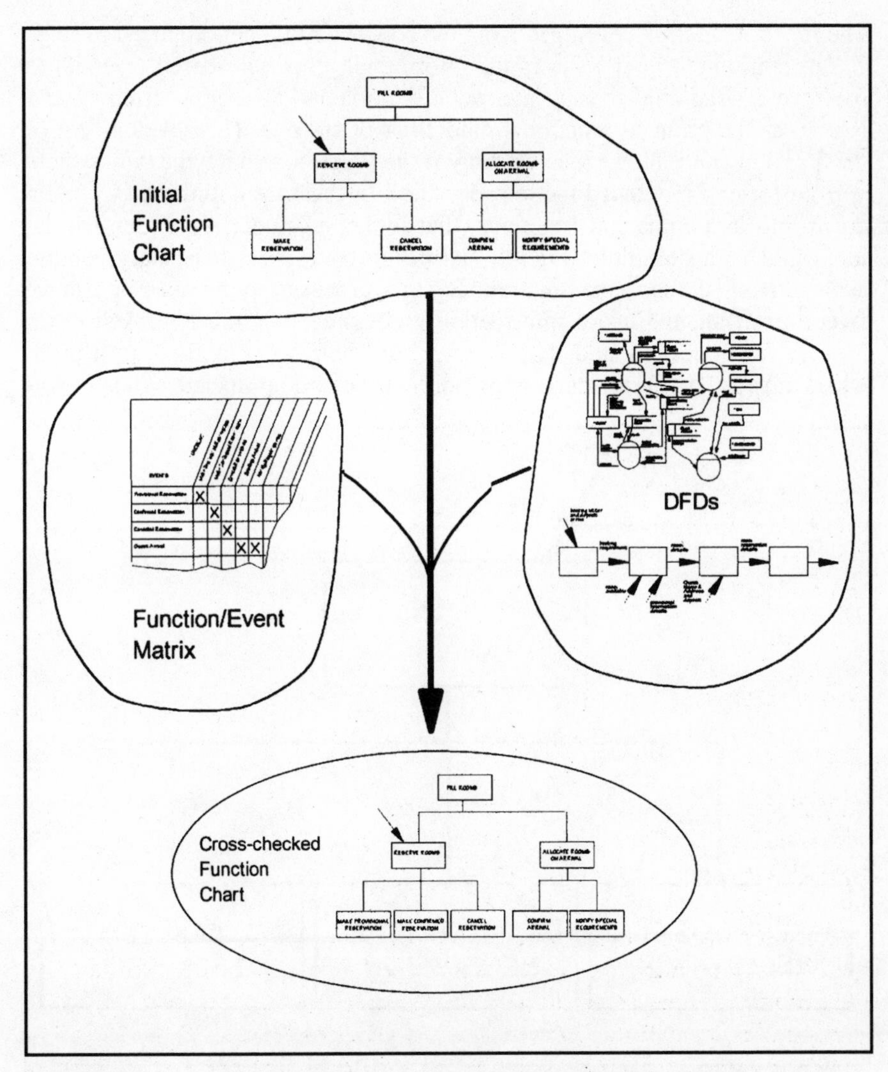

Figure 4.5 Function Chart, Data Flow Diagrams and the Function/Event Matrix as used in Multiview

4.3.1 Functional Model

The functional model comprises a hierarchical set of function charts which go one step further than the conceptual model by taking out the physical aspects of the system such as references to documents and physical goods, etc. As a cross-check upon the functional model Multiview uses DFDs and a Function/Event Matrix (see Figure 4.5), both of which will be discussed below after the discussion of Function Charts.

4.3.1.1 Function Chart

The function chart is developed from the conceptual model produced in stage 1. The conceptual model shows the information flow considered essential to meet the desired objectives of the root definition and this output from stage 1 is used as the input to functional modelling of stage 2. The development of the function chart allows the detailing of the functions which the new system will perform. The main function identified in the root definition is broken down into its component functions, that is the sub-functions which can be identified as a complete but indivisible unit which is necessary for the achievement of the main function. By this process, known as **Functional Decomposition**, the major information processing function is broken down into its sub-functions and they in turn into their sub-functions to a point where no further useful decomposition can be accomplished. (Avison and

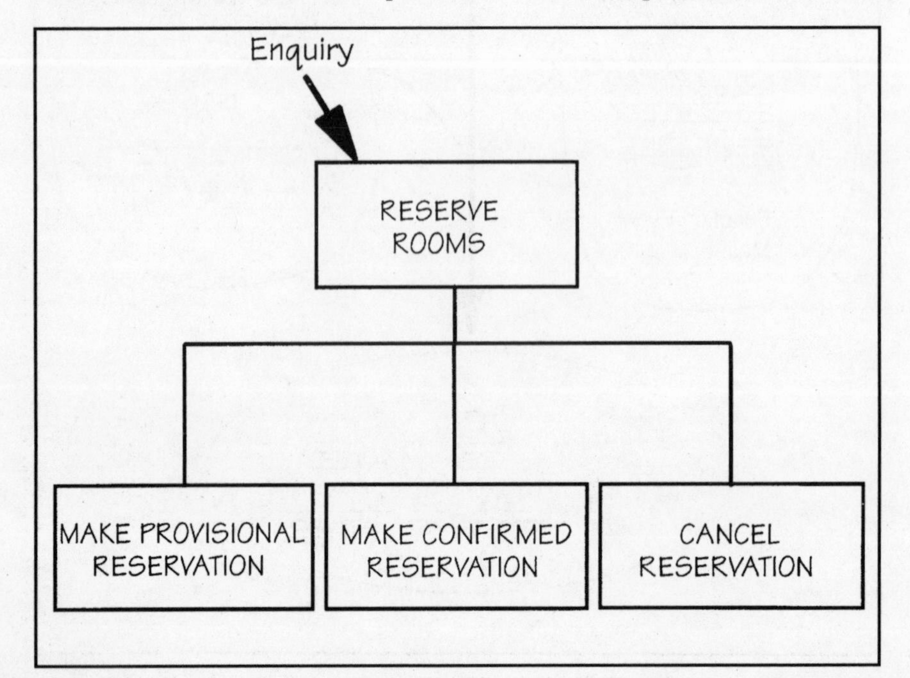

Figure 4.6 Function Chart for Sub-function Reserve Rooms – Midlinks Motel

Wood-Harper (1990) suggest that four of five levels of decomposition will usually be sufficient.) Hence a hierarchy of sub-functions is established. These functions and sub-functions are represented on a chart akin to the charts used to represent a management organisation structure. The sequence in which the functions are written down in the chart in no way implies any ordering in which the functions are performed (unlike Jackson Diagrams where sequence is represented). Multiview allows for the production of a Corporate Functional Model detailing **all** the functions of an organisation at a high level. This is equivalent in part to the Information Strategy Planning stage planning of Information Engineering (see Chapter 6).

Where an existing system, albeit with problems, is in place the functional model is more easily developed. Here the users have a clearer idea of how things work, having already faced and more than likely overcome (or even circumvented!) many of the practical problems of the day-to-day work pattern. Where the system is new, unclear views of how the system will function inevitably leads to deficiencies in the functional model which will be uncovered later. This will of course necessitate review of the model to eliminate such deficiencies as they come to light. The **triggers** or **events** which prompt a particular function can also be represented on the function diagram as a named arrow which points to the function which this named event triggers. The function chart 'Reserve Rooms' for the hotel case study triggered by the event 'Enquiry' is shown in Figure 4.6.

It is, however, more usual to show the events which trigger functions on Data Flow Diagrams, since many of the triggers are the completion of a particular function and the output of information which is required to trigger another function (internal trigger).

4.3.1.2 Data-flow Diagrams (DFDs)

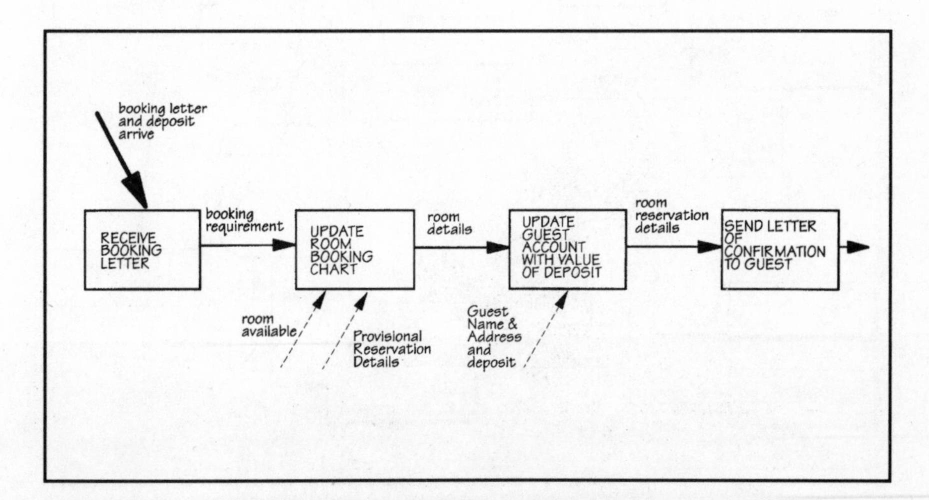

Figure 4.7 Simplified DFD for Make Confirmed Reservation – Midlinks Motel

In contrast to Function Charts, a DFD does show temporal relationships between functions and the flow of information shown by data flows, both by flows connecting processes (for example an invoice cannot be paid until invoice details have been entered) and by the addition of events or triggers to the diagrams, as Multiview allows.

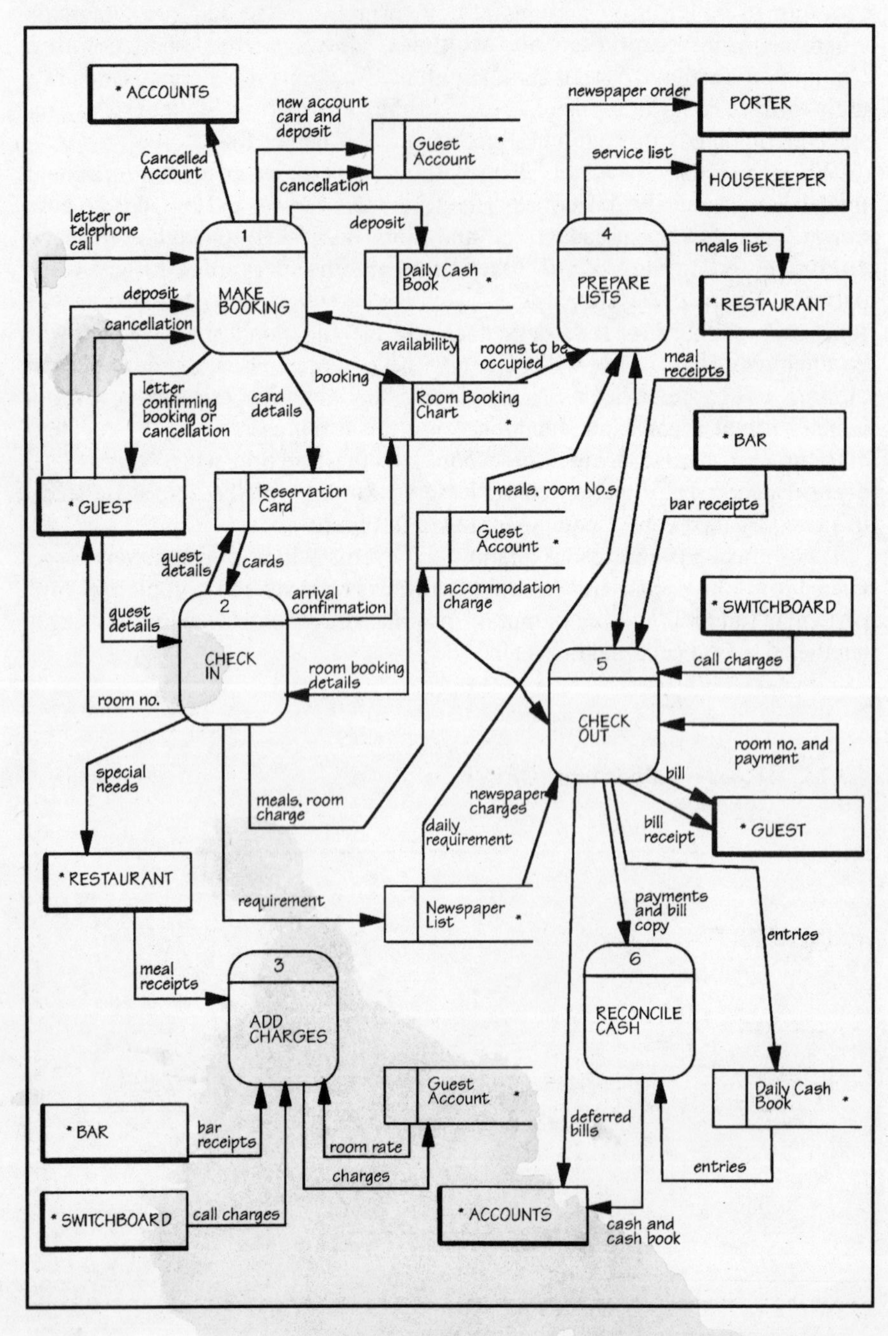

Figure 4.8 Sophisticated Level 1 DFD for Front Desk System – Midlinks Motel

A more complete account of Data Flow Diagramming is given in Chapter 3. It should, however, be noted that Multiview identifies two types of DFD: a simplified form and a more complex form.

The **simplified form** consists of hard boxes for named functions (processes), data flows as arrows between functions and triggers (events) as arrows pointing to the function which they trigger. It is derived by reference to the lowest level functions on the Function Chart. Such a DFD for the sub-functions of 'Make Confirmed Reservation' for Midlinks Motel is shown in Figure 4.7.

The **complex form** also details data flows and processes with the addition of data stores and external entities. The events which trigger functions relate to both external entities (external trigger) and the data flows within the system (internal trigger). Here soft boxes (a rectangle with rounded corners) are used to enclose the processes. The process box has a small section partitioned off at the top of the soft box which is used as a reference to identify the process, for example, in the data dictionary. The external entities are shown by a rectangle with a thickened left and bottom side. Data stores are shown as open-ended rectangles with a vertical bar sectioning off an area which can be used for a reference number to identify the store (in the data dictionary). Asterisks on the diagram indicate that the entity or store appears more than once on the diagram this being necessary to clarify the diagram by reducing the degree to which the flows cross over each other. From an overview DFD, through a series of levelling procedures the processes are decomposed to a greater level of detail with consistency checks along the way to ensure that data flows from higher level diagrams are present on the lower ones. The Level 1 Front Desk System DFD of the current physical system for Midlinks Motel is shown in Figure 4.8.

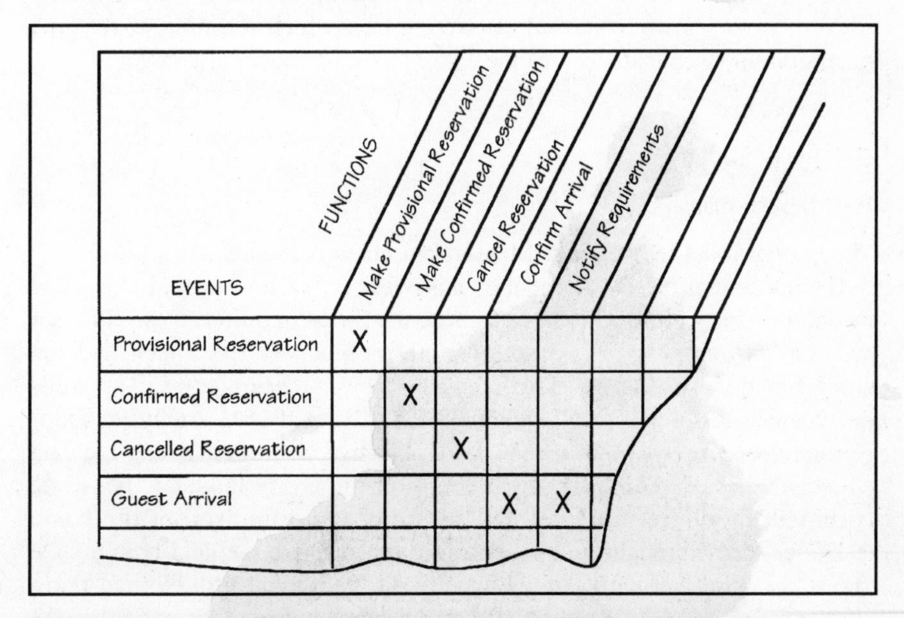

EVENTS / FUNCTIONS	Make Provisional Reservation	Make Confirmed Reservation	Cancel Reservation	Confirm Arrival	Notify Requirements
Provisional Reservation	X				
Confirmed Reservation		X			
Cancelled Reservation			X		
Guest Arrival				X	X

Figure 4.9 Function/Event Matrix – Midlinks Motel (Part Complete)

The DFDs act as a check upon the function chart uncovering any entries which are missing from the function chart. The hierarchy of subfunctions together with DFDs, which show the time-sequence of events and information flow, are produced as input for stage 3, Analysis and Design of the Socio-technical system.

4.3.1.3 Function/Event Matrix

A Function/Event Matrix, showing which events trigger which functions, can be used as a further check on the functional model. Part of such a matrix for Midlinks Motel is displayed in Figure 4.9.

4.3.2 Entity Model

If a sound analysis of human activity has been performed in Stage 1 then the analyst, in the light of the root definition, should be able to identify the entities needed for the system to fulfil the main function embodied in the root definition. The entities are the things about which the organisation needs to keep information in order to support the functions identified in the functional model. Each entity has a 'type' associated with it, which means that every occurrence (record) of the entity consists of the same data elements (also known as fields or attributes). The Entity Model consists of named entities enclosed within soft boxes with the relationship between those entities, where present, shown by named connecting lines which according to the type of relationship terminate against the entity boxes in one of three ways:

- One to one;
- One to many;
- Many to many.

The Entity Model for Midlinks Motel is presented in Figure 4.10.

Having drawn the entity model, Multiview considers ways to check the model for 'correctness'. Firstly users are asked what questions they will want the system to answer. The responses allow the analyst to check that the entity/functional models are sufficient to answer these queries. The well-documented technique of **Normalisation** (Relational Data Analysis, Third Normal Form Analysis) provides a very objective way of analysing the data collected such that direct comparison with the entity model is possible. Although there are more refined versions of normalisation the basic technique leads through three rules to the point where the data is in a state known as Third Normal Form. (Boyce-Codd, fourth and fifth Normal Form are the others.) The technique is described more fully in Chapter 3 but the three basic rules are:

- remove all repeating groups;
- remove all data items that are not dependent on an whole key;
- remove all non-key data items that are not dependent on the primary key.

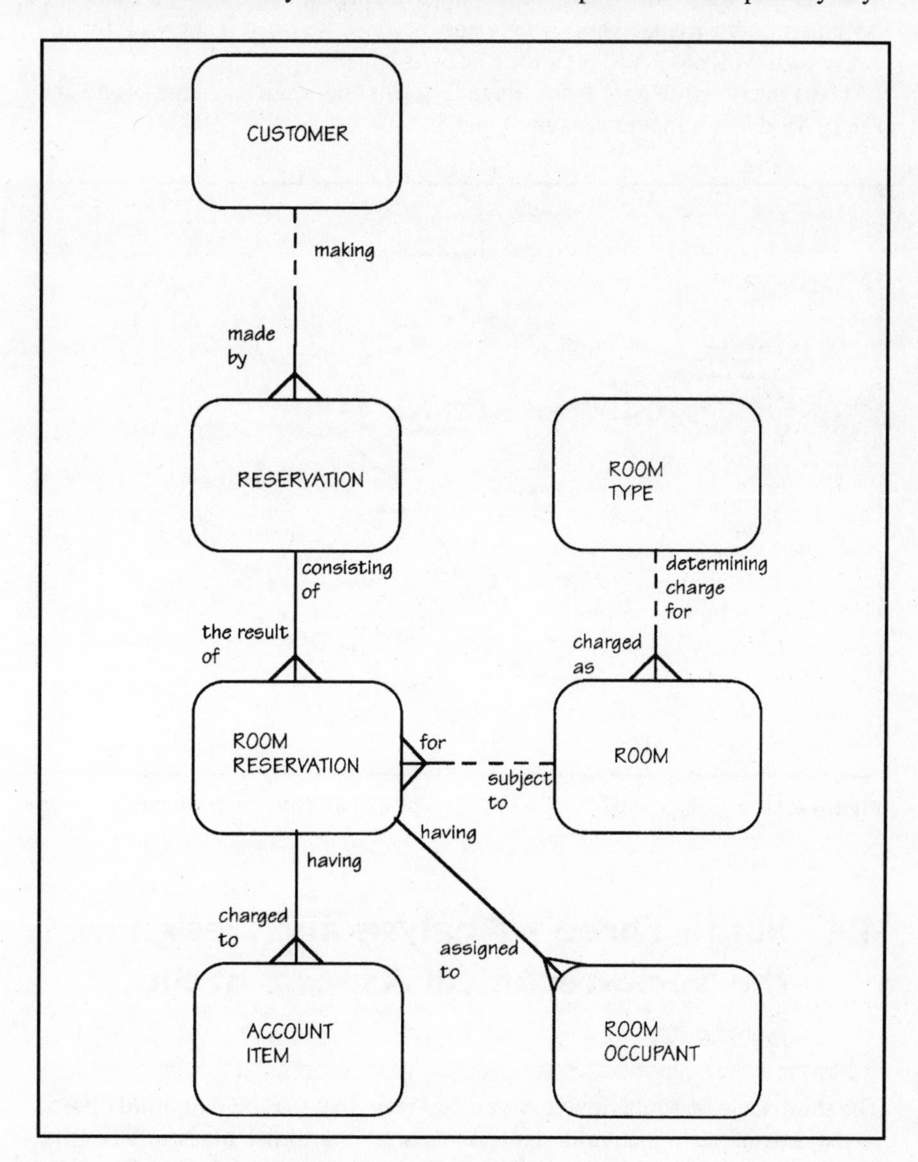

Figure 4.10 Entity Model for Midlinks Motel

4.3.2.1 Entity Life Cycle

Entity Life Cycles show the changed states that an entity passes through within its lifetime in the system, e.g. for the entry GUEST, valid states may be guest *enquirer*, guest *booked*, guest *checked-out*. These changes in state are triggered by events and are effected by functions. The state at which the entity exits the system is crucial since an entity should not be locked into the system forever. The Entity Life Cycle diagram used is of the type documented by Rock-Evans (1981) and is illustrated in Figure 4.11 for the entity ROOM RESERVATION for Midlinks Motel.

The Functional Model from Stage 2 forms the input to Stage 3 and the Entity Models are input to Stages 4 and 5.

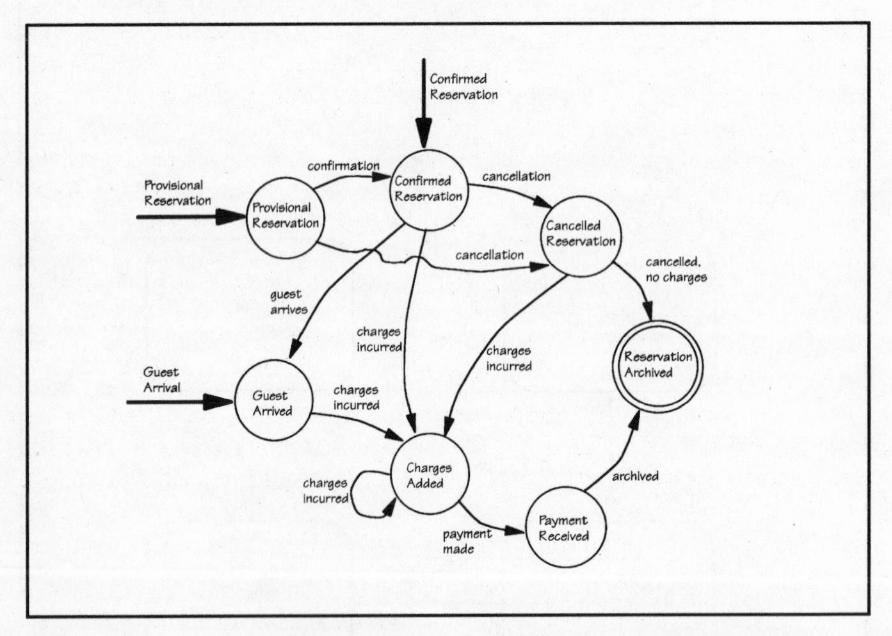

Figure 4.11 Entity Life Cycle for ROOM RESERVATION – Midlinks Motel

4.4 Stage Three – Analysis and Design of the Socio-technical Aspects of the System

The third stage of Multiview is developed from the work of Mumford (1985) in the area of participative design which is central to her method, **E**ffective **T**echnical and **H**uman **I**mplementation of **C**omputer-based **S**ystems (**ETHICS**). Mumford identifies various types of participative design which are distinguished by the user's level of involvement and responsibility in the decision-making process.

A system may technically be capable of achieving the functions for which it was designed but may still fail simply because the users do not accept its imposition upon them. They have to be taken along with or even drive it and understand the benefits that it will give to **them**. The users must have a 'vested interest' in the new system and in making it successful! The following anecdotal story relating to a laboratory system illustrates the point:

The laboratory consisted of several departments (different scientific disciplines) and provided a diagnostic service to all other departments (external to the laboratory) in the organisation. When tests were requested, the sample came to a central laboratory reception. Reception recorded receipt of it and sent the sample and its accompanying multi-part request form to the appropriate laboratory department. On completion of the test the laboratory would send a copy of the results to reception for filing. Reception would often have to answer queries in relation to these. The Accounts Department wished to introduce a computerised departmental budgeting system with inter-department cross-charging. The outcome of the system would be used to determine the budget for each laboratory department for the next year. Reception were charged with the responsibility of using the new computer system to enter the data in addition to continuing the previous paper-based system. They would gain no benefit from it, would have no enquiry access and no extra staff would be allocated to cope with the extra work.

Shortage of time and the increased pressure from the additional workload had the following effects. The paper records which could not be filed until the data had been entered into the computer system began to accumulate. To be able to answer enquiries related to these meant that the backlog of forms had to be sorted twice daily. The clerks continued to enter data to the computer system, but with an ever increasing backlog. The fact was that once the data was entered into the computer the clerks were never able to retrieve it for any purpose. It was only used by the Accounts Department to generate a report of number of tests and total test costs plus inter-department charge information. The result was that:

- there was an increase in days lost through illness by Reception staff;
- the information that was generated for Accounts, such as the number of tests per month was shown to be in error by the laboratory departments' individual paper records which they still kept;
- some results were never entered in to the Accounting System since the accuracy of the Accounting System had neither a negative or positive impact upon the Reception staff; failure to file the paper test results had a negative impact upon Reception staff and therefore filing took priority over data entry.

Precisely to prevent situations such as this the rationale behind Stage 3 of Multiview is that the users must be allowed to play an active role in the conception, design and development of the system which will impact upon their work flow/satisfaction/achievement/role. The view is that a contented work force will utilise the system to its full potential, swimming downstream rather than upstream actively confronting the system (and finding ways to circumvent it!). They are then far more likely to operate the system successfully if they feel the system **belongs** to and benefits them.

Obviously it is not always possible, nor for that matter economic, to design the ideal system that the users would like and the analyst is therefore faced with the task of balancing the requirements of the people on the one hand with the organisation's requirements (and often limitations) on the other.

The analyst must identify *social objectives* with their *social alternatives* to meet those objectives. A social objective of the project may be, for example, to bring all previously distributed purchasing decisions under the control of one person. A social alternative could be to create a new department, and redeploy staff to it. The alternative may involve a change of job role or responsibilities for a user or group of users, or sometimes the distribution of one manager's empire between other managers. The analyst must, in parallel with the above, identify the alternative *technical objectives* with their associated *technical options*. A matrix of social and technical alternatives is created such that each social alternative is paired with a technical alternative. From this, the socio-technical combinations are ranked as to how well they achieve the stated objectives in the light of other considerations such as cost/design limitations. Future Analysis (Land, 1982) may also be used to predict the expected life of the system and the system environment so that ranking of the socio-technical alternatives can take account of the impact of environmental changes upon the system.

This stage then produces:

- computer task requirements, the computer role set and people tasks which are the products passed to the next stage;
- the non-computer role set, people tasks and social aspects which are passed for action out of the method. These may result in organisational changes, for example.

4.5 Stage Four – Design of the Human–Computer Interface (Man–Machine Interface/User Interface)

This stage is concerned with how the user will interact with the computer system. This interface must not be so complex that the user avoids using the system nor should it be inflexibly structured such that when the user becomes more experienced the interface 'gets in the way' slowing down and frustrating the human activity.

The entity model from Stage 2 and computer tasks, role set and people tasks from Stage 3 are fed in to the design of the Human–Computer Interface. If an on-line system is adopted then the dialogue which the user will have with the computer must be considered whereas a batch system will only require the user consideration of when and how the data is input and when the output is required and in what form. Where the user interfaces directly with the system, such features as arrangement of data on-screen, on-line help, error checking on data entry and the consequent handling of such errors and ease/method of data entry must be considered (e.g. all numeric fields of information are arranged together where possible to speed up entry via the keyboard numeric pad). Having established the human–computer interface, the attention is drawn to the technical aspects of achieving this interface. The output from this stage to stage five is these technical requirements. In fact a major output of the method is the human– computer interface, in particular the specific dialogues tailored to various categories of user with their different levels of experience.

4.6 Stage Five – Design of the Technical Aspects

Multiview approaches the technical design of the system by using the following products from previous stages as inputs to stage five:

- Entity Model (from stage two);

- Computer Tasks (from stage three);

- Technical Requirements (from stage four).

Since attention has been given to social and technical objectives together with a suitable human–computer interface, the design of the technical aspects has already been pro-forma moulded to take account of human factors. In stage 5, the analyst is concerned with the technical aspects consistent with a complete and **efficient** system specification.

Multiview defines the technical design stage to comprise seven areas for the design effort to focus upon:

- the **application** which is how the functions from the function hierarchy will be designed such that the system will do everything that the user has agreed it should do. These will form the basis for the programs which will be developed either manually or in an automated way;
- **information retrieval** which considers how information retrieval will be achieved. Information Retrieval is considered separately in Multiview since it fundamentally differs from other aspects of the system by not changing any information which is stored, e.g. in our hotel case study 'Room Availability' enquiry, although it may lead to a booking, does not *per se* alter the stored data. Design for information retrieval takes account of standard reports with code written to implement them and 'ad hoc' queries which could be handled for example by trained users of Structured Query Language (SQL). Another factor which influences the design here is the level of user experience and skill in relation to computer-based systems. A menu driven enquiry procedure, with the availability of standard reports, may be more appropriate for the less technically oriented users than a totally do it yourself query facility. The design of this area is developed from:

- the **entity model**;
- the **socio-technical stage** (stage three) where the people who were going to use the system were identified and characterised;
- the **dialogue design** (stage four), where the human–computer interface was considered;
- the **database** where database refers to the non-specific collection of information (rather than a Database Application) however it is organised. The Entity Model together with Entity Attributes are used to structure the data in a way which will facilitate the applications and information retrieval;
- **database maintenance** which lays down how data will be updated to accommodate insertion, deletion and amendment of records within the stored data;
- **control** which is part of the wider subject of Computer Security. Specifically this design area of Multiview addresses how errors introduced by the user, the programmer or by hardware failure are to be trapped and handled. Included within these are such things as error checking on data entry, software failure and hardware failure;
- **recovery** is the 'what to do in the event of system failure' aspect of the design. It is naturally linked to the control area of the design and considers such things as programmed recovery from a control error pick-up. In these circumstances the procedures for recovery should be laid down in a recovery procedure manual which is formulated as part of the design process;
- **monitoring** whereby management information can be obtained detailing who has been doing what with what and for how long. Tied into this are log-on procedures which may be associated with charging out the use of computer time and security aspects of control.

Ultimately acceptance testing criteria are established to ensure that the socio-technical objectives, the human–computer interface requirements and the technical design objectives detailed above are met by the delivered product, i.e. the users' requirements are met and the organisation's objectives achieved. Inevitably maintenance will be needed to fix program errors, incorporate minor changes or additional features. This will require backtracking into Multiview, the extent of this backtracking depending upon the degree of change which is needed.

4.7 Summary

This chapter has examined Soft Systems Method within the framework of Multiview Method. SSM differs from the commercially available structured methods since it is directed toward the investigation of a problem situation and proposal of a course of action which will improve the problem situation. In this context SSM focuses on Human Activity Systems from a holistic viewpoint such that a system is considered within the context of the wider system which is all components of the organisation. By using rich pictures, root definitions (fully formed using CATWOE) and conceptual models, the comparison of the real world with the conceptual is possible. In this way feasible changes which will improve the problem area can be identified and implemented. Multiview then follows with 'hard' techniques of information analysis as used in other methods. By combining alternative social options with alternative technical options for the new system, a selected socio-technical alternative can be arrived at using the ETHICS method of Mumford. Following design of the Human–Computer Interface to match various levels of user experience, the design of technical aspects of the system is completed under the seven design areas listed.

Multiview is not prescriptive about the techniques to be used in the analysis and design process, and advocates a contingency approach. The analyst should use whatever techniques are considered to be useful within the circumstances of the project. This could be viewed as a weakness, since the correct use of techniques relies on the experience of the analyst, and there is no framework to ensure that a sufficient set of techniques is used for a full definition of the problem area. The strength of Multiview in relation to other methods is in its inclusion of the human aspects, which are often belittled by the event, data and function-driven system specifications of those other methods.

5

Structured Systems Analysis and Design Method Version 4 – SSADM (V4)

5.1 Overview

SSADM is a systematic approach to the analysis and design of Information Technology (IT) applications. It was commissioned by the government of the United Kingdom in the early 1980s and has been developed under the control of the government IT advisory body, the government centre for information systems (CCTA), to the current version, Version 4. SSADM is the IT systems development method recommended for use in UK government departments and the method most widely used in the UK, both in the public and private sector, accounting for some 40 per cent of the methods marketplace (Euromethod Phase 3a Information Pack 1993).

SSADM aims to help the IT project team to accurately analyse the requirement for an IT system to support an organisation's IT strategy, and to design and specify an IT system to cost-effectively meet that requirement. The method pulls together a large number of tried and tested techniques in a manner which renders them compatible with each other, and provides a well-documented framework within which they sit. SSADM does not claim to be a universal method, encompassing every aspect of (IT) or for that matter a panacea for every IT problem. Most importantly SSADM does not claim to be a substitute for common sense or skill. SSADM is carefully placed within the systems development life cycle to cover just systems analysis and design, and not other project issues such as project management or quality assurance, although it provides good interfaces to these areas. It does not attempt to cover Information Strategy Planning or Construction, Testing and Implementation of the eventual system. The techniques within SSADM are intended to meet the needs of the practitioner in defining functional and information requirements for a wide variety of Information Systems (IS) applications.

A fundamental principle underlying SSADM is that the system belongs to users and hence their participation in the development process is essential. They are consulted during fact finding, and the products of the various stages of SSADM are discussed with the users to confirm completeness, correctness and understanding. Some of the products can only be derived with the active assistance of the users (e.g. Function Definitions). User management makes the major decisions during Information System development and one phase only proceeds to the next when users have accepted the products of the current one.

SSADM (V4) is in the public domain and is fully documented in a well-structured four-volume set of manuals. Additionally, several guides are available or being developed under the auspices of the CCTA. These aim to expand upon the core information held in the manuals and assist in adapting the standard framework to suit a variety of different situations, for example where the target environment for the developing system will have a graphical user interface or where development is required to be object-oriented. Other guides will cover areas such as project management, quality assurance, capacity planning. The continued enhancement and development of SSADM is assured by the existence of an active International User Group, by the stewardship of the CCTA and the Design Authority Board and by the involvement of SSADM with Euromethod. SSADM is committed to gradually align itself with the rationale of Euromethod, upon which it also has a major influence.

In this text, we can only give a flavour of the richness of such an extensively documented method. Thus, we have tried to highlight the major techniques and their interrelationships and the underlying structure of the method, SSADM Version 4.

5.2 The Three-view Model of SSADM

SSADM adopts three perspective views with the potential for cross-checking between them, in order to highlight errors early in the development of a new system. These views are directed towards:

- **Functions** Data Flow Modelling yields a processing/data flow view. Function Definition builds upon this, expressing required system processing in user-defined groupings;
- **Events** Entity/Event modelling through Entity Life Histories and Effect Correspondence Diagrams provides an event-oriented view;
- **Data** Logical Data Modelling provides a data/relationship view.

5.3 The SSADM Framework – The Structural Model

The Framework of SSADM is shown in the Structural Model given in Figure 5.1, which closely follows that presented in the SSADM (V4) Reference Manual. The diagram has an information highway (Information and Control)

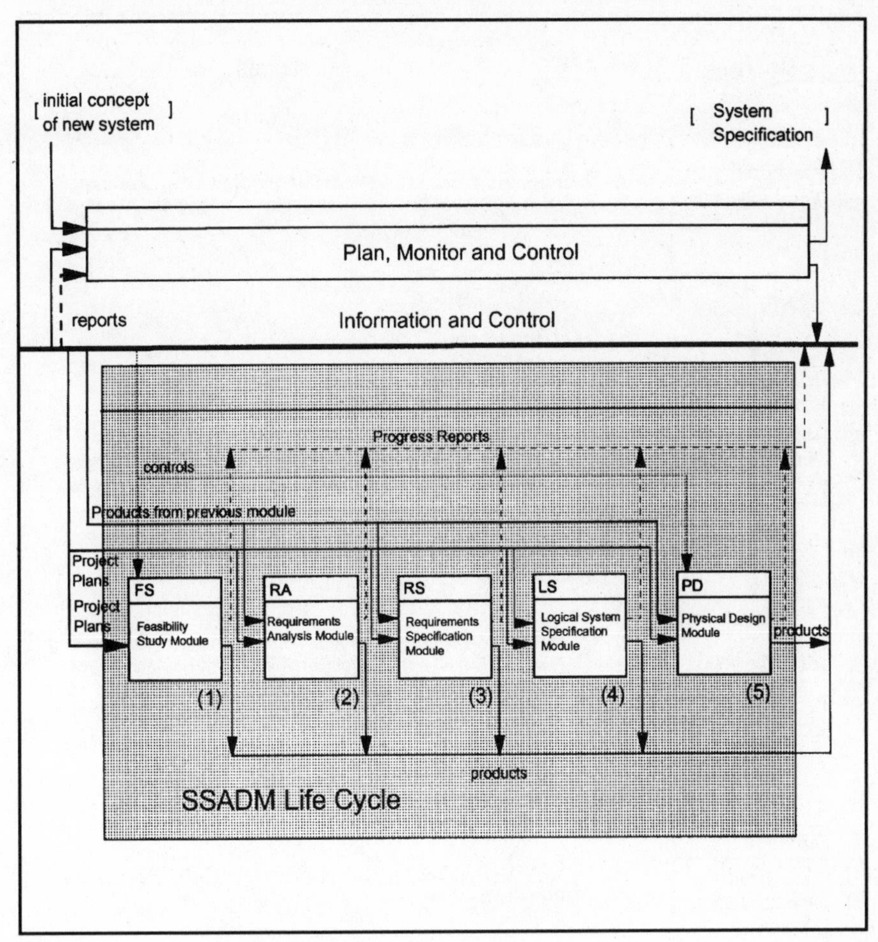

Figure 5.1 The Structural Model for SSADM Version 4 (after SSADM (V4) Reference Manual)

which can be assumed to carry all products produced by the modules within SSADM. It also shows the separation of the technical processes, which are the province of SSADM, from the management processes (Plan, Monitor and Control) which are outside the scope of SSADM, although maintaining an essential communication path with it.

SSADM (V4) comprises five modules which form the central core of the method. Modularisation was introduced to be consistent with current thinking on Project Management with emphasis being placed on the delivery of clearly defined products at the end of each module.

The core modules are shown again in Figure 5.2 which also indicates the techniques to be used within individual modules.

The structure of SSADM (V4) is represented in Figure 5.3 at an overview level of detail, showing the Modules, Stages, Steps and Tasks which form the framework of the method. For further detail of the steps and tasks the reader is referred to the SSADM (V4) Reference Manual.

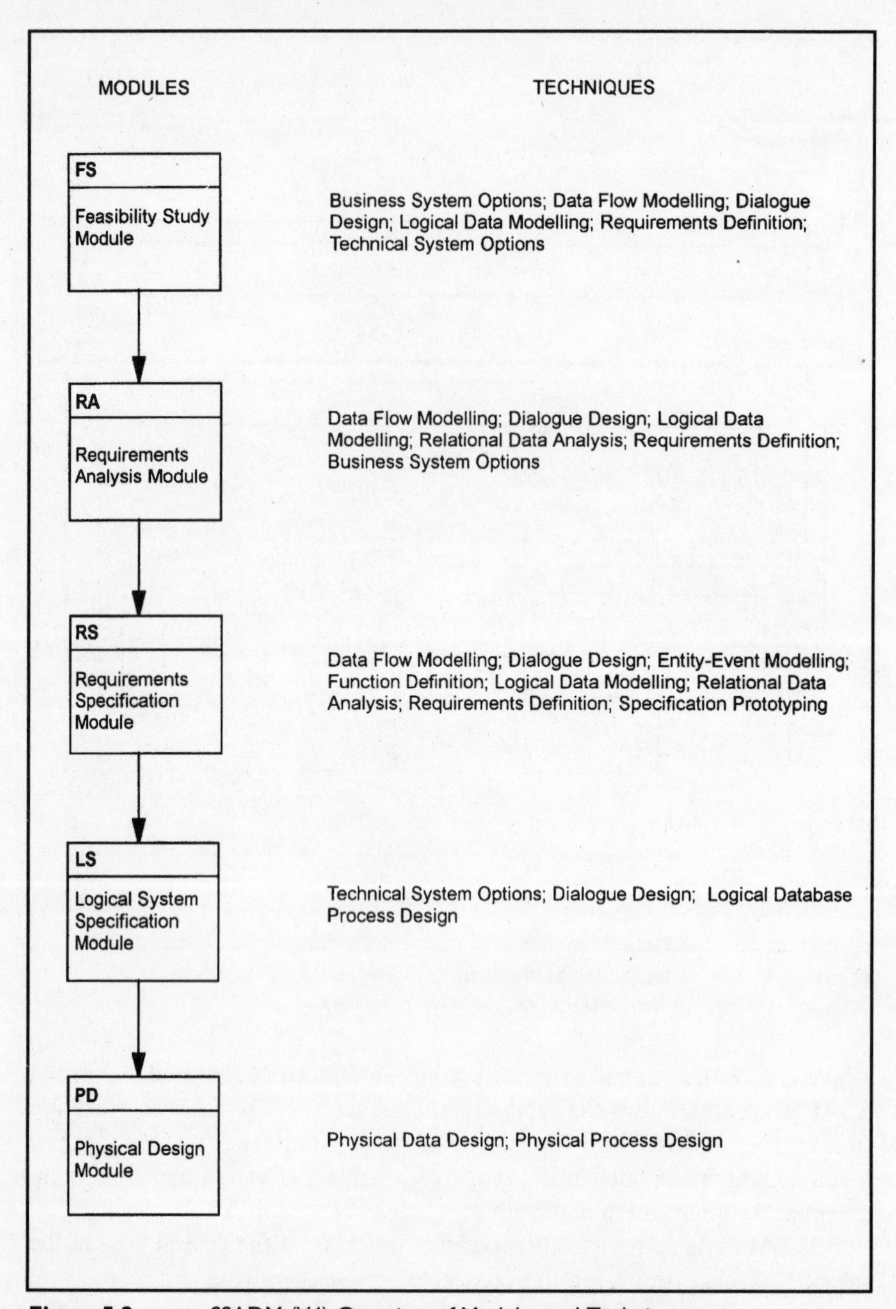

Figure 5.2 SSADM (V4) Overview of Modules and Techniques

Whilst the framework follows a predominantly linear flow from current physical system through current and required logical models to required physical system, it is not necessarily implied that steps should be followed in this linear fashion. Many steps and tasks are iterative or are more practically addressed in parallel with others. The contents and techniques of each module are described in turn below.

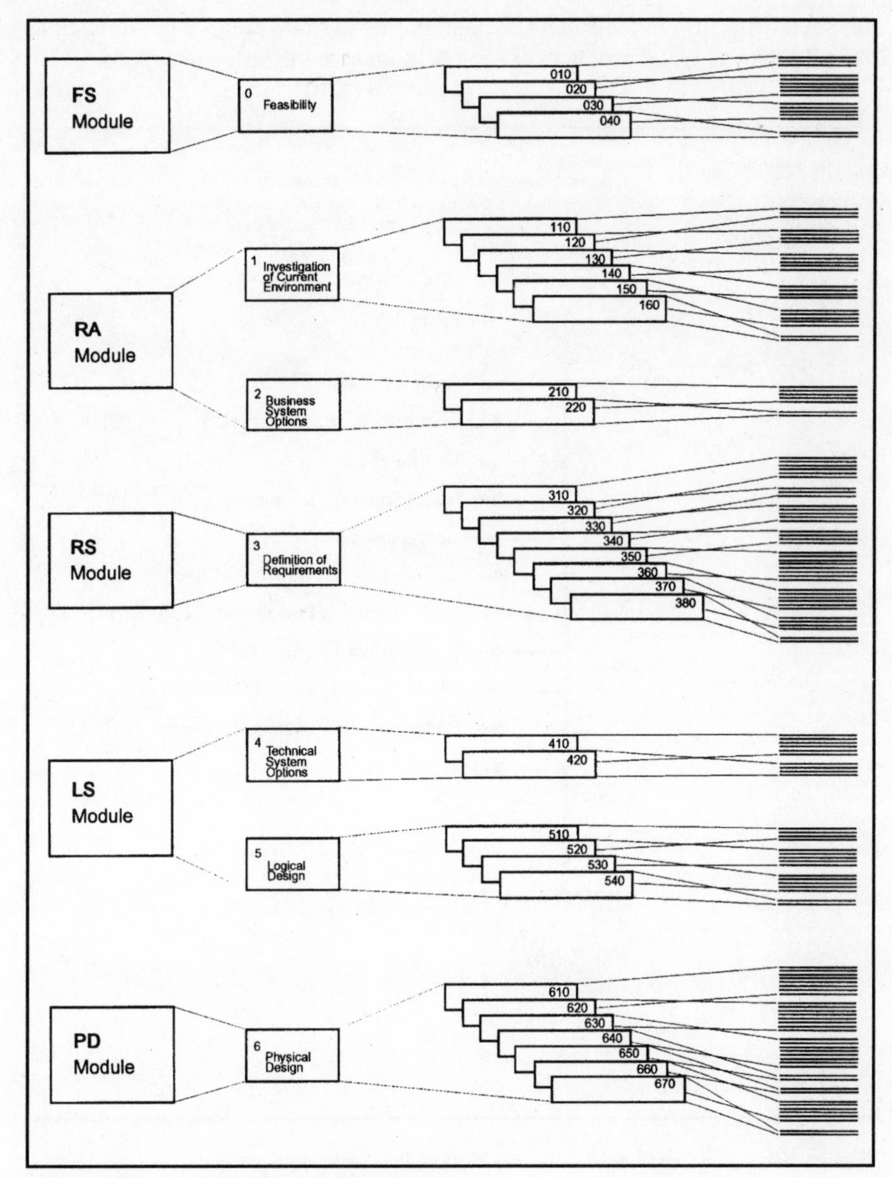

Figure 5.3 SSADM (V4) Framework

5.4 Feasibility Study (FS) Module

In response to a Project Initiation Document (PID), a Feasibility Study may be initiated. The PID is based on management information compiled outside the SSADM project. It results from strategic and tactical planning decisions of the organisation and should incorporate the business case for the project, its objectives, terms of reference, resources and responsibilities. (A more detailed list of its contents can be found in the SSADM Reference Manual, as for all products.) The products of the FS Module are detailed in Figure 5.4.

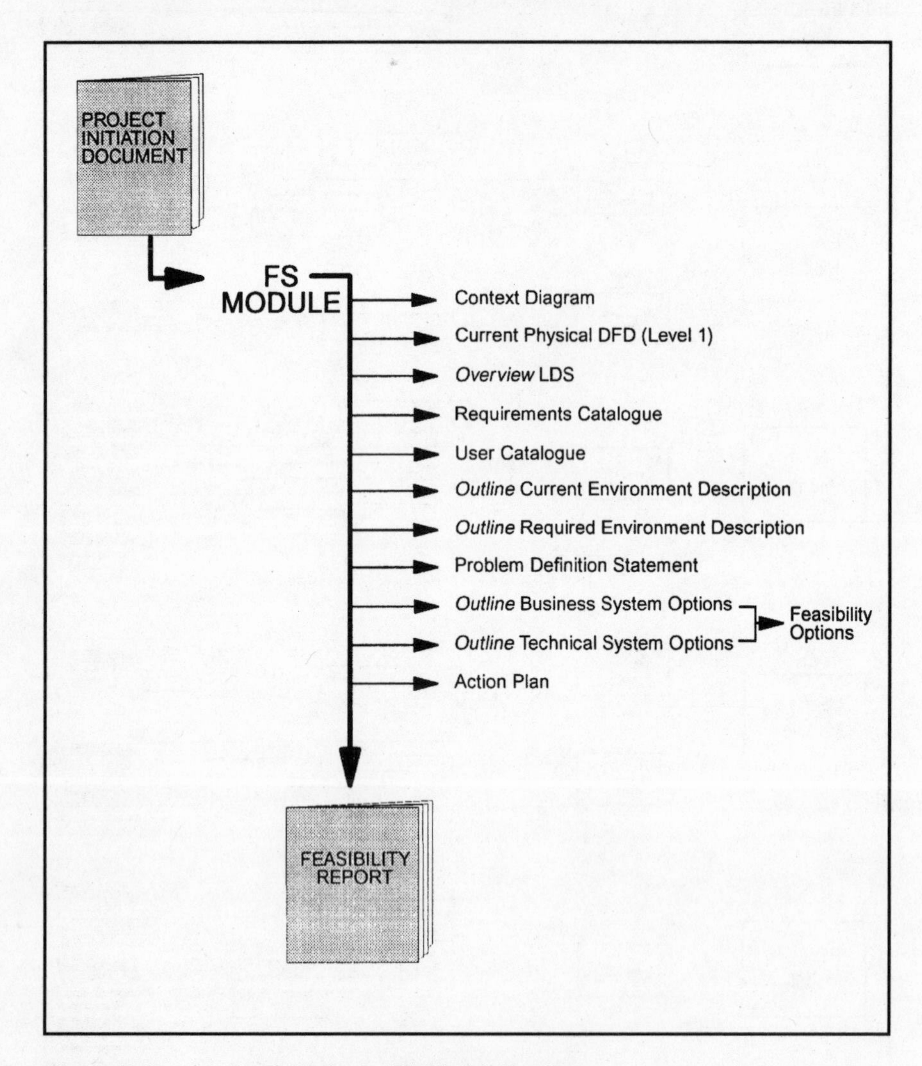

Figure 5.4 Feasibility Study (FS) Module Products

The decision to undertake a feasibility study (or not) is governed by the nature of the project. For low-cost, low-risk projects or where a corporate strategic plan has already embraced the work of an FS module, a feasibility study is not indicated. However large, high risk or politically sensitive projects must be the subject of a Feasibility Study. SSADM (V4) acknowledges that the activities of a Feasibility Study are increasingly being incorporated in Information Strategy Planning although guidelines for module FS are detailed, including an outline structure for the resulting Feasibility Report. If the FS module is not undertaken, the PID will be input to the Requirements Analysis (RA) module.

For our case study, Midlinks Motel, a Feasibility Study is probably not indicated on the basis of project size and also because the hotel must have a Front Desk system which interfaces to the Back Office Accounts system: feasibility and an outline direction have already been proscribed by management.

5.5 Requirements Analysis (RA) Module

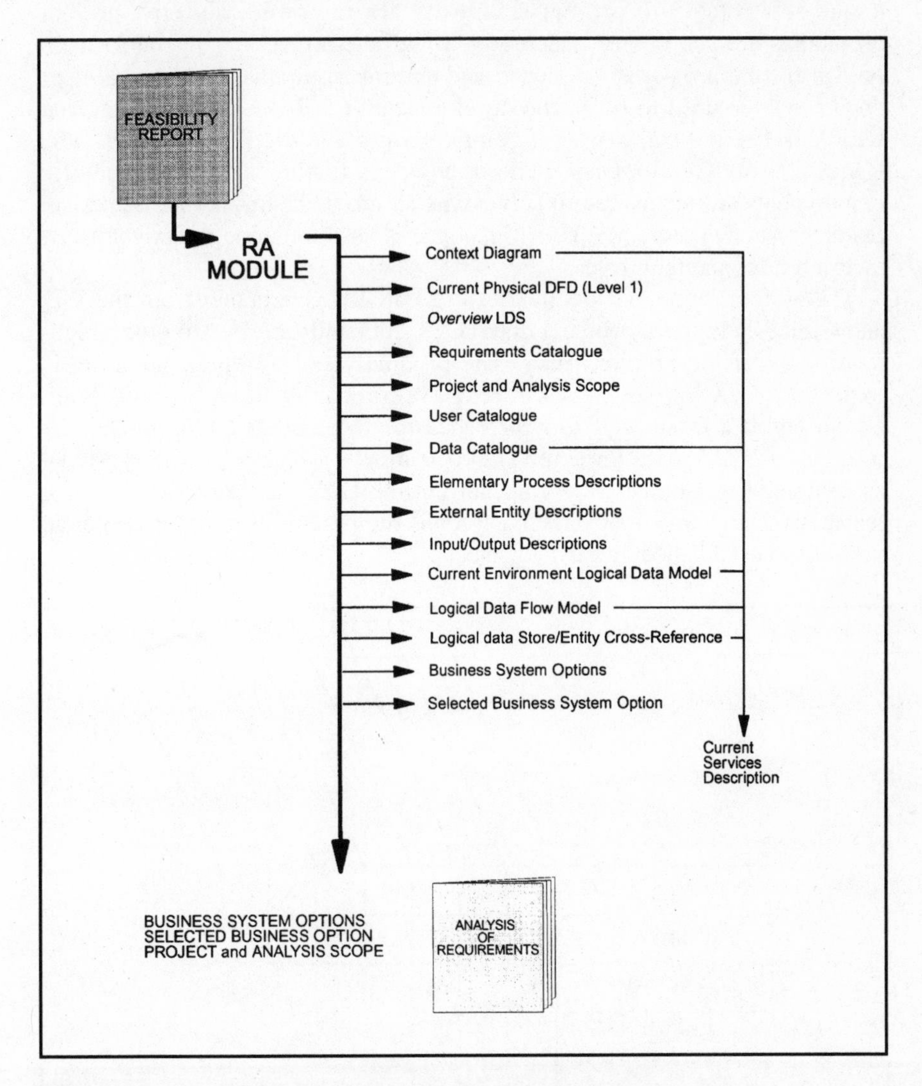

Figure 5.5 Requirements Analysis (RA) Module Products

Requirements Analysis is concerned with information gathering, in order to fully understand the business area which is the subject of the proposed system. It also identifies the 'stake-holders' of the system and the business objectives which the system must support. Business models of the existing system (both manual and computer based activities) are produced and the future business requirements are identified in terms of functionality needed and data to be held. A set of business options is presented to the users and sponsors of the system, detailing the scope of the proposed system. A major decision point at the end of the module determines which option is to be developed further. The products of the Requirements Analysis (RA) Module are identified in Figure 5.5.

At the start of this module, the Project Initiation Document and the documents related to project feasibility are reviewed, and any project difficulties are determined and resolved with the project management team. A plan for the project is developed and system requirements from the input documents are used to begin the development of a Requirements Catalogue which is the central store of requirements for the new system. The Requirements Catalogue will be updated as further and more specific requirements are uncovered to solve existing problems or provide additional features. Another store, the User Catalogue is used to define the target users, their job titles and their roles.

The system scope and boundaries are initially ascertained from the PID and feasibility report. A context diagram is, optionally, created to give a high-level view of the area of study. The production of a context diagram is explained in Chapter 3. A context diagram may have already been established if a feasibility study preceded the RA module. Additionally, the overview Logical Data Structure (LDS) is drawn, which is an Entity Model as described in Chapter 3. Again, an outline LDS may already exist if a feasibility study was done previously or if the organisation has a corporate data model established by earlier projects.

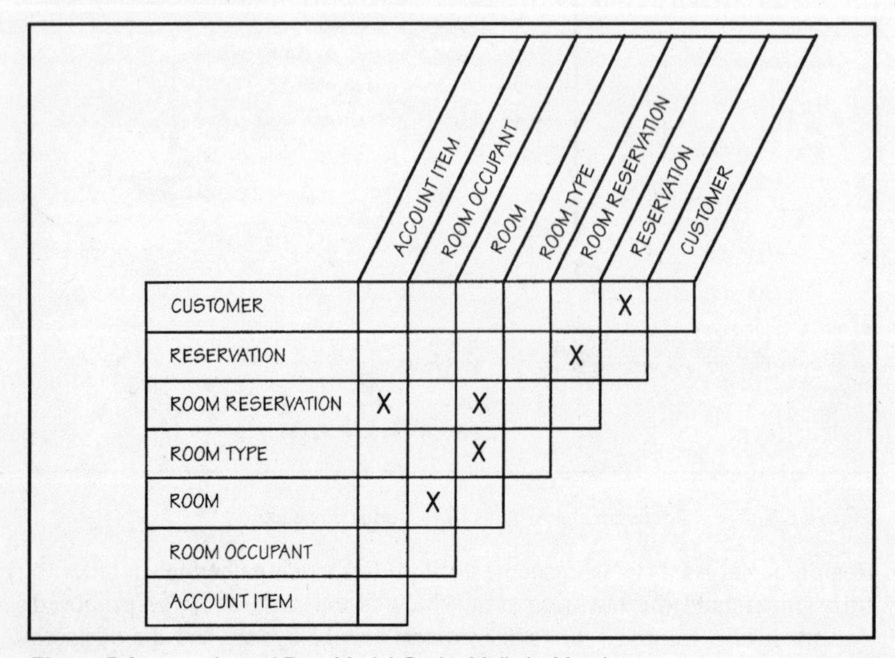

Figure 5.6 Logical Data Model Grid – Midlinks Motel

5.5.1 The Logical Data Model (LDM)

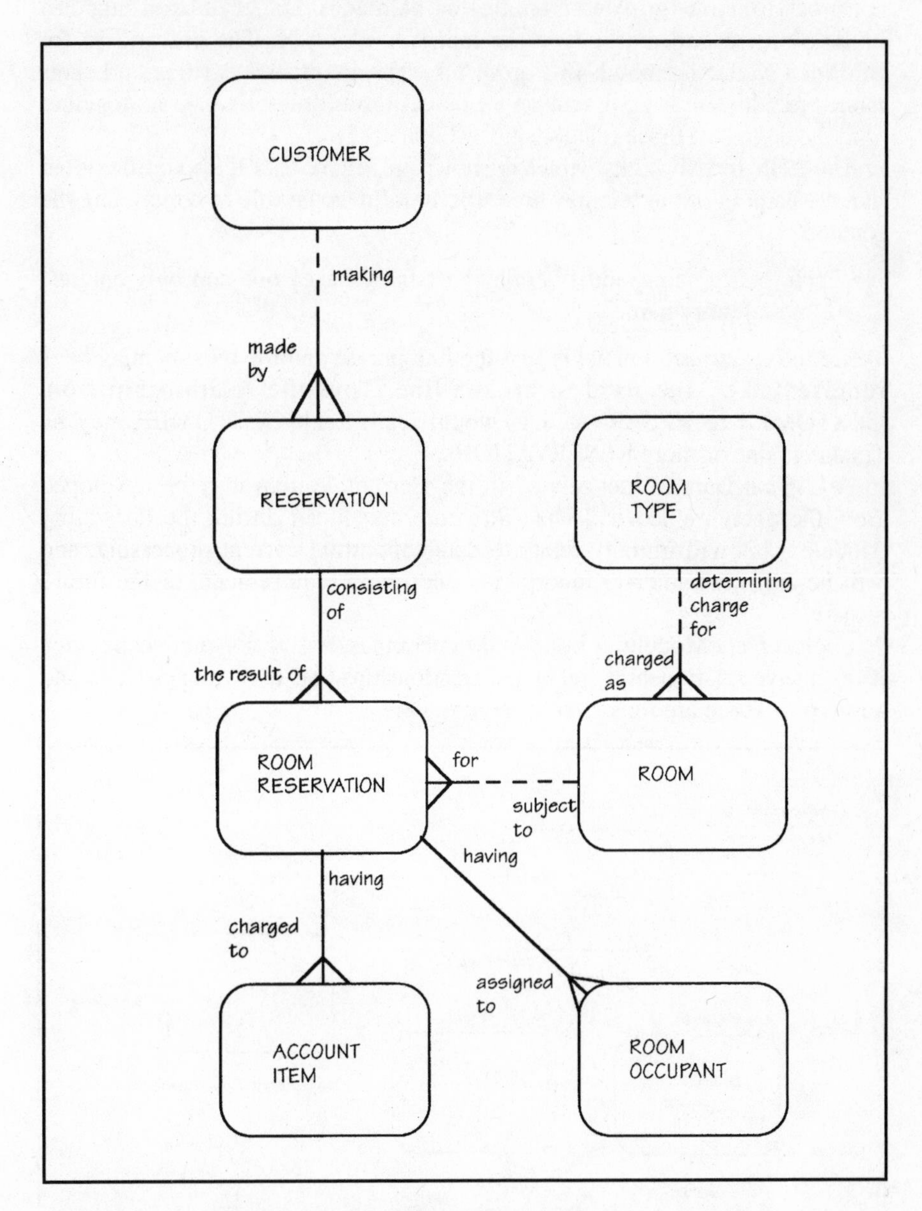

Figure 5.7 Logical Data Structure – Midlinks Motel

The Logical Data Model (LDM) is an Entity Model, as described in Chapter 3. The LDM consists of a Logical Data Structure diagram (LDS) plus supporting documentation about entities, attributes and relationships. By reference to the Midlinks Motel interview notes (see Appendix A) which describe the contents of the user documents (index card, booking chart, etc.), entities are identified for the Front Desk system. In order to ensure that the possible relationships of all pairings of entities are considered, it is useful to construct a matrix (grid) with entities on both axes. Direct relationships can be determined and marked on the matrix with an X. The entity grid for Midlinks Motel is shown in Figure 5.6. The identified entities and their named relationships are drawn on a Logical Data Structure diagram together with the degree of those relationships. (Figure 5.7)

The LDS for Midlinks Motel is shown in Figure 5.7. It should be noted that the naming of the relationships strictly follows the rule of completing the sentence:

- Each [entity name] must be/may be {link phrase} one and only one/one or more [entity name];

where the relationship name is **just** the link phrase and the must be/may be is represented by the solid or broken line. Thus the relationship from CUSTOMER to RESERVATION would read: Each CUSTOMER may be {making} one or more RESERVATIONs.

A Logical Data Model related to the Current System may be developed from the overview Logical Data Structure developed during the Feasibility Module. This will initially relate to data supporting current processing, and will be enhanced later to incorporate additional requirements of the future system.

Logical Data Modelling in SSADM encompasses additional concepts such as exclusive relationships, recursive relationships and entity super-types and sub-types. These are illustrated in Figure 5.8.

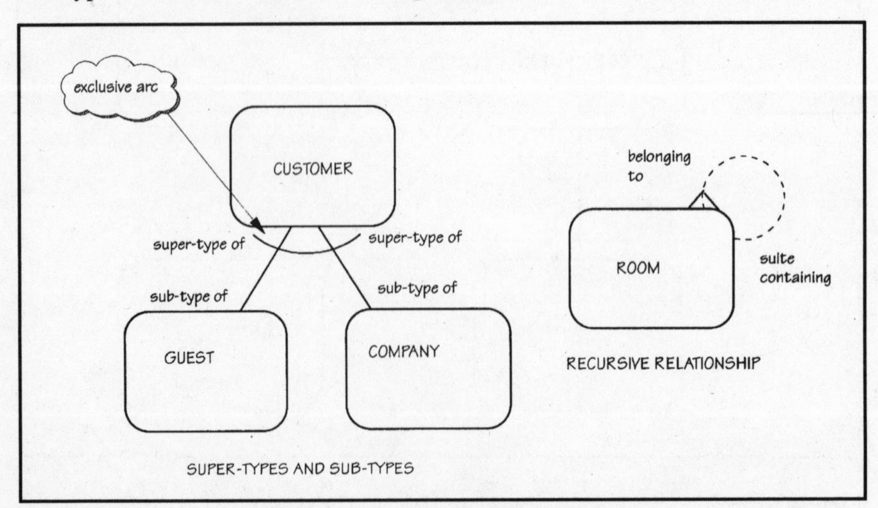

Figure 5.8 Examples of exclusive arc, super-typing and recursive relationship

5.5.2 Data Flow Modelling

The Data Flow Model (DFM) consists of a set of Data Flow Diagrams (DFDs) plus supporting documentation about the components of the DFDs.

An initial DFD (level 1) is developed showing the Current Physical system. The level 2 DFD for the Current System Physical process 'Make Booking' (originally identified within the level 1 DFD) within the Midlinks Motel Case Study is shown in Figure 5.9. This and the Context Diagram (if drawn) will be updated through discussion with users and investigation of document flow around the current system. Further levels of DFD are developed as each process is decomposed.

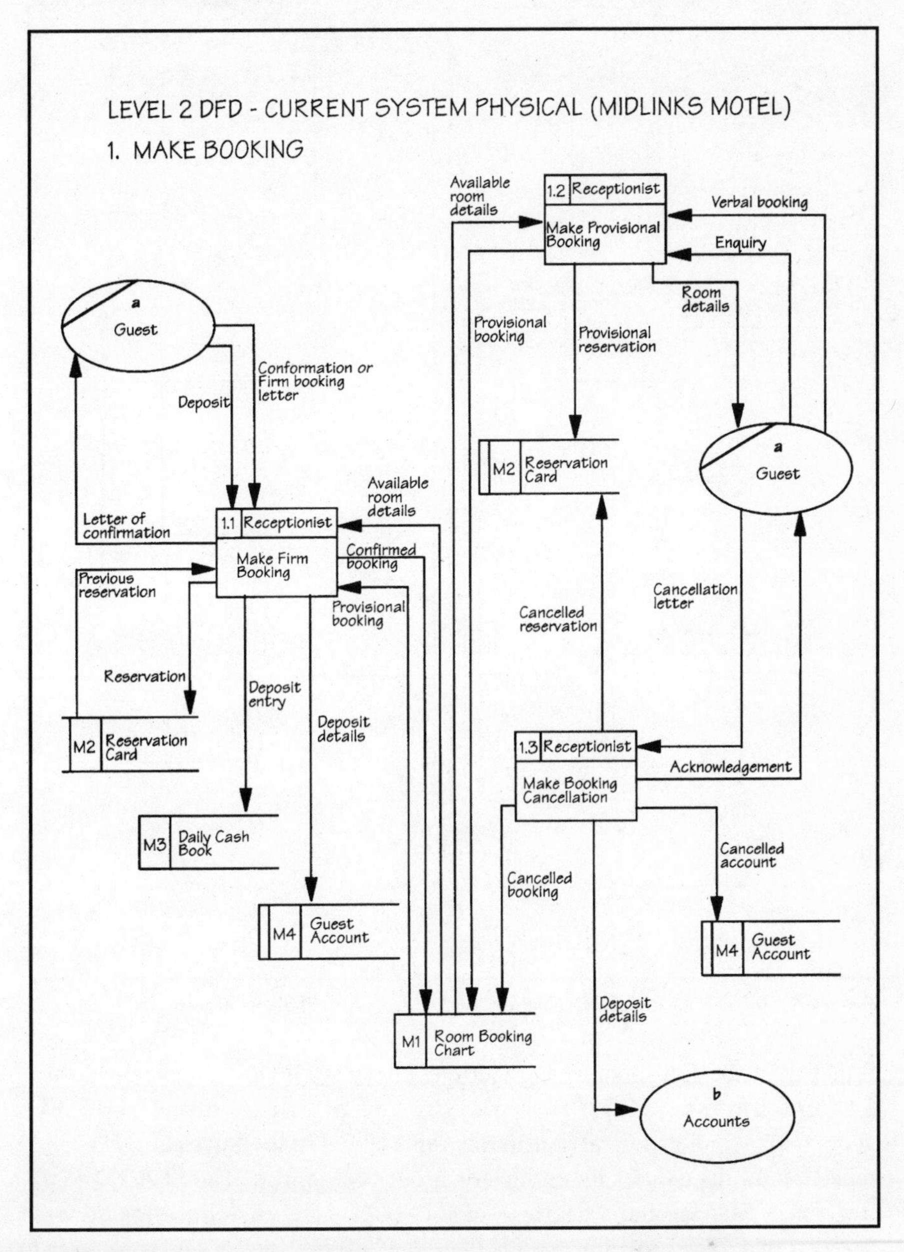

Figure 5.9 Level 2 DFD Current System Physical – Midlinks Motel

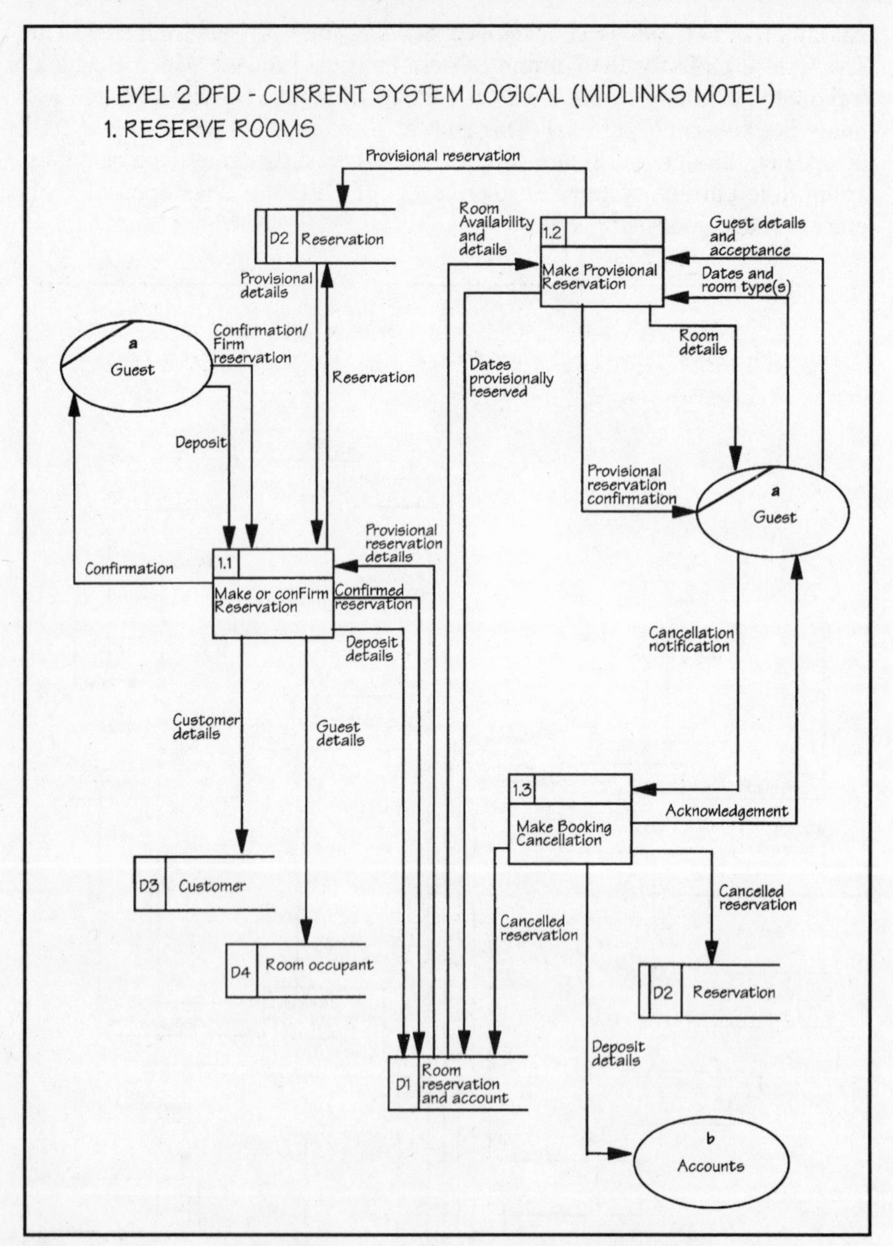

Figure 5.10 Level 2 DFD Current System Logical – Midlinks Motel

Still within the RA module, the Current Physical DFDs are developed into Current Logical DFDs by removing the physical, organisational, geographical and historical constraints represented in the Physical DFDs and replacing all data stores by entities or groups of entities. The SSADM (V4) Reference Manual and Chapter 3 of this book give guidelines for how to perform this transformation. In this way the Current Logical DFDs represent a level of abstraction, showing what is being done in the current system without details of how, when, where and by whom. This Current Logical DFM will be formally developed into the Required Logical DFM at the start of the Requirements Specification (RS) module after the chosen Business System Option has been identified. In practice, outline Required Logical DFDs may be developed during the RA module to form part of the Business Systems Options presentation, as a graphical way of presenting the options to the user community.

Figure 5.10 is the Existing System Logical level 2 DFD derived by extraction from the level 2 DFD Existing System Physical DFD shown in Figure 5.9 for Midlinks Motel. The data stores are now the entities from the LDS, grouped in a user-understandable way. SSADM provides for the documentation of these groupings.

In parallel with the development and subsequent updating of LDS and DFDs a Data Catalogue is established and progressively updated as the central repository for all descriptive information about items of data. Elementary Process Descriptions, I/O Descriptions and External Entity Descriptions are also developed. The LDS plus supporting documentation constitutes the Logical Data Model (LDM). Similarly, the DFDs plus supporting documentation form the Data Flow Model (DFM).

The examination of the current system and its deficiencies is concluded by cross checking the DFM and LDM against each other and the context diagram as well as reviewing the Requirements Catalogue with the users.

5.5.3 Requirements Definition

The steps related to building the DFM and LDM proceed in parallel with investigation and definition of requirements. The technique for requirements definition aims to:

- identify requirements to meet the needs of the users and the business as a whole;
- describe requirements in quantifiable terms;
- provide a basis for future decisions regarding the new system;
- contribute to a complete accurate requirements specification;
- focus analysis on requirements of the future system.

Requirements must be defined in functional and non-functional terms. In essence, the functional requirement describes what is required and the non-functional requirement states how well, how quickly, accurately, etc. this is required to be done. For example, looking again at the hotel scenario, one requirement might be:

- Functional: given a guest's request, the clerk must be able to determine room availability by room number, room type, room charge and date and allocate an appropriate room for each day of the requested period.

- Non-functional: the clerk must be able to ascertain room availability within 3 seconds of the request on 95% of occasions and have allocated an appropriate room for the requested period within two minutes of the request in 75% of cases.

Some non-functional requirements, particularly those concerned with security, reliability and response time, may be system-wide.

The requirements must be added to the Requirements Catalogue. SSADM provides a form-layout detailing the aspects of requirements to be documented. Requirements must be iteratively refined throughout the RA and RS modules to ensure that they are Specific, Measurable, Accurate, Realistic and Traceable to the originator. Although it is not found in the SSADM manual, the acronym 'SMART' has been found useful by the authors in defining requirements to SSADM Requirements Catalogue standards.

5.5.4 Business Systems Options

The second stage, within the RA Module, concerns the creation of several (up to six) possible Business System Options (BSOs) which will usually range from an option with the minimum mandatory requirements for the new system to one having all the 'bells and whistles'. These options are described in terms of what the system does, its boundary and the inputs and outputs. This description may be diagrammatic or narrative or a combination of both, as deemed appropriate for the audience. Following discussion with the users the list of options is pared down to two or three possibilities for further detailed description and for cost/benefit analysis. The implications to the organisation of each option are also discussed. Consideration of these options by the project board (the steering group required by the project management method) and other interested parties, as necessary, results in a Selected Business System Option to be carried forward into the RS Module. The final choice need not necessarily be one of the options put forward but may be a modification of one or even a hybrid of them. The selected option must fit the overall objectives and constraints (costs, timescales, etc.) imposed by the PID. The reasons for selection of the chosen option and for the rejection of the others are documented within the Selected BSO.

5.5.5 Completion of the RA module

The product of the RA module is the Analysis of Requirements, a formal product comprising:

- Current Services Description which consists of the Data Catalogue, Current Environment LDM, Context Diagram, Logical DFM (the physical DFM being only a means to an end), and a Logical Data Store to Entity Cross Reference;

- User Catalogue;
- Requirements Catalogue;
- Selected BSO.

These are carried forward, following management approval, to form the initial working documents for the Requirements Specification Module.

5.6 Requirements Specification (RS) Module

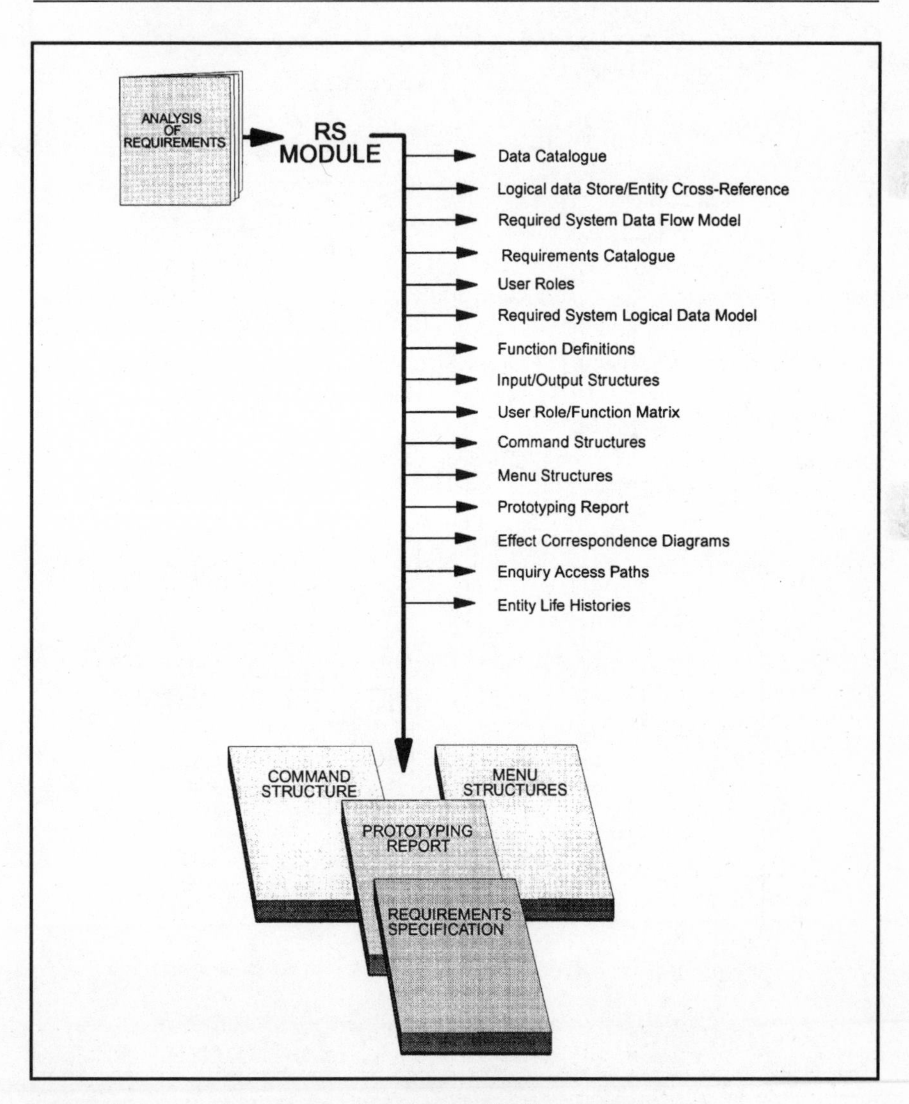

Figure 5.11 Requirements Specification(RS) Module Products

The objective of Requirements Specification is to take the Analysis of Requirements and review it in the light of the Selected BSO in order to add detail to the requirements of exact data, functions and events to be present in the future system. The end product is a Requirements Specification which is sufficiently detailed to form the basis for contractual agreement with a supplier of software, or to allow an in-house development team to proceed into design with a precise description of what is required. This module generates and updates various products which are noted in Figure 5.11.

5.6.1 Required System DFM

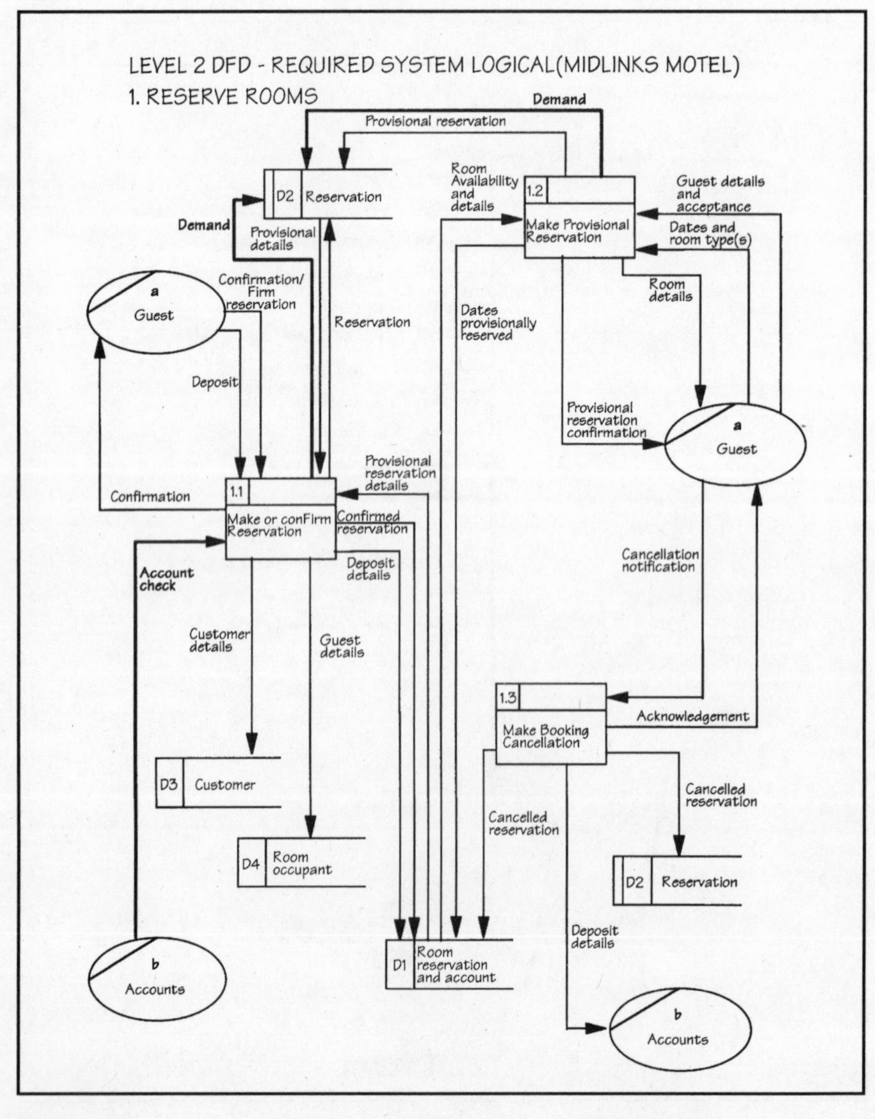

Figure 5.12 Level 2 DFD Required System Logical – Midlinks Motel

By reference to the Selected BSO, the Logical Data Flow Diagrams are modified, adding further business processes which were identified and eliminating processes which are not part of the Selected BSO. A level 2 diagram for the required system for Midlinks Motel is presented in Figure 5.12. Additional flows required in comparison with the current logical system have been emboldened for ease of identification. All levels of DFDs are amended to maintain consistency within them and the set now becomes the Required System Logical Data-flow Model. Elementary Process Descriptions for all new lowest level processes must also be created at this point and other supporting documentation maintained.

5.6.2 Required LDM

Similarly, the Logical Data Model is transformed in line with the Selected BSO into the Required System Logical Data Model by removing or adding entities, relationships and attributes as defined by the Selected BSO.

5.6.3 Requirements Catalogue

At this point other requirements (from the Requirements Catalogue) such as security and back-up requirements are overlaid upon the LDM. Requirements incorporated in the DFM and the LDM are cross-referenced to the Requirements Catalogue, which is updated.

5.6.4 Function Definition and I/O Structures

Functions are a central feature of SSADM (V4). Functions are distinct areas of processing, as defined by the user, which support the required system. Function Definition is not a precise technique, but rather the result of discussion and agreement with users as to what constitute separate units of processing to be performed at one time. It is useful to think ahead and equate functions with items which will appear on the users' menus, as function definition has great impact on the eventual dialogues of the future system. In the hotel scenario, functions such as: Reserve Rooms, Check-out Guest would probably be chosen by the users as discrete and complete tasks which they need to perform.

Consistency with the Required System DFDs is ensured by checking that each bottom-level process is assigned to at least one function. At this point the first steps are taken toward the development of Entity Life Histories (ELHs) and Effect Correspondence Diagrams (ECDs) by noting the **events** associated with each update function. Function Definition is revisited when ELHs and ECDs have been drawn and also as a result of Specification Prototyping (described below) because these invariably uncover further detail which must be taken account of in Function Definition. Volumes and service-level requirements are also laid down and documented as part of Function Definition.

For each function defined, an **Input/Output Structure Diagram** and **I/O Structure Description** are created, detailing the constituent data items of the inputs and outputs in the structure which the dialogue is required to follow. The data items within the I/O Structures can be checked against the data flows wherever a lowest-level data flow crosses the system boundary on the Required System DFM. At this point no error conditions are shown in the I/O Structure. These will be carried forward to Dialogue Design in the Logical System Specification (LS) Module. Figure 5.13 shows an I/O Structure Diagram for the Midlinks Hotel. I/O Structure Diagrams are drawn to Jackson structure diagram standards.

Jackson structure diagrams have three basic constructs: sequence, selection and iteration. These are represented as shown in Figure 5.13. Further conventions related to parallelism and to 'quit and resume' changes to the sequence are described in relation to Entity Life Histories later in this chapter.

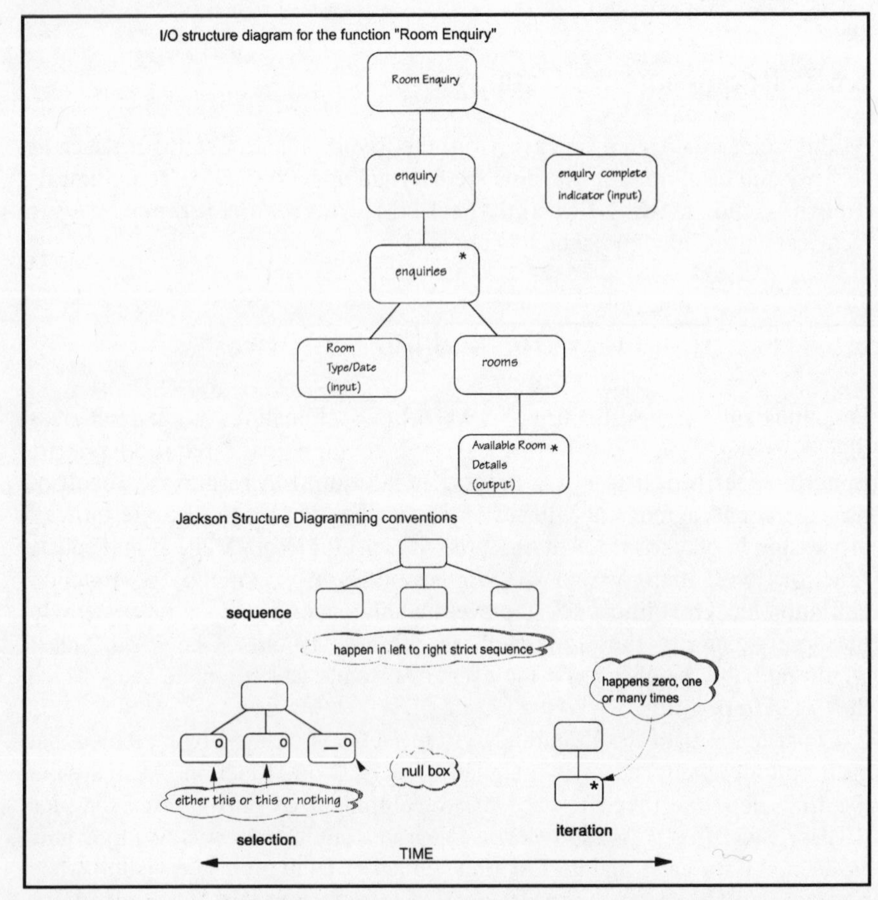

Figure 5.13 I/O Structure Diagram for the Function 'Room Enquiry' plus Jackson Structure Diagramming Conventions

In order to refine further the LDM for the required system, the I/O Structures and Descriptions related to functions are used as a source for Relational Data Analysis (RDA) to generate data models which can be compared with the LDS. This process follows the rules for normalisation described in Chapter 3. Any discrepancies are resolved by consideration of processing requirements together with user consultation and the Required System LDM is adjusted as necessary.

5.6.5 Specification Prototyping

Specification Prototyping in SSADM (V4) is the checking with users of the feasibility and detail of the critical functions which they have defined and of the outline dialogues produced from the I/O structures. Dialogues are not formally developed until the Logical System Specification (LS) Module, but this is one case where the apparent linear structure of SSADM would not be followed in practice, and a 'preview' of the critical dialogues to be prototyped would be developed during the RS module.

Specification Prototyping may be used particularly if the system is high risk, high cost or where there are many possible solutions to be evaluated by the users. Prototyping of the Requirements Specification is used to demonstrate to users that their requirements for the system have been fully understood and to generate feedback from them upon the style of the interface/menus/ command structures. Prototyping would concentrate on the dialogues identified by the users as critical to the success of the system. Iteration of the cycle of demonstration/prototype/interface/redesign leads to improvement of the Requirements Specification and updating of the Requirements Catalogue. One major benefit of prototyping is that the users are able to see what their system will look like (after all the interface and not the underlying processing **is** the system as far as they are concerned). Care must be taken, however, since showing a screen-based prototype can give the false impression that the system is almost finished! Additionally, prototyping may take place before the TSO has been agreed, and thus the final platform for the system will not have been fully established. An outline TSO, agreed at Feasibility Study time, if done, should have established the broad picture of the technical environment.

5.6.6 Entity/Event Modelling (ELHs and ECDs)

Now, holding a clearer definition of Requirements Specification, it is possible to delve more deeply into the processing outlined in the Function Definitions. **Update** events are identified to be used in Entity/Event Modelling. This forms the event-oriented view of the required system. Within Entity/Event modelling a matrix is constructed with the entities along one axis and the events along the other. Completeness can be confirmed by checking that every event affects one or more entities, every entity has an event which creates it and every entity has an event which deletes it. This matrix is the key to the development of Entity Life Histories (ELHs) and Effect Correspondence Diagrams (ECDs). The part-complete Entity/Event Matrix for Midlinks Motel is shown in Figure 5.14. Characteristically within the matrix three types of effects of the events upon the entities are represented: **C**reation, **M**odification and **D**eletion. The *columns* in the matrix show the events which affect each entity and are used to derive an Entity Life History for each entity whereas the *rows* show all the effects caused by the occurrence of a particular event which is then used to develop an Effect Correspondence Diagram for each event.

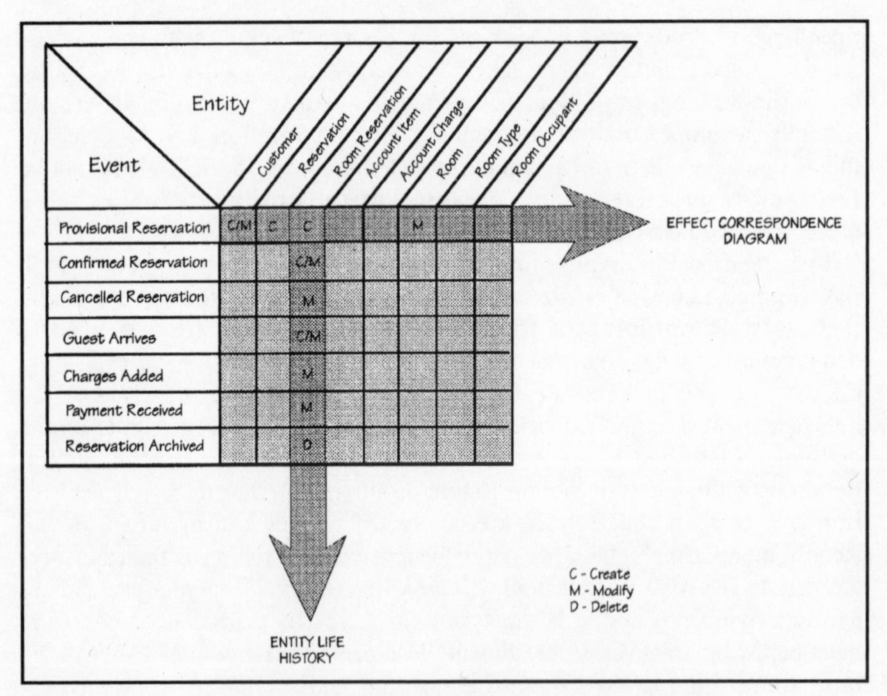

Figure 5.14 Entity/Event Matrix (part complete) – Midlinks Motel

5.6.6.1 Entity Life History (ELH)

Focusing attention upon the completed column for the entity 'Room Reservation' for Midlinks Motel, the ELH for Room Reservation is developed below in Figure 5.15. ELHs use a Jackson-like structure with sequence, selection and iteration constructs, as discussed in Chapter 3. These are augmented with parallel constructs (where sequence cannot be imposed) and 'Quit and Resume' notation. Quit and Resume are used to show that the processing which follows an event may not be the next event implied by the Jackson diagram sequence. An associated number is used to indicate which

Resume partners any particular Quit situation. These constructs are illustrated in Figure 5.15. Quit and Resume should be used sparingly. Rather like the 'go to' construct in programming, overuse leads to confusion. Its use should be restricted to cases where processing may *either* follow the normal pattern of the Jackson structure or under specific conditions may need to quit from that event and resume at another part of the structure (a *conditional quit*). To use a quit to imply a mandatory change of sequence is widely agreed to be a crime worthy of capital punishment!

The entity which is the subject of the ELH is drawn at the top of the diagram and, by considering the events which create, modify and delete it, the ELH diagram can be drawn. The diagram for the normal (most common) life-sequence for one occurrence of an entity should be worked out first, without quits and resumes and these used for abnormal events (e.g. sudden death within the system of the entity occurrence) if unavoidable. Operations related to specific aspects of processing are attached to the events and an operations list detailing those operations is appended and subsequently updated. The valid types of operation are detailed within the SSADM (V4) Reference Manual but some examples are given in Figure 5.15. The Room Reservation ELH also contains State Indicators, which are the unboxed numbers beneath the effect boxes. These are shown here but are produced as part of the LS module, and will be mentioned again in the section below dealing with that module. Their inclusion here serves to remind the reader that the modular breakdown of SSADM does not always imply a rigid modular isolation.

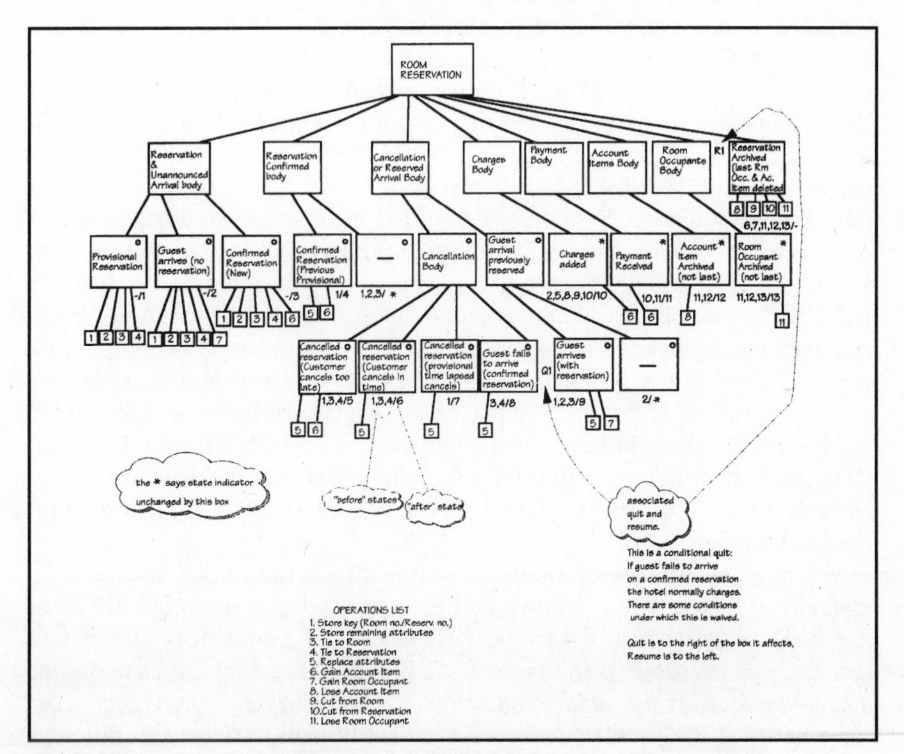

Figure 5.15 Entity Life History for Room Reservation – Midlinks Motel

In creating the ELH, it is likely that new events not foreseen in the Entity/Event matrix will be discovered. Archive Account Item and Archive Room Occupant are examples of this. These must be added to the matrix and their effects on other entities considered. They will also have an impact on Function Definition and the I/O Structures.

5.6.6.2 Effect Correspondence Diagrams

Consideration now returns to the Entity/Event matrix (Figure 5.14) and to the row for the event 'Provisional Reservation' to derive the Effect Correspondence Diagram (ECD) for Provisional Reservation. ECDs are intended to show all the entities affected by a particular event and, in essence, to show the access path for obtaining data and updating entity occurrences for that event. They can be checked by navigation through the LDS, but are also directly derivable from the effect boxes on the ELHs.

To construct an ECD, soft boxes (representing *effects*) are drawn for each entity which is affected by the event under examination and Jackson notation is used to indicate where an entity is affected iteratively (i.e. several occurrences of the same entity are accessed) or where an entity occurrence may be affected in two or more mutually exclusive ways (selection construct). Where an entity is affected iteratively a new box is introduced and annotated 'Set of' <entity>. Finally, one-to-one correspondences are indicated with a line with an arrowhead at both ends. Where the one-to-one correspondence is between two entities (rather than a *set of* box and an entity) one entity occurrence of each entity is read or updated. The ECD for Provisional Reservation for Midlinks Motel is shown in Figure 5.16. In Figure 5.16, one occurrence of CUSTOMER is read/updated and one occurrence of RESERVATION is created. Where the correspondence arrow joins an entity with a 'Set of' box, the first entity is read/updated once in relation to potentially many occurrences of the second entity. In Figure 5.16, a RESERVATION is created, and a set of possibly many ROOM RESERVATIONs is also created. The ECD also shows the input attributes required by the event coming in to the entry point entity which in this particular case may be an Account No. or New Customer Details, the processing of these alternative inputs being slightly different and accounted for by the selection boxes (*Current Account* CUSTOMER and *Transient* CUSTOMER) shown on the diagram. The ECD is later used in the LS Module to develop the Update Process Model (UPM) for Provisional Reservation.

Where an occurrence of an entity will be affected in one of two or more mutually exclusive ways, this would have appeared as a **qualified effect** on the ELH. Reservation Confirmed (New) and Reservation Confirmed (Previously Provisional), shown on the ELH in Figure 5.15 are examples of qualified effects of the same event. Where more than one occurrence of the same entity will be updated, we have an entity being affected in different **Entity Roles**. This would happen if, for example we transferred a guest from one room to another. The ROOM entity would have two affected occurrences: the room transferred from and the room transferred to. For further discussion of these features, the reader is referred to the SSADM manual.

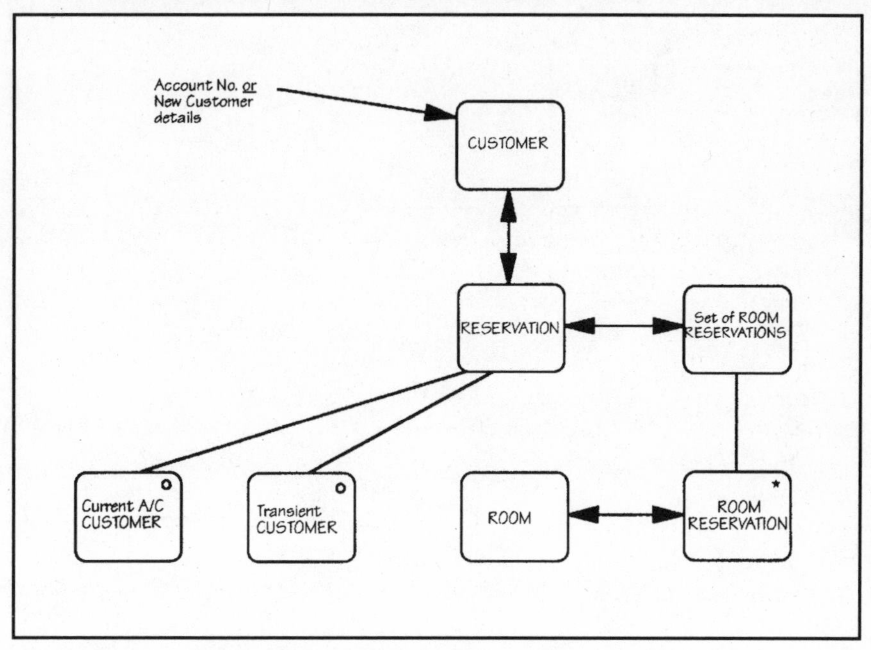

Figure 5.16 Effect Correspondence Diagram for Provisional Reservation – Midlinks Motel

5.6.7 Enquiry Access Paths (EAPs)

Attention is now given to Enquiry Functions which were initially identified by discussion with the users and subsequently included in the Requirements Catalogue. At this point an Enquiry Access Path (EAP) for each enquiry is developed ready to be used to define an Enquiry Process Model in the LS module. An EAP is, as implied by the name, an access path through the LDS to obtain data to satisfy a specific enquiry. By reference to the Required LDM and the enquiry, the entities necessary to gather all the required information are selected for the basic structure. The data items which invoke the enquiry, via the entity which forms the entry point of the enquiry are determined with due reference to the I/O structure diagram, Relational Data Analysis and Entity Description documentation. The development of the EAP then follows a similar pattern to that described above for ECDs with accesses being represented in this case by single-headed arrows and the familiar Jackson-like structure appearing again. Figure 5.17 shows the EAP for the Midlinks Motel enquiry, 'Room Enquiry'. This has input data of several room types and dates for which availability is sought. The I/O Structure diagram was presented in Figure 5.13.

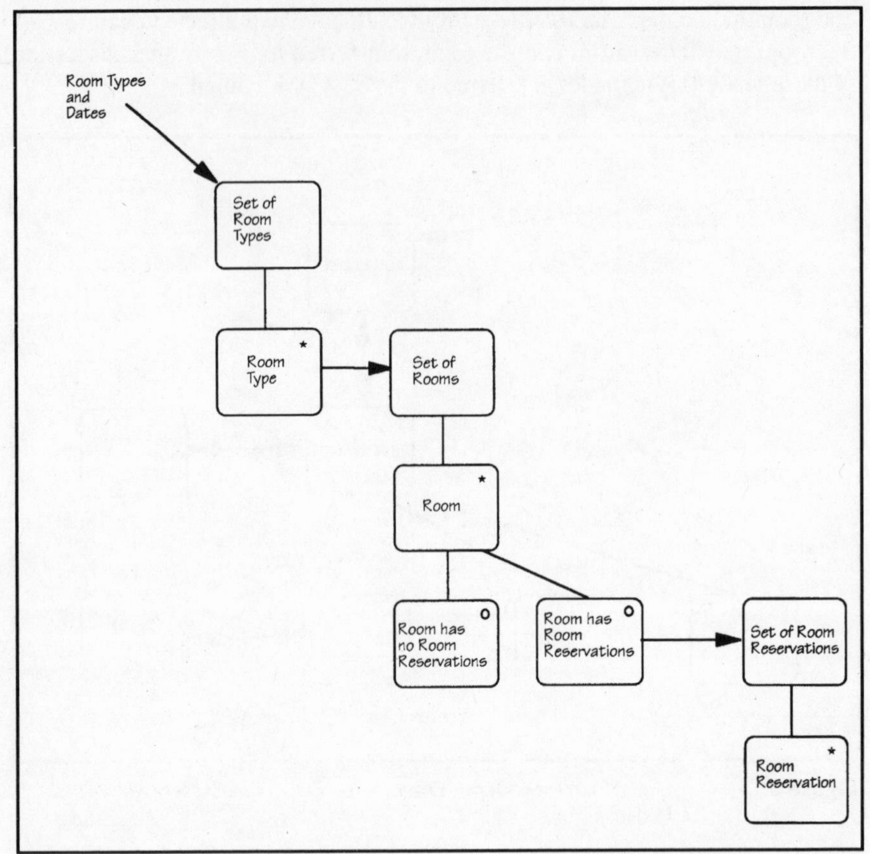

Figure 5.17 Enquiry Access Path, Room Availability Enquiry – Midlinks Motel

5.6.8 Concluding the Requirements Specification Module

Any implications of the development of ELHs, ECDs and EAPs upon other SSADM products must be considered with potential updating of the Requirements Catalogue, the LDM and Function Definitions. At this point the Required System LDM is modified to hold entity/relationship volumes so that **First Cut Physical Data Design** can be undertaken within the PD module.

Finally System Objectives are revisited and refined or confirmed. The Requirements Catalogue and other products of the module are reviewed in order to ensure that the Function Definitions and the Required System LDM are a complete statement of the requirements which can be carried forward into the next module to form a sound foundation for the new system. The resulting Requirements Specification product may form the basis for contractual agreement with an outside software supplier or with an in-house development team. It is also the 'sign-off' document for the users' acceptance of the future system.

5.7 Logical System Specification (LS) Module

The LS Module comprises two parallel stages: Technical System Options (TSO) and Logical Design (LD). The intermediary and major products of the LS module are detailed in Figure 5.18.

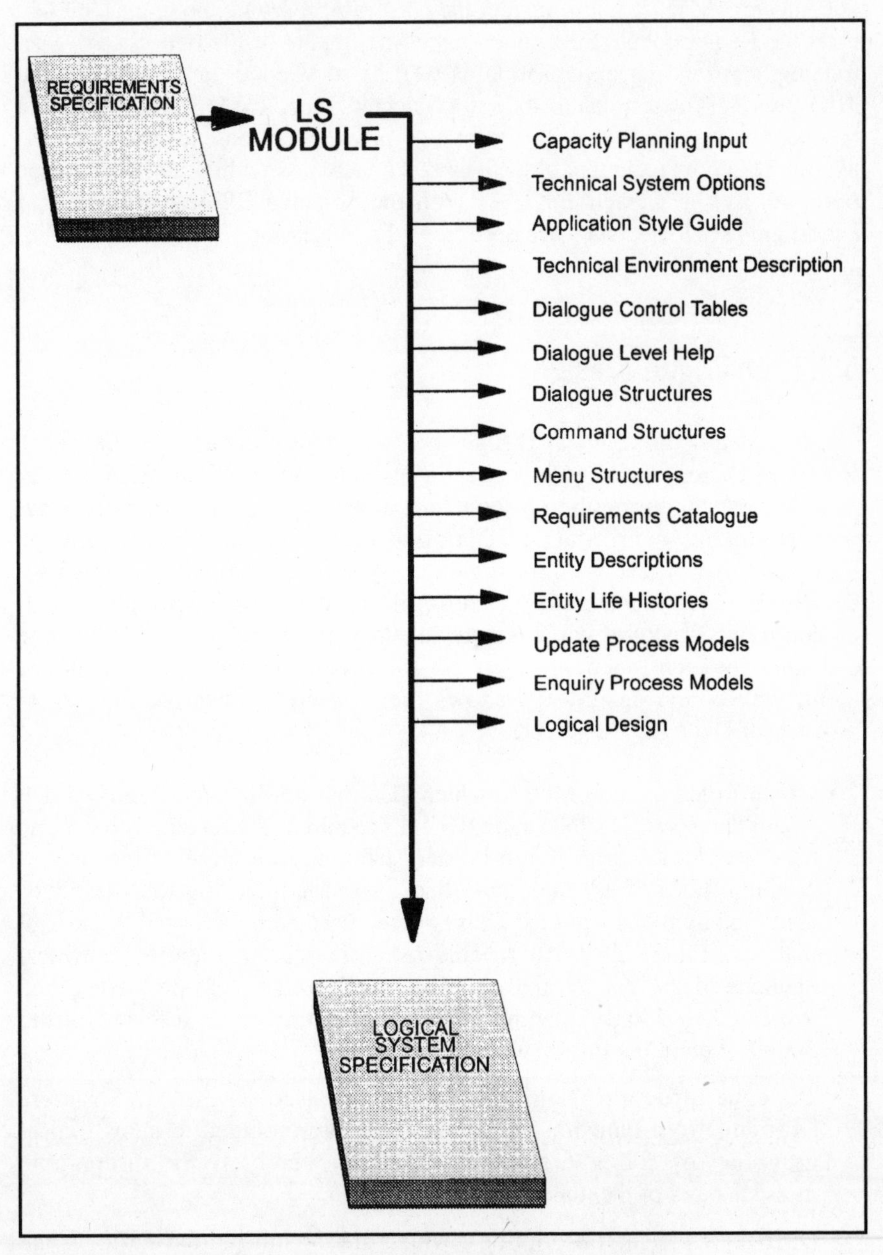

Figure 5.18 Logical System Specification(LS) Module Products

5.7.1 Technical System Options

Technical System Options (TSO) is similar in procedure to Business System Options discussed previously but is concerned with **how** the Required System is implemented and not purely with **what** the system will do (BSO). Up to six technical options are identified which fall within the constraints of the Requirements Catalogue, the Selected BSO and any other strategy documents. These options consider the hardware, software, how the TSO will meet the Requirements Specification and the impact of the option upon user training, staffing and organisation as well as cost/benefit analysis. Potential TSOs are presented to users so that a Selected Technical System Option can be chosen from them or a hybrid of two or more options developed as the chosen TSO. The technical objectives and constraints within the PID must be observed by the chosen TSO. As with the Selected BSO the reasons for choosing the Selected TSO are documented at this point.

5.7.2 Dialogue Design

The first step of the Logical Design (LD) Stage is to define user dialogues. However, Dialogue Design is an activity which started way back in the RA module (or even during the FS module if a feasibility study was performed) with Dialogue Identification. Dialogue Identification begins with the identification of target users for the future system and the building of a User Catalogue. Function Definition, during Requirements Specification (RS) produced I/O Structures which are picked up again in the LD stage as outline dialogues for each function. Figure 5.19 illustrates the products of dialogue identification and design and shows their interrelationships. The major products of Dialogue Design are:

- a User Role/Function Matrix, which identifies all dialogues required (i.e. where the X's or C's fall) and all menus required. A user role is a generic title for a collection of job holders who share a large proportion of common tasks. Each user role has a menu containing the dialogues identified by the X's or C's. These are represented by a row of the central matrix in Figure 5.19. An **X** indicates a dialogue. **C** indicates a critical dialogue of the system and one which will probably be prototyped and subjected to detailed timing and storage estimation exercises during optimisation in the Physical Design module;
- Dialogue Structure Diagrams, which are based on the I/O Structure Diagrams from Function Definition. They are modified to show logical groupings of dialogue elements (LGDEs) which are the inseparable message pairs of the dialogue;
- Dialogue Element Descriptions, which are documentation defining the data items within the dialogue;

- Dialogue Control Tables, identifying the navigation paths within the dialogue;
- Menu Structures, which are Jackson-type diagrams indicating the hierarchy of dialogues within each menu;
- Command Structures, which are documentation indicating the paths that can be taken when a dialogue ends;
- Dialogue Level Help, which defines contextual, navigational and job-related help to be provided for each dialogue.

Dialogue design does not attempt to provide the design for physical screens as this is hardware and software dependent. The dialogues produced are logical but may be fed into specification prototyping for refinement.

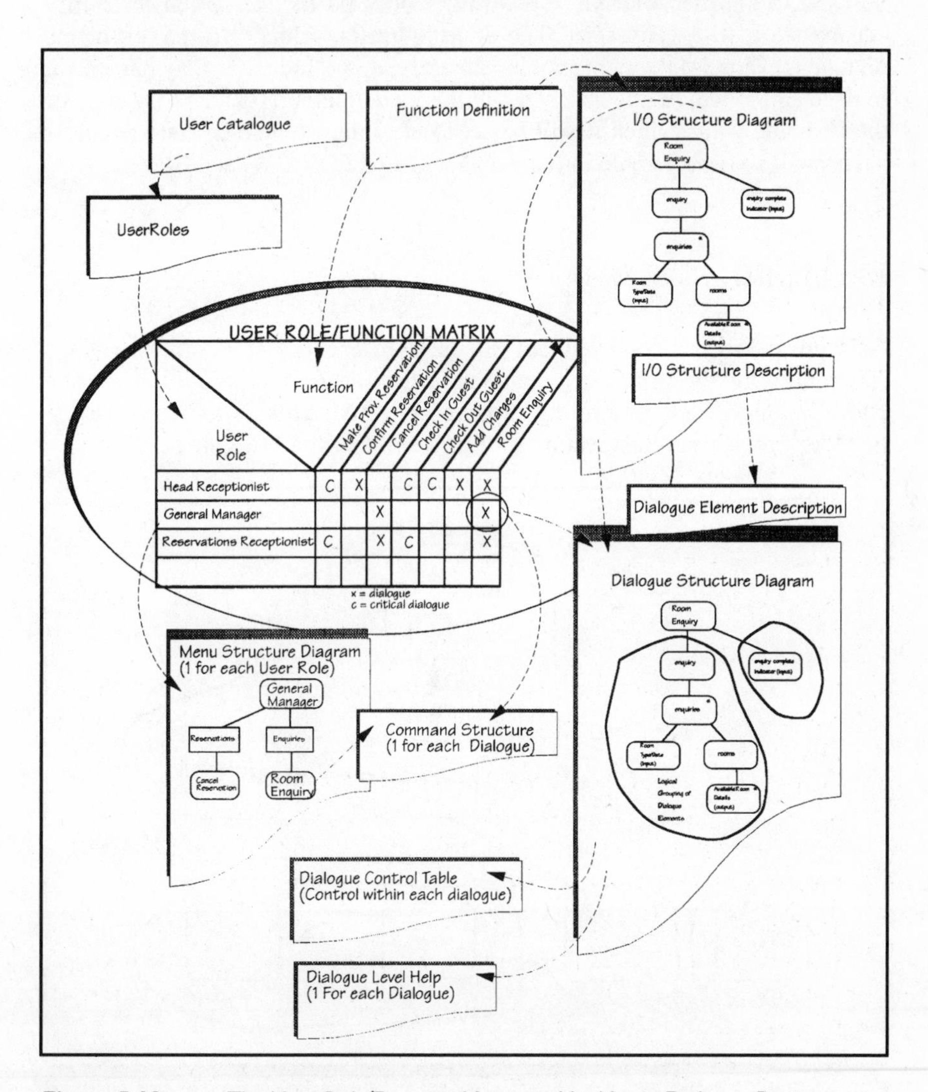

Figure 5.19 The User Role/Function Matrix and Its Use in Dialogue Design

5.7.3 Update Process Models and Enquiry Process Models

Logical Design progresses the development of the functional components of the Requirements Specification toward implementable pieces of processing. Update and enquiry functions are each considered separately and Update Process Models (UPMs) and Enquiry Process Models (EPMs) are developed for each of them. Examples of Update and Enquiry Process Models are presented in Figures 5.20 to 5.25.

It is also at this point that the ELHs are allocated State Indicator values as shown in the ELH for Room Reservation, Figure 5.15. These indicators are attributes of the entity which are updated by events and which therefore can be used to confirm that any updating is only performed when the entity occurrence is in a valid state. The State Indicator value(s) to the left of the oblique (/) show(s) the permissible value(s) that the indicator may have prior to the event to which it is attached and the State Indicator value to the right of the (/) is the value which it will be set to after the event. The state indicator values will be incorporated into the operations on the UPMs and EPMs.

5.7.3.1 Update Process Model

An Update Process Model is developed from each Effect Correspondence Diagram by firstly grouping the effects of the event which are in one to one correspondence, and encircling those correspondences as indicated in the ECD for Provisional Reservation for the Midlinks Motel (Figure 5.20).

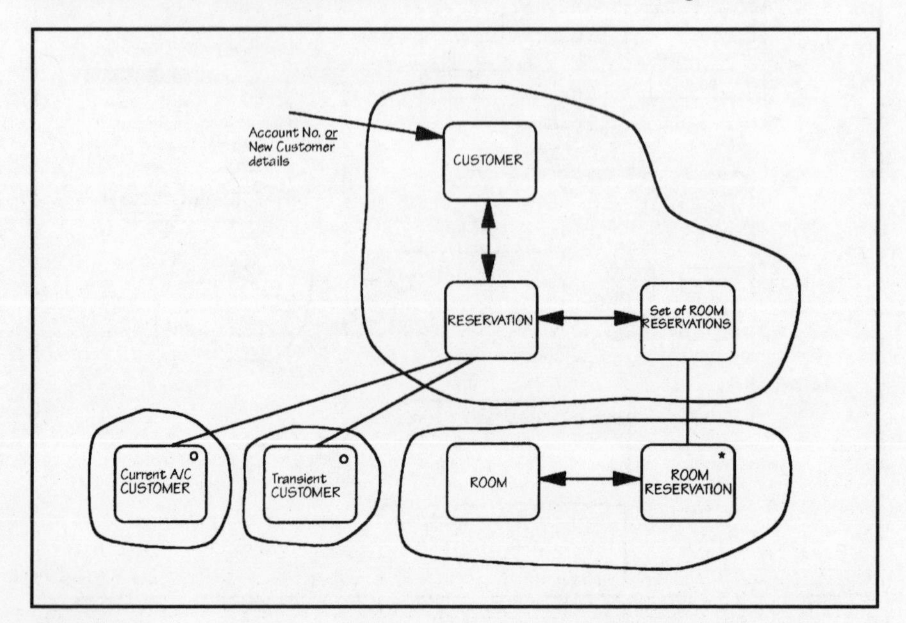

Figure 5.20 Effect Correspondence Diagram, with Process Model Grouping

Each of these groups of processing forms a discrete unit which will become a named process in the Update Process Model. Initial operations, already identified in the production of ELHs, are copied to the UPM and to these may be added reads, writes and creates so that processing is more fully specified on the UPM. Each discrete unit from the ECD is drawn as a process box on the Update Process Model, which uses Jackson notation. Additional boxes may be added to conform to Jackson Diagramming rules. Finally operations (including the operations list) and wording to express the conditions for selection/iteration constructs are added to the Update Process Model. State indicator settings from the ELHs are included in the operations, and a 'fail if' operation added to accommodate conditions where the state indicators are found to be invalid. The UPM Provisional Reservation for Midlinks Motel is represented in Figure 5.21.

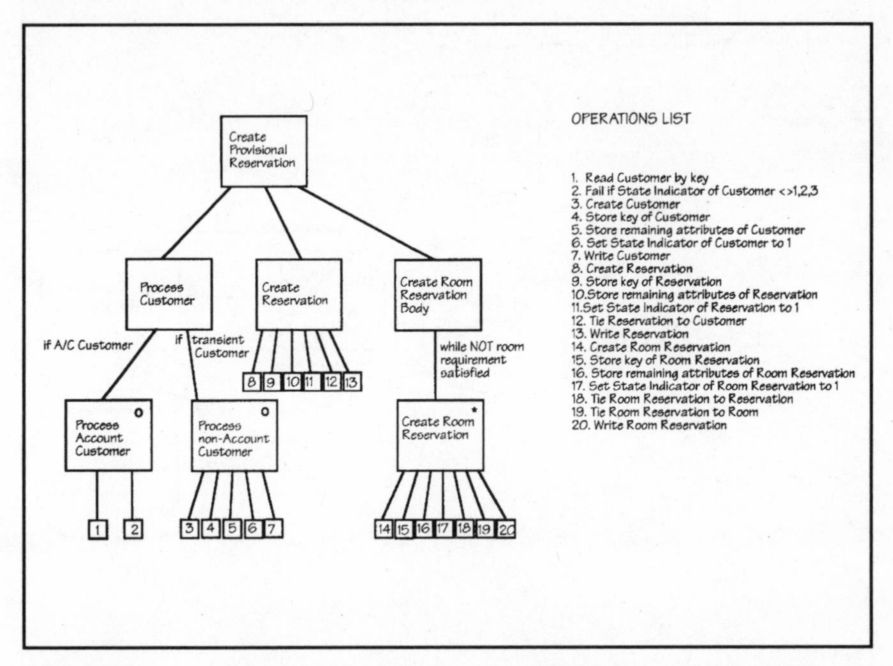

Figure 5.21 Update Process Model Provisional Reservation – Midlinks Motel

The UPM only moves through to the PD module after user and analyst have worked through the model to confirm that what is required is accurately described by the model.

5.7.3.2 Enquiry Process Model

The process for development of Enquiry Process Models (EPMs) is similar to that for Update Process Models, described above. However, additional attention has to be paid to the sequence and structure of the required output. For this purpose, the I/O Structure for the enquiry is used.

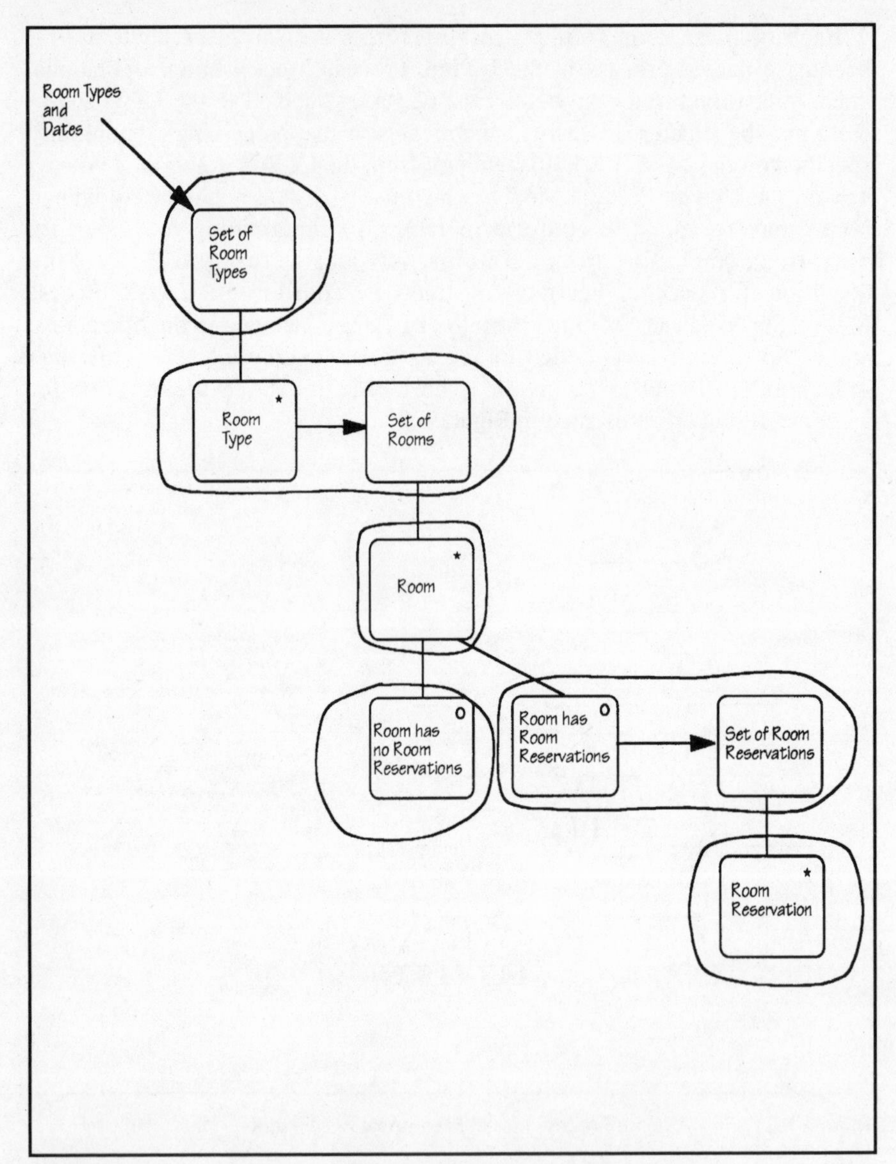

Figure 5.22 Enquiry Access Path, with Process Model Grouping

Elements of the EAP are grouped where accesses (correspondences) are shown by a single-headed arrow. These discrete groups will form boxes in a Jackson-style Input Structure Diagram. This is matched against the Output Structure, derived from the I/O Structure diagram with inputs removed. This process is illustrated in Figure 5.23.

A combination of these two structures allows the initial EPM to be built, as shown in Figure 5.24. The structure again is of Jackson type with the introduction of any further structure boxes being made to comply with Jackson diagramming conventions. The final EPM for the Room Enquiry from Midlinks Motel is drawn in Figure 5.25.

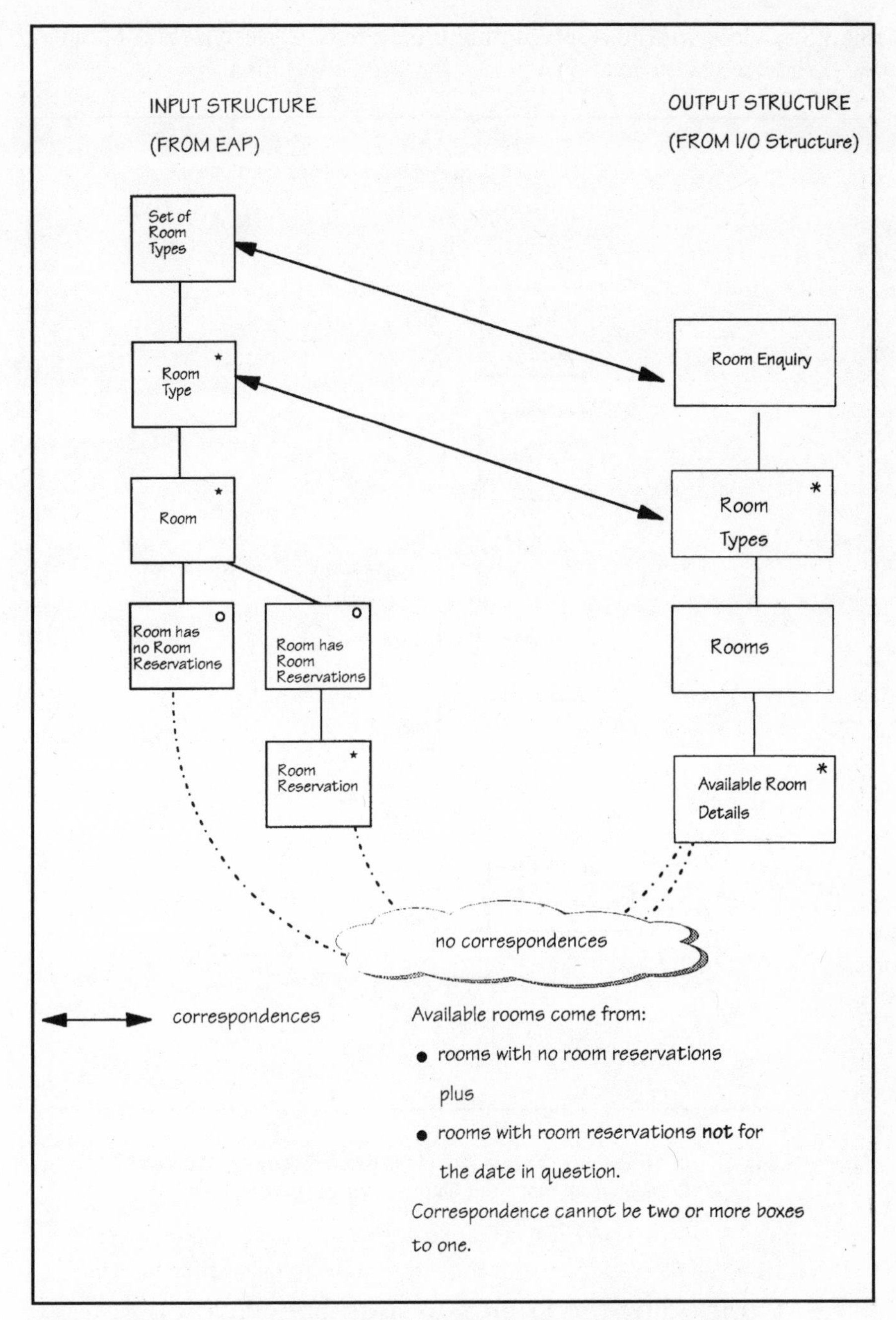

Figure 5.23 Matching of Input and Output Structures of Room Enquiry

Incompatibilities between the input and output structures can occur if, for example, the groupings in which data is captured do not match the groupings in which it is required to be displayed. This is known as a *structure clash* and can create the requirement for additional processing (e.g. sorting). Structure clashes are further defined in (Jackson 1975). Structure clashes identified in the LS module will be resolved during the Physical Design (PD) Module, where any necessary additional processing will be specified.

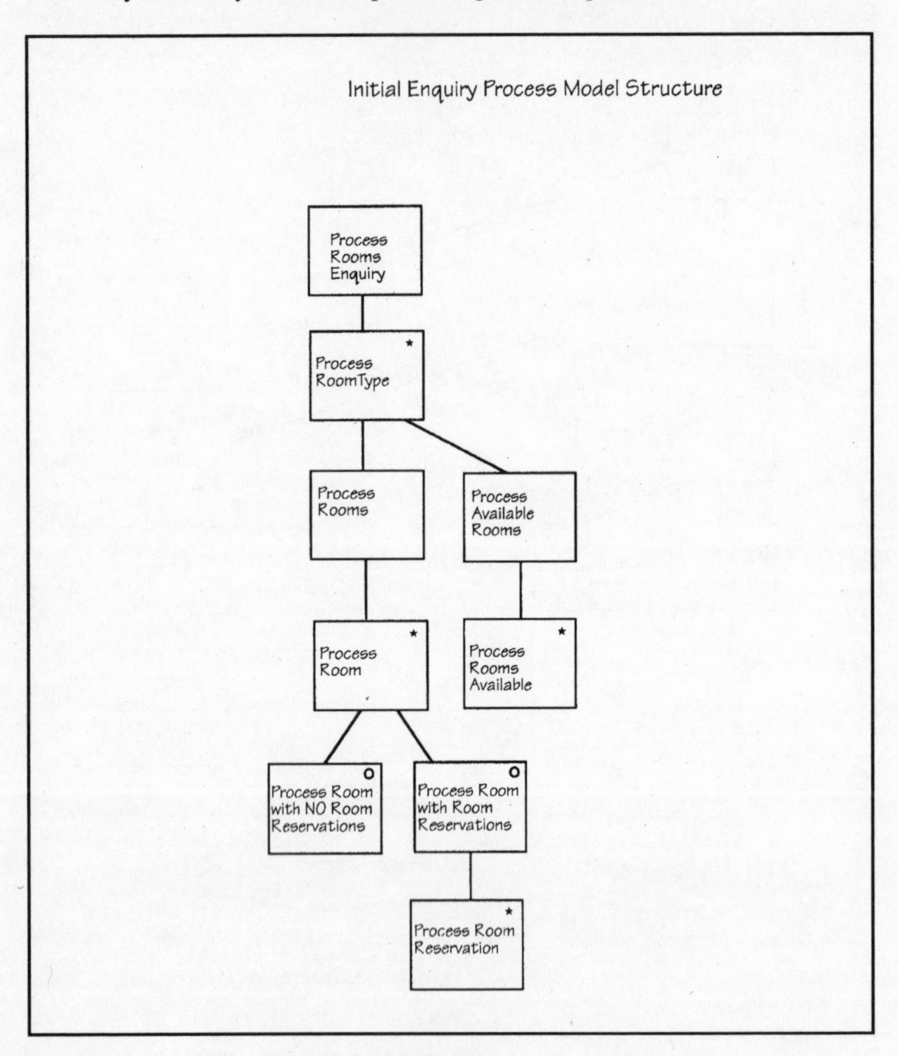

Figure 5.24 Initial Enquiry Process Model for Room Enquiry – Midlinks Motel. Derived from Input and Output Structures combined

5.7.4 Concluding the Logical System Specification Module

The EPMs and UPMs are taken forward to the Physical Design module to form the basic units of processing. Also essential to this process are the Function Definitions and the products of Dialogue Design. The LDM will give the structure on which the Physical Data Design will be built. The selected TSO gives the Technical Environment Description and physical constraints. The Requirements Catalogue remains the central reference point for functional and non-functional requirements, including objectives for reliability, performance, etc.

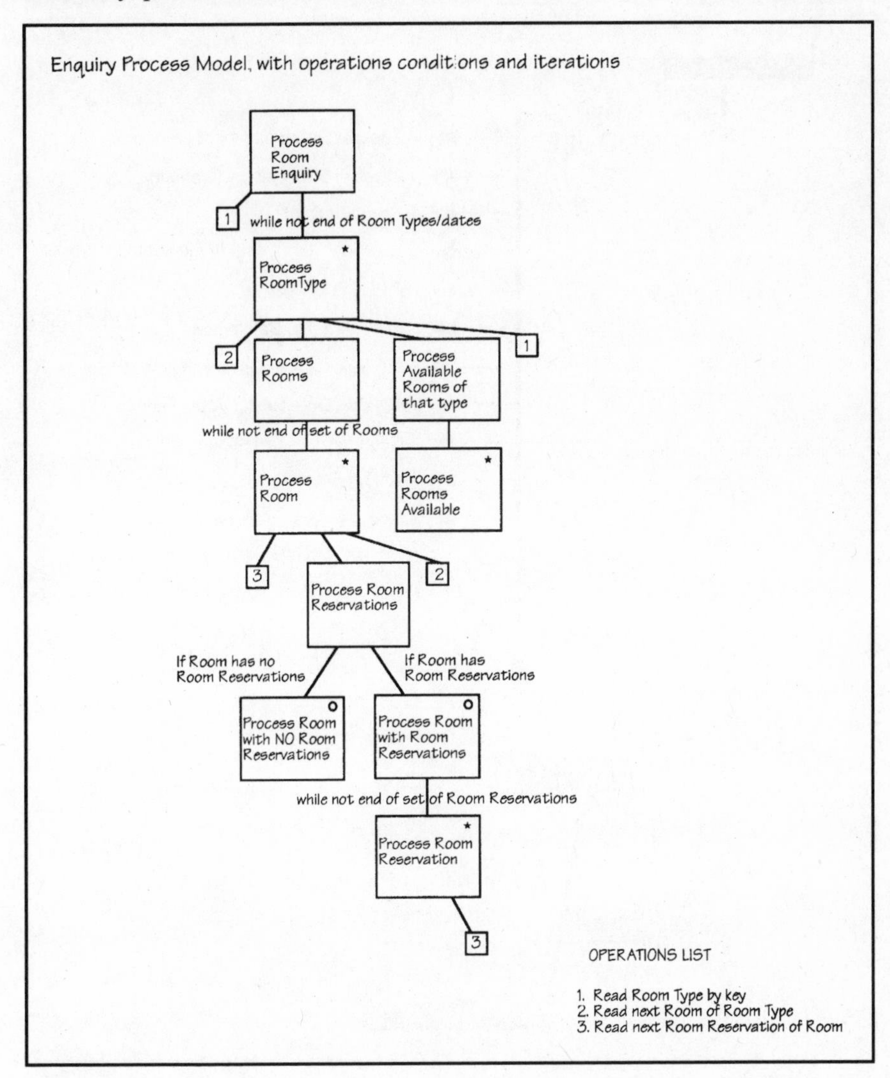

Figure 5.25 Enquiry Process Model for Room Enquiry with Iterations and Conditions

5.8 Physical Design (PD) Module

The Physical Design Module specifies the physical data, processes and I/Os within the framework of the chosen physical environment and any other constraining factors. The PD module is summarised in Figure 5.26.

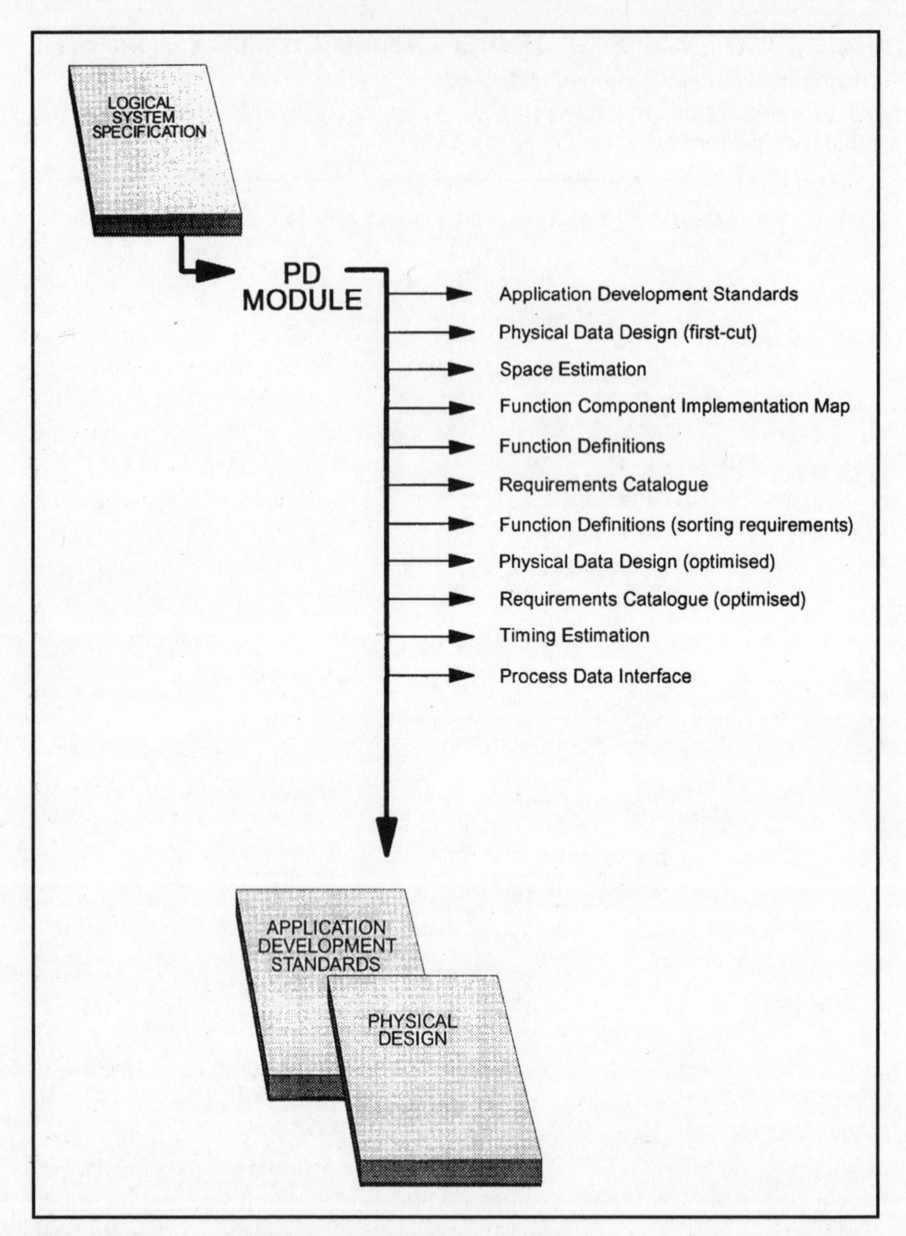

Figure 5.26 Physical Design (PD) Module Products

The purpose of the Physical Design (PD) module (from the SSADM (V4) Reference Manual) is:

- to specify the physical data, processes, inputs and outputs using the language and features of the chosen physical environment, incorporating installation standards.

At the end of the PD module, everything should be in place specifying precisely how the application is to be constructed. The integrity of the

Logical Design is preserved as far as possible within constraints and requirements of performance, storage, etc.

The PD module steps are outlined below.

5.8.1 Prepare for Physical Design

The first activities in the PD module are concerned with planning the approach to physical design (the physical design strategy) and investigating and documenting the performance features and requirements of the target hardware and support software (operating system, DBMSs, programming languages, screen handlers, etc.) The physical design strategy will be determined largely within the Technical Environment Description (TED) which was produced during the TSO stage. The organisation's Installation Standards (the Installation Style Guide) and standards agreed for this application (the Application Style Guide) will provide necessary guidelines.

5.8.2 Create a First-cut Physical Data Design.

In the light of the target Database Management System the Required System LDM is converted into a First-cut Physical Data Design using a number of guidelines presented in the SSADM manual, and based on experience and good practice.

5.8.3 Create a Function Component Implementation Map

In this step, we detail how functions will be implemented and produce a Function Component Implementation Map (FCIM). The FCIM is built to show how all logical processing will be physically implemented, e.g. C++, Visual Basic, COBOL, 4GL and the groupings in which they will be physically constructed. The functions defined in Function Definition are the basis for physical process design together with the EPMs, UPMs and the dialogues specified in Dialogue Design. The UPMs constitute 'success units' which must succeed or fail in their entirety.

5.8.4 Optimise the Physical Data Design

The first-cut design is evaluated in the light of volumetrics to estimate whether space and performance objectives can be met. A cycle of design, test, optimise, review is followed until a satisfactory design is reached.

5.8.5 Complete Function Specification

In this step, we complete the specification of functions, combining the logical processes into physical programs or run-units. Functions requiring procedural code are further specified and structure clashes discovered in the LS module are resolved. The FCIM is updated to reflect any additional processing specified.

5.8.6 Consolidate the Process Data Interface (PDI)

This step completes the definition of the mapping between physical data structure and the logical data structure as seen by programs. The FCIM has a view of data as structured in the LDM. The Process Data Interface (PDI) specifies the mapping between physical and logical views of the data. This mapping may be handled by the database management system. Alternatively, it may require code construction and if so this must be specified. The aim is to prevent the processing from having to change in line with the physical data structure, or the data structure from having to mirror what the processing expects from the LDM when this is not able to meet performance or storage objectives. By preserving data and function independence in this way we can retain the benefits of the logical design (which has low coupling between processing units and high cohesion within them (Yourdon & Constantine 1979)) and thus produce a flexible, maintainable system.

5.8.7 Assemble the Physical Design

The products of the PD Module are finally assembled for management decision and, subject to user agreement, hand-over to the construction team. The output should be sufficient for construction to begin without further specification.

5.9 Summary of SSADM (V4) and Onward to SSADM (V4.2)

At the beginning of this chapter we described, at a high level, the framework within which SSADM (V4) operates. This framework was then examined in more detail, interspersed with specific techniques at the relevant points. We have in no way attempted to be exhaustive in our coverage and for this the reader is advised to refer to the SSADM (V4) Manual. We will, however, conclude this chapter by summarising the method and its major products in an informal way, to show where these products are produced and used and to attempt to show the interactions between products, where one product is produced as the result of another. This summary is shown in Figure 5.27.

The structural model of SSADM (V4) is tried, tested and well-established. However, there has recently been (May 1995) a release of SSADM beyond V4. Subsequent releases will be generally termed SSADM V4+. This release is specifically identified as SSADM V4.2. It represents a revised view, in line with Euromethod and present-day implementation requirements. This view is now overlaid on to V4. It does not replace or invalidate SSADM (V4) but rather changes the perspective and adds flexibility. The next paragraph considers SSADM (V4.2) in more detail.

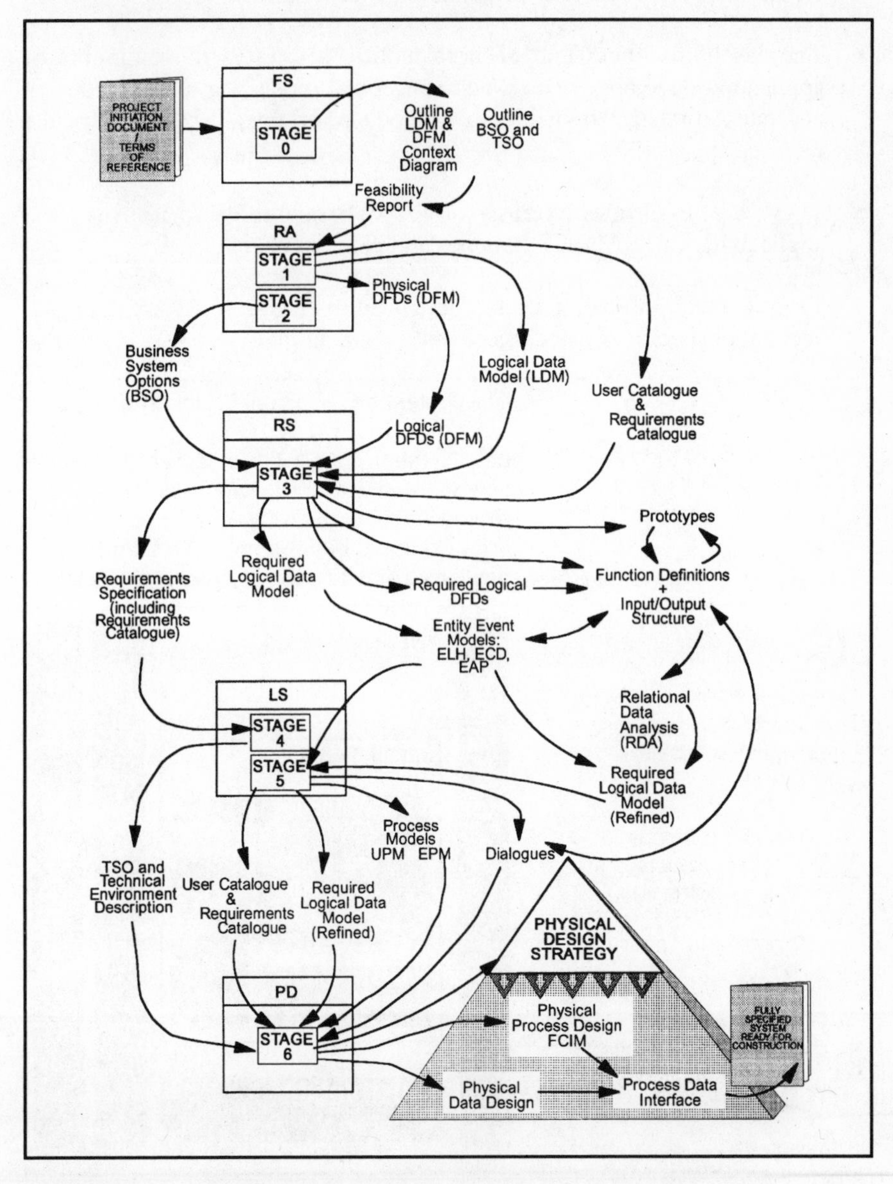

Figure 5.27 Summary of the Stages and Major Products of SSADM (V4)

5.10 SSADM Version 4+

During the production of this book, CCTA announced the launch of Version 4.2 of SSADM4+. This is a summary of the significant developments in the new version:

- SSADM4+ Version 4.2 has reduced emphasis on the structural model (which still provides a default starting point) and increased emphasis on the System Development Template (SDT) which re-positions SSADM within an overall template for application development.
- The 3-schema specification architecture separates the elements of application development between conceptual modelling, external design and internal design to enable a flexible and portable specification of the required application.
- There are technique extensions and enhancements to improve the applicability of SSADM techniques and deliverables.
- There is much more emphasis on the need to customise SSADM to take account of specific project objectives and situational factors.

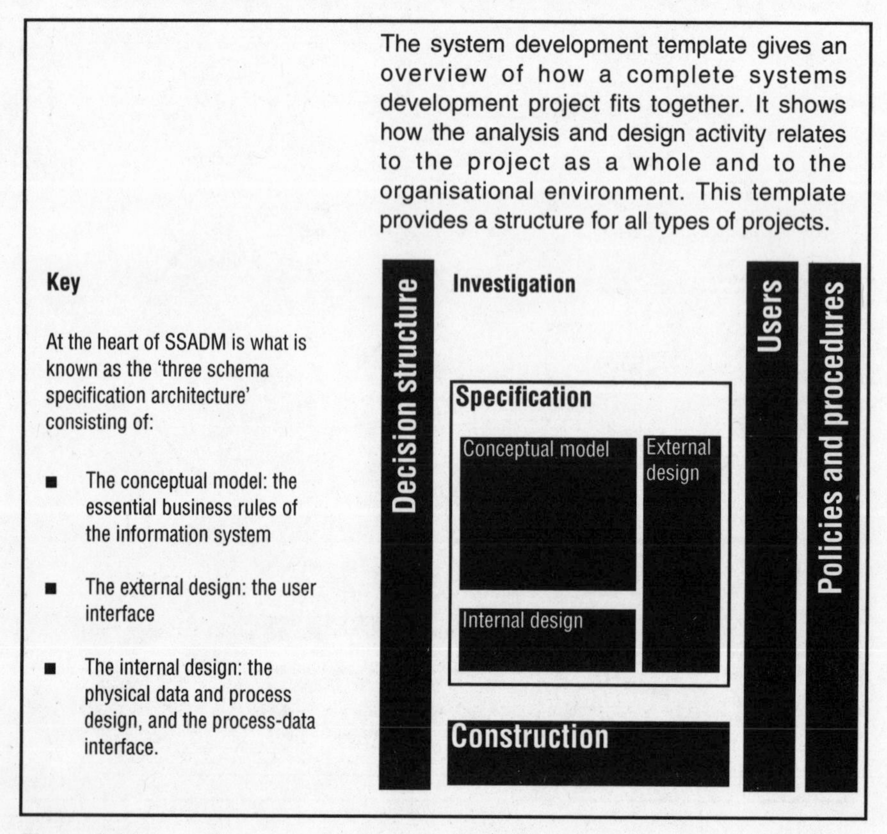

Figure 5.28 The System Development Template

- Business Activity Modelling and Work Practice Modelling interfaces are included. The former extends the front-end of SSADM to allow integration with non-IT business requirements and process improvement issues. The latter allows integration with organisational and job design issues. The extensions enable a more complete mapping between SSADM and Euromethod which is strongly influenced by systemic methods such as Soft Systems. Although SSADM4+ Version 4.2 provides examples of how Soft Systems can be used to provide these two extensions, the broad principles are described in general terms to enable other methods to be used.

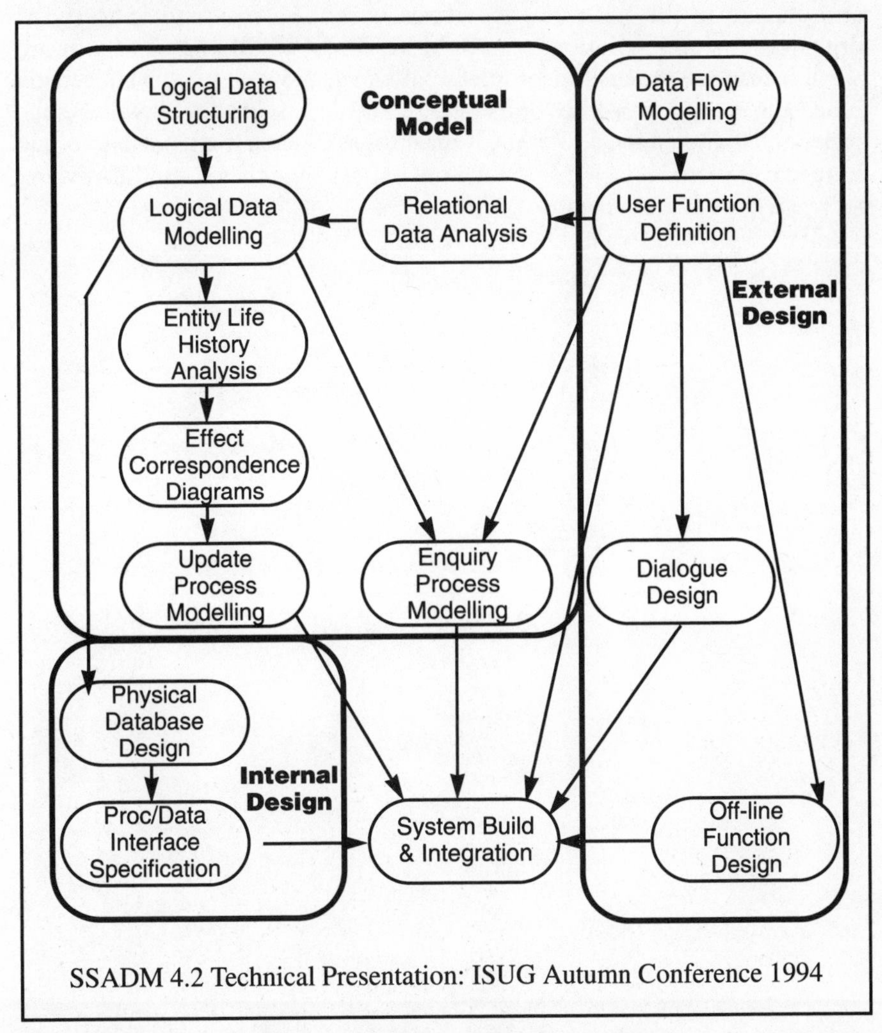

Figure 5.29 SSADM (V4) Techniques in the 3-Schema Specification Architecture

- Note that the term SSADM4+ Version 4.2 refers not only to the enhancement to SSADM but also to the integration with surrounding ISE Library volumes which have resulted from a major CCTA research and development programme. The issues examined include distributed system design, client server design, achieving reuse using object orientation and application partitioning.

SSADM4+ Version 4.2 was published by NCC Blackwell in May 1995.

5.11 Summary

SSADM is a well-documented method that is subject to continuing review and change in the light of experience of its use in a wide variety of organisations and situations. SSADM Version 4+ will encompass future version changes, the first major release of which is Version 4.2. This version is accompanied by a new set of materials. However, it does not invalidate the material within Version 4. Rather, it repackages it into a more flexible format to render it more suitable for use with a variety of situations and hardware/software target environments.

6

Information Engineering (IE)

6.1 Overview

Before looking in detail at Information Engineering we shall consider its background, its objectives and underlying principles.

6.1.1 Background

Information Engineering first emerged in the late 1970s and early 1980s, from the co-operation of Finkelstein and Martin. The name 'Information Engineering' was coined to reflect the fact that the approach was modelled upon engineering principles. The method is of a somewhat varied nature as, from those early beginnings, a number of divergent interpretations of the method have developed to the present day. Indeed with the introduction of a professional qualification of proficiency in Information Engineering, administered by the Information Systems Examinations Board (ISEB), it has been acknowledged that there are variations which are accepted as the Information Engineering Method. Within the context of this text we shall therefore offer a generic version of IE, rather than that of any one specific vendor, in order to present the method to the reader.

6.1.2 Objectives and Principles

Information Engineering has two basic objectives:

- to develop integrated systems which support real business needs defined by the business objectives and strategies;
- to deliver information systems which meet the needs of the business at the *time of delivery* but in a framework which allows flexibility for future change.

Fundamentally underlying IE is the assertion that **data** is central to any information system and that while the values of the data may change the basic structure of and types of data within an organisation's information system will remain reasonably stable. What will change is **what is done with that data,** that is the processing which operates upon the data. The approach is top-down with attention focused firmly upon the needs of the business.

The key principles which underpin the method are:

- business orientation;
- architectural approach;
- multiple development paths;
- staged approach;
- user involvement;
- graphical representation of deliverables;
- CASE tool support;
- investment protection.

These are discussed in more detail below.

6.1.2.1 Business Orientation

Models of the business, which support the business objectives, strategies and management information needs, form the basis upon which systems are developed. By focusing upon the business needs, the constraints which the existing systems and existing structure of the organisation would otherwise impose upon the development of new systems are excluded.

6.1.2.2 Architectural Approach

Organisation-wide architectures of business activities and data form a framework for systems development. This framework affords a number of benefits. Firstly, the systems boundaries and interdependencies are defined (scoped), clarifying the scope of a project. Secondly, critical systems are identified and can thus be developed first (prioritised). Thirdly, by encouraging a full understanding of the system size, a system can be developed in its entirety or sub-divided into smaller systems. Finally, borne out of the 'top-down' approach, the systems which are developed can be formed in a co-ordinated manner with cross-system integration and consistency built in.

6.1.2.3 Multiple Development Paths

IE has a number of development paths each of which is appropriate for a particular circumstance. These are illustrated in Figure 6.1 and are discussed in overview detail below:

- **Information Engineering** (complete path). This will form the bulk of this chapter and is covered below;

- **Rapid Application Development** (RAD) forms an accelerated route aimed at development of systems which are of limited scope. RAD involves specified stages of Requirements Planning, User Design and Rapid Construction stand-alone systems can be accommodated. Where a small system is to be developed as part of an integrated system an outline Business Area Analysis is followed by User Design and Rapid Construction. Essential to the RAD approach is the use of an integrated CASE tool, with code-generation capabilities.

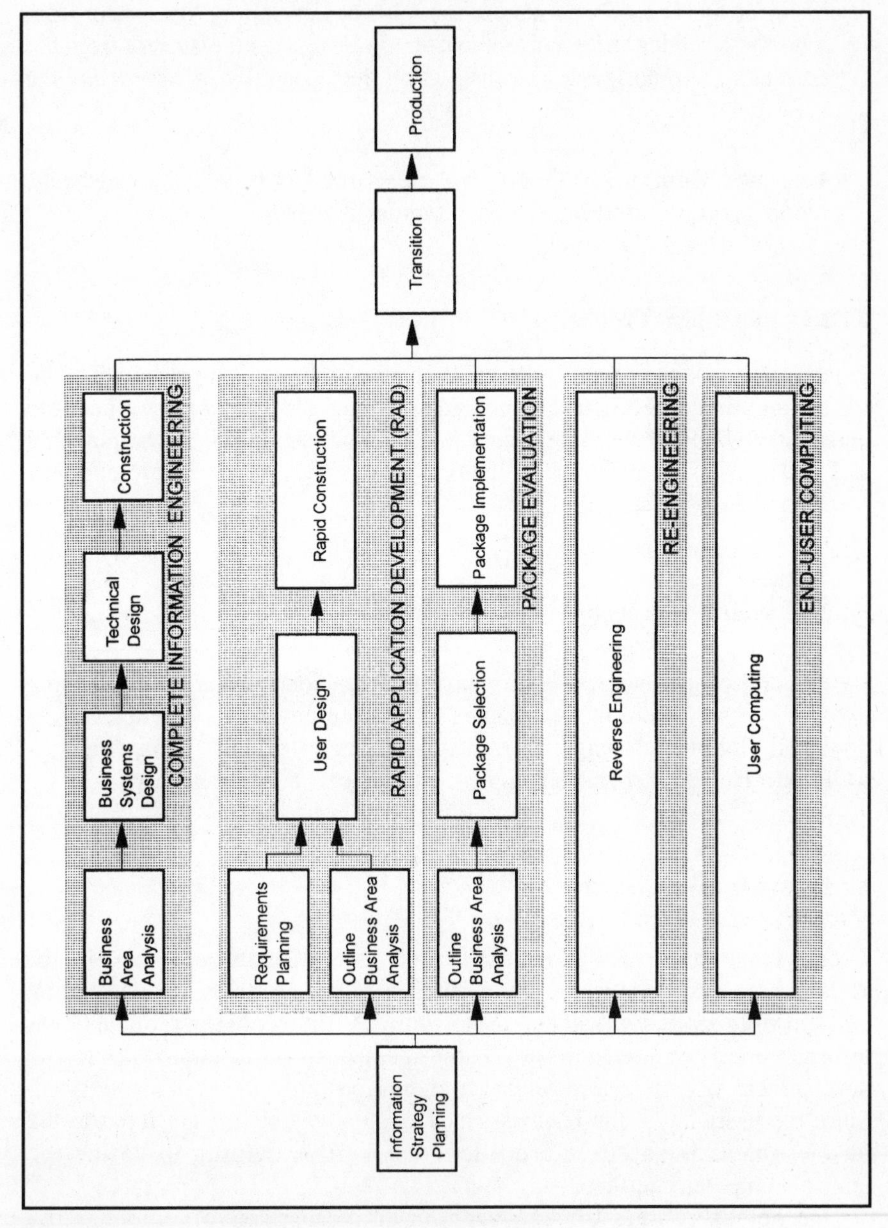

Figure 6.1 Multiple Development Paths Within IE

- **Package Evaluation** is supported by IE. An outline Business Area Analysis is carried out to develop Entity and Process Models against which the available packages can be compared. Other stages from full IE can then be used to design tailored aspects for the system;

- **Re-engineering** is a path acknowledged by IE in order to take account of the investment which organisations have made in their existing systems. This path uses a 'bottom-up' approach to determine the start point for re-engineering. The start point could involve *re-developing* the system from a business model of the current system, *re-designing* the system from the current system design or re-constructing the system by *re-structuring* the code and data for the current design;

- **End-user Computing** describes the use of tools such as spreadsheets, report generators and fourth generation languages (4GLs).

6.1.2.4 Investment Protection

Existing Systems may well represent a considerable investment on behalf of the organisation and it is understandable that they may not wish (or need) to replace the whole of that investment. By acknowledging this in the planning and analysis stages, IE identifies the options to retain, replace or re-engineer parts of the existing system.

6.1.2.5 Staged Approach

By defining project stages, within each development path, IE identifies the tasks to be completed, the techniques to be used and the deliverables to be produced within each stage. In turn this facilitates project management as well as developing a consistent approach to systems development.

6.1.2.6 User Involvement

User involvement throughout the project is considered essential in IE. Although analysts contribute their skills in planning, analysing, designing and building systems it is the users who are the repository of business information. They are therefore in the ideal position to contribute to the development of architectures and to assign priorities to the systems required. Since the users know the business, they can make a significant input to the business models and user procedures as well as helping to define the human–computer interface.

One potent vehicle for encouraging active user involvement is the establishment of **User Workshops** where users are guided through planned steps to deliver high quality results in a shorter time-scale than traditional analysis and methods. Workshops reduce user dissatisfaction by enhancing the analyst's understanding of the business and of the users' requirements whilst also generating, in the user community, a feeling of involvement and responsibility for the system. Workshops may be used for strategic planning, analysis and very commonly for design, where it is known as Joint Application Design (JAD).

6.1.2.7 Graphical Representation of Deliverables

Information Engineering is based upon the belief that diagramming is the most effective way of communication between analysts and users and, to that end, diagrams are central to the products within IE.

6.1.2.8 CASE Tool Support

From its early inception, IE was directed toward CASE automated development. Although it is possible to apply Information Engineering manually, the benefits of CASE technology, through the embedded IE techniques and deliverables, must not be overlooked or underestimated. The increased rigour of CASE tools in cross-validation leads to improved quality within the analysis and thereby the design. Automation of screen design, code generation and re-use of existing design components leads to greater productivity.

6.2 Information Engineering

This section covers the complete path known as Information Engineering as originally defined in the Information Engineering Method. It encompasses seven stages and is directed toward the production of integrated systems. Each of these stages will be discussed below.

6.2.1 The Background to an Information Engineering Project

The pre-cursor to Information Engineering is the existence of a **Business Strategy Plan** (BSP) which is derived via Business Strategy Planning at senior management level with final approval of the Business Plan at Board level.

Within the Business Strategy Plan there will be 'Mission and Purpose' statements both for the business overall and also for functional units within the business. This will identify where the business is now and where it will and should be in the future.

In pursuit of the organisation's mission, the BSP will focus on key concerns and issues defined by top management. These are called **Critical Success Factors** (CSFs) and are the areas in which the business *must* succeed since poor performance in them would threaten the future development or even the very survival of the business. Areas are highlighted for the development of both broad long-term achievements (**Objectives**) and short-term specific targets to be reached by a specific point in time (**Goals**). These objectives and goals address the CSFs of the business.

The Business Strategy Plan must also specify:

- **how to measure improved performance** in these areas;
- **what level of performance** is to be achieved;
- **the time-scale** over which the improvement will be achieved.

Having determined what the business needs to achieve and how to measure performance, the BSP must specify the policies and strategies of the organisation. These strategies will define how the objectives and goals, embodied in the policies, will be met. Additionally, the tactical steps, incorporating the specifics of implementation of the strategies, will be laid down. The Information System(s) forms *one* of the tactical tools available to the organisation. IE addresses the development of the Information System(s) taking the products of BSP as its start point.

IE addresses the Business Life Cycle (see Chapter 1) from **Information Strategy Planning** through to Implementation taking the products of BSP as its inputs. The seven stages of IE are shown in the overview diagram in Figure 6.2 and each of these stages is further discussed below.

6.3 Information Strategy Planning (ISP)

Information Strategy Planning bridges the gap between the BSP and the development of Information Systems by identifying strategies to fulfil the range of **information needs** of the whole organisation (or a clearly scoped area of it) in support of its business aims. ISP requires clarification of the priority of the components of the BSP which the Information Systems will support. Indeed, ISP will almost certainly feed back into the BSP for clarification. The ISP can be used to:

- evaluate how technology can impact upon the business strategy;
- identify data as a corporate resource with opportunities for data-sharing;

- define a business plan of the business systems to meet the information needs, prioritising the systems to be developed;
- evaluate the effectiveness of the current systems and define a migration path to the target systems;
- define a technical strategy to make best use of information technology;

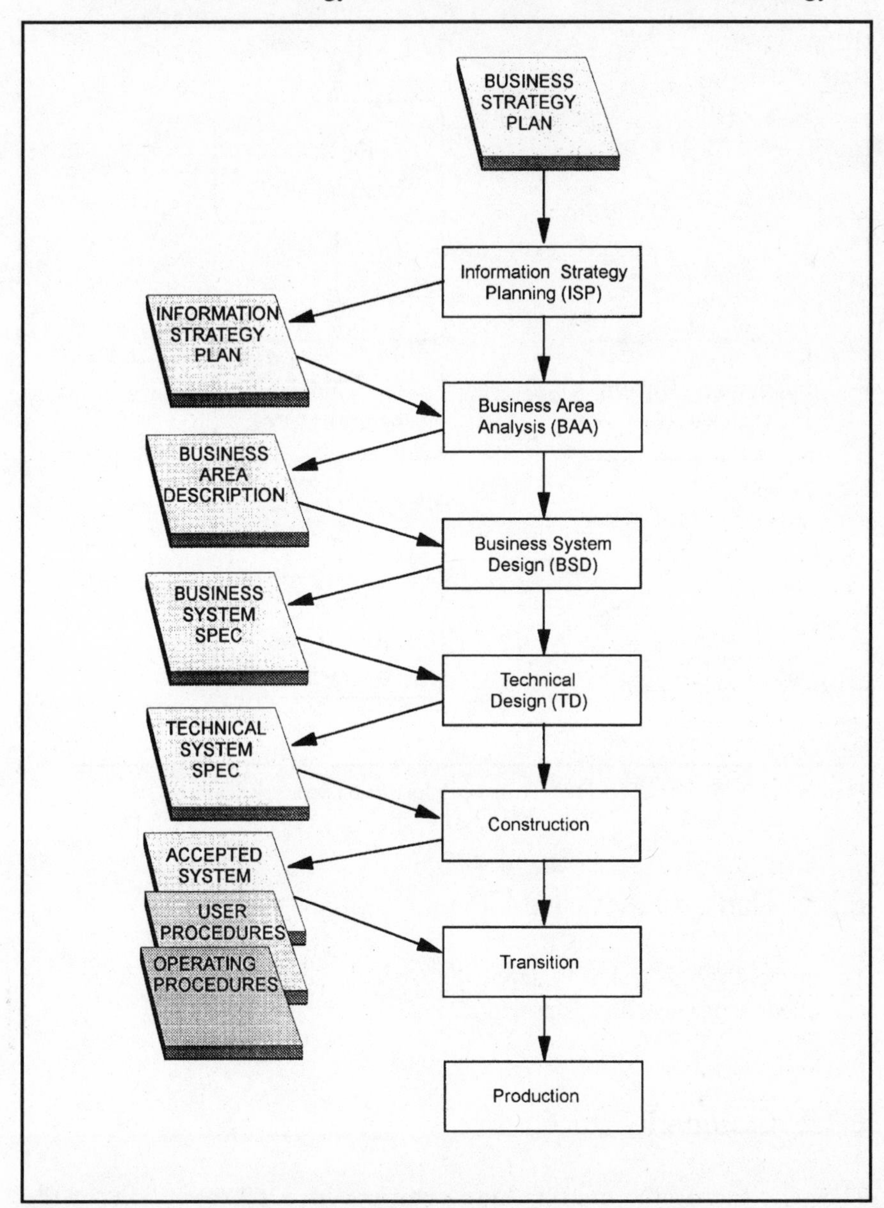

Figure 6.2 Full Information Engineering Overview

ISP consists of five major tasks which are shown in Figure 6.3 and which are described below.

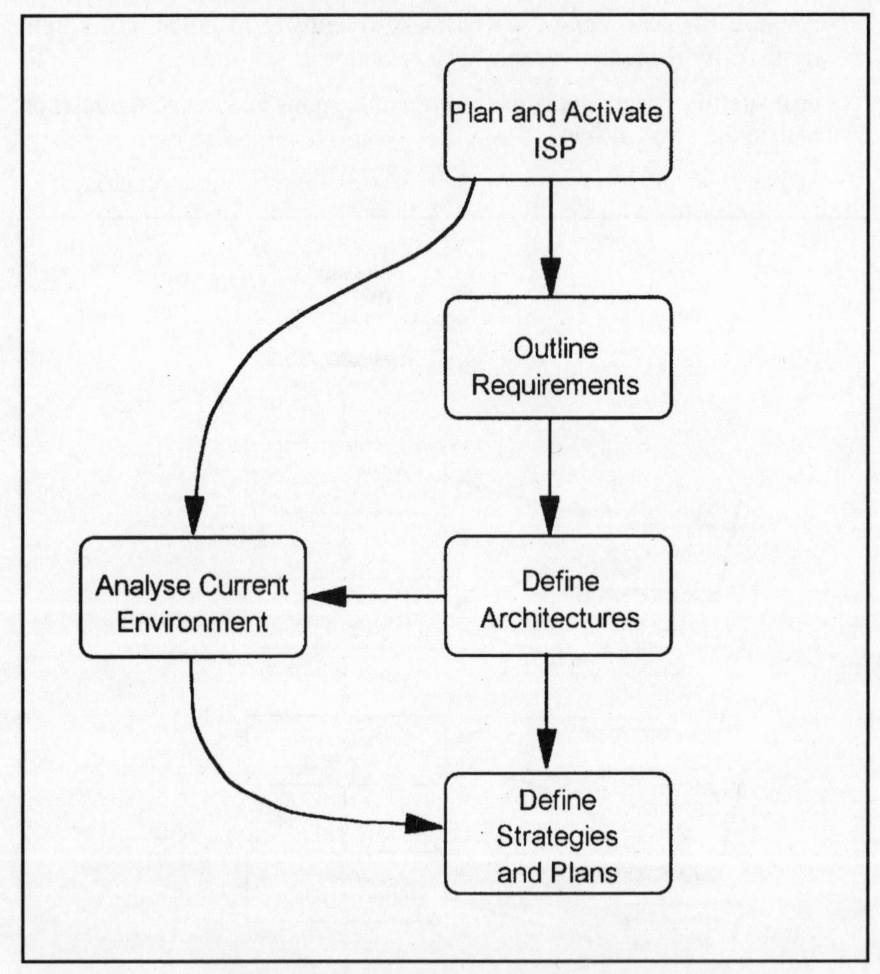

Figure 6.3 The Five Major Tasks of Information Strategy Planning

6.3.1 Plan and Activate ISP

The purpose of this task is to define the scope, the project team resources and the schedule of work for the project.

6.3.2 Outline Requirements

The task facing the analyst here is to produce a broad view of the organisation by analysing the Business Strategy Plan whilst considering the potential impact which IT will have on the strategy. The activities and information needs, which will support the business objectives, are also defined at this point.

6.3.3 Analyse the Current Environment

The current information environment is examined and scrutinised to ascertain how well the current systems meets the organisation's needs by comparison with the architectures. An inventory of current systems and data stores, is built with the current technical environment also forming a focus for analysis.

6.3.4 Define the Architectures

The central task of ISP is to build an **architectural framework** for future developments of Information Systems to meet the organisation's information requirements. There are three architectures to be developed:

- **Information Architecture** which is a *high level view* of the data, activities and their interaction across the organisation. The information needed is defined by modelling the data in a high-level entity relationship model and the activities in a high-level activity model. The interaction between these is analysed to understand which activities use which data;

- **Business System Architecture** which pinpoints probable business systems and data stores needed to support the Information Architecture. The task is to take the activities and data described within the Information Architecture and group them together into business systems within business areas. By reference to the current environment and the business priorities, the requirement for new systems is ranked to focus the next stage on the most urgent areas of change for the business;

- **Technical Architecture** which describes the broad hardware, software and communications environment to support the Business System Architecture, comparing this to the current technical environment. The focus is upon types of product rather than vendor specific products.

6.3.5 Define the Strategies and Plans

At this point the alternative strategies for implementing the architectures are evaluated to allow planning of subsequent projects which may, for example, be:

- Business Area Analysis Projects;

- RAD Projects;

- User Computing Projects;
- Technical Projects;
- Organisational Projects.

The recommended strategies and plans are then presented to Senior Management for approval.

6.3.6 Deliverables of ISP

The deliverables from the ISP stage are shown in Figure 6.4.

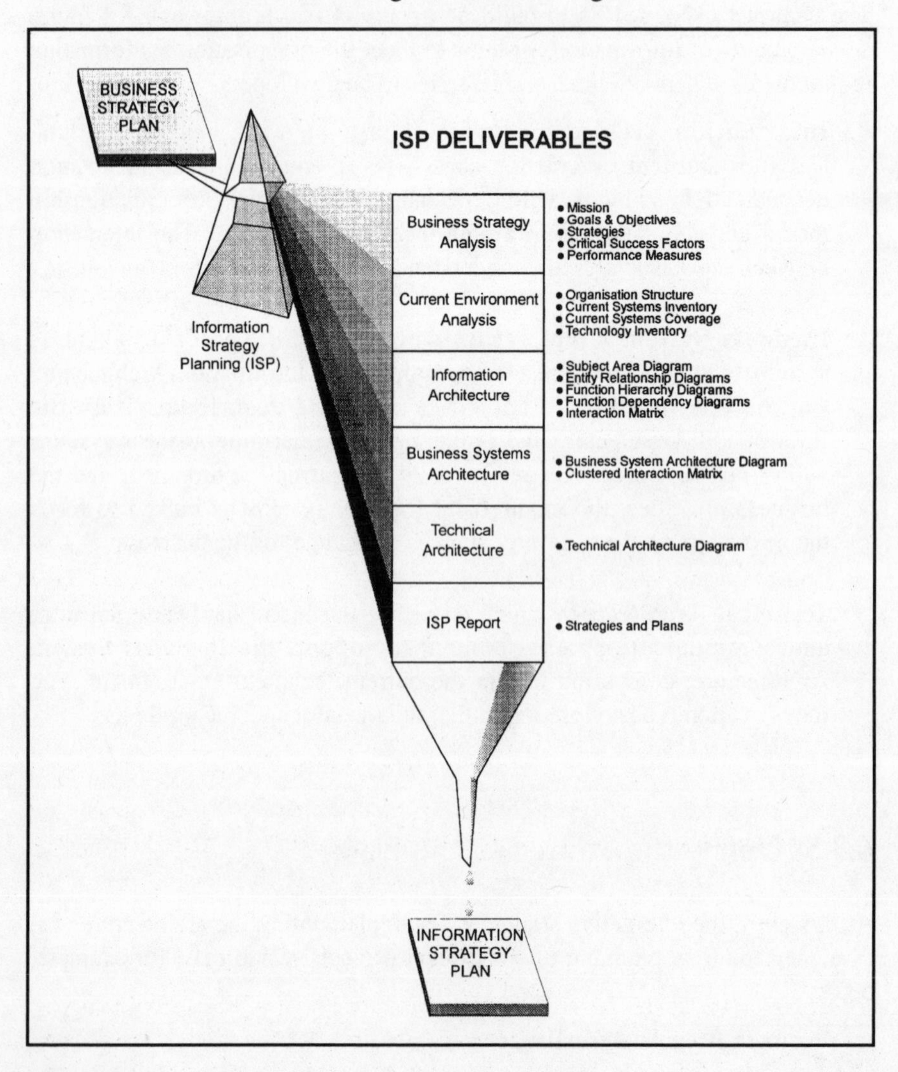

Figure 6.4 Information Strategy Planning Deliverables

One ISP can, and usually will, result in many separate BAA projects. The deliverables from ISP, formulated as the Information Strategy Plan are carried forward as input to the next stage: Business Area Analysis.

6.4 Business Area Analysis (BAA)

Business Area Analysis takes an area of the business, identified within the ISP, and makes a detailed analysis of it from the perspective of **what** is done, not how or who does it. To be successful, however, it requires the users to play a crucial role, that of information providers both in terms of their job function and their information requirements. The BAA seeks to define to greater detail the business activities (Business Functions and Business Processes), the entities and the activity/entity interactions which were first discussed as part of the business area in the ISP. It, therefore, focuses on the part of the Information Architecture which is of specific relevance to the particular business area which is the subject of the BAA.

The major purpose of BAA is:

- to model the selected business area by defining the business activities, the data necessary to support the information needs and the interaction of these. The models are used to record, communicate and confirm understanding of the business area and are the basis for system design and automated system generation;
- to identify, scope and prioritise systems to support the business area. In BAA the initial view of the system, which was identified in ISP, is further refined.

The Major Tasks of BAA are shown in Figure 6.5 and are expanded below. The reader should note that there are occasions when BAA is sub-divided into *Outline* Business Area Analysis and *Detailed* Business Area Analysis. This allows focus to be placed upon the detail of priority areas whilst maintaining an outline understanding of the total BAA project area.

6.4.1 Plan and Activate the BAA Project

As in the ISP stage, we are again concerned with defining the scope, the resources, organisation and the schedule of work but now for the BAA project.

6.4.2 Gather Information

Information about the business area is initially collected by examination of current documents and then by interviewing the users and/or by conducting user workshops. By homing in on the each of the user's activities, the analyst is able to elicit:

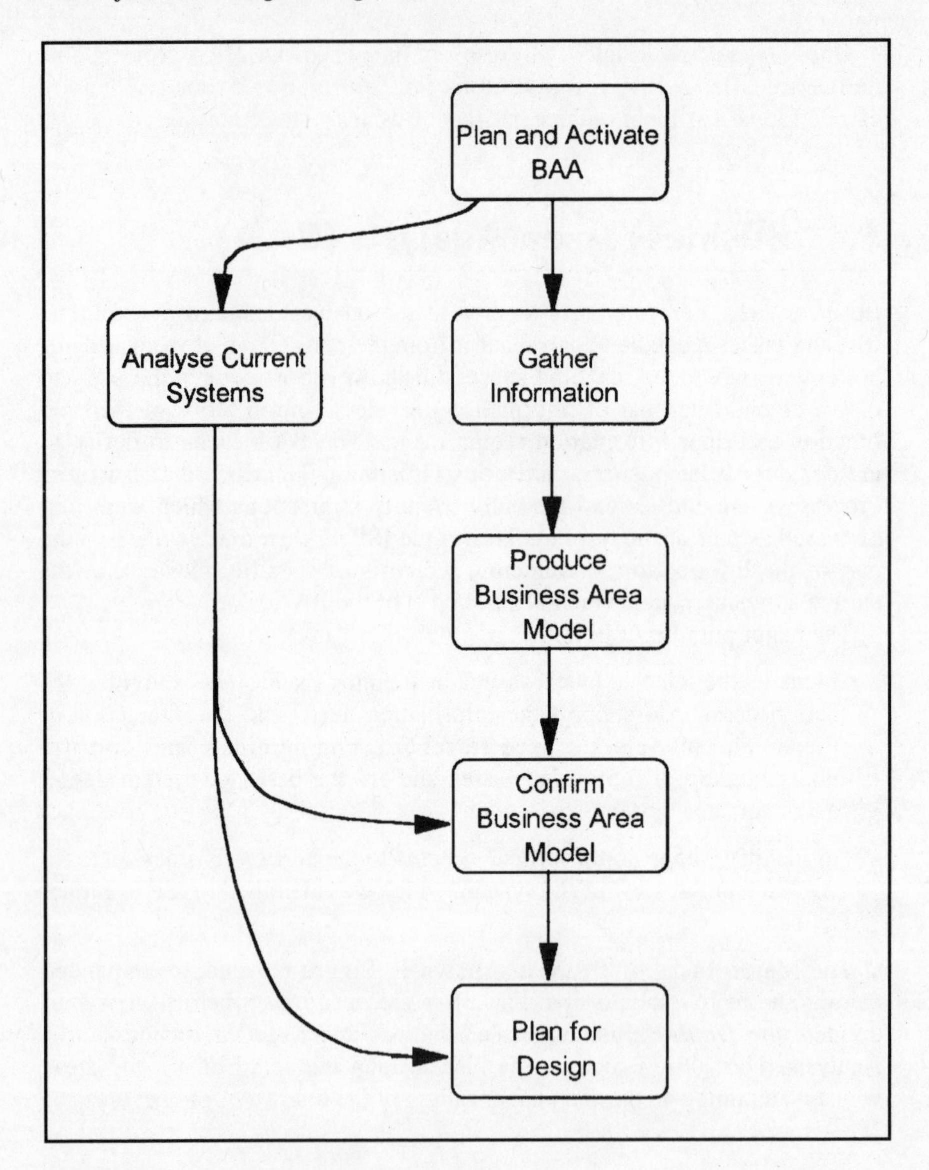

Figure 6.5 Business Area Analysis Tasks

- the purpose of the activity;
- the performance measure;
- the information needed;
- the information created;
- the actions performed.

The analysis is concerned with what must be done rather than how it is currently done, e.g., 'make provisional reservation' not 'record provisional reservation on room-booking chart'.

6.4.3 Produce Business Area Model

The development of the Business Area Model begins with the definition of two models in parallel with each other: the **Entity Model** and the **Process Model.**

6.4.3.1 Entity Model

The information produced and needed by an activity provides a source of information for Entity Analysis, which commences with the identification of entity types. The analysis of the business area builds a list of entity types which can then be subject to Entity Relationship Diagramming. Within IE, the diagramming conventions represent entities as hard boxes with relationships having optionality marked on the relationship line by a circle and singular mandatory relationships marked by a short bar across the line. The Entity Relationship Diagram (ERD) for Midlinks Motel using this notation is shown in Figure 6.6.

Further refinement of entities is accomplished by defining them in terms of:

- the **attributes** which characterise them. IE categorises attributes as *basic* (e.g. surname), *derived* (e.g. age) or *designed* (patient no.) with mandatory or optional requirement status. Multi-valued ('repeating groups') attributes are removed from being one of the attributes of the entity and are made the entity occurrences of a newly created entity with a new relationship being established between the original entity and the new entity, e.g. a CUSTOMER entity may initially be thought to have a 'reservation' as one attribute. However, the customer may make several reservations at the time of booking and, therefore, there would be several reservation numbers (multi-valued) associated with that customer. Consequently, a new entity RESERVATION is required (reflected in Figure 6.6);
- an **entity identifier** which is an attribute or combination of attributes which uniquely identifies one entity occurrence from any other;
- **volumetrics** indicating the minimum, maximum and average number of occurrences that may exist at a time together with any seasonal trends, etc.

6.4.3.2 Process Model

Whilst considering the data aspects of the Business Area Model, IE also focuses upon the activities of the Business Area which were first considered in the Information Architecture of the Information Strategy Planning. The Process Model, which was at a high level during ISP, has further detail added to it in BAA, defining the lowest level processes.

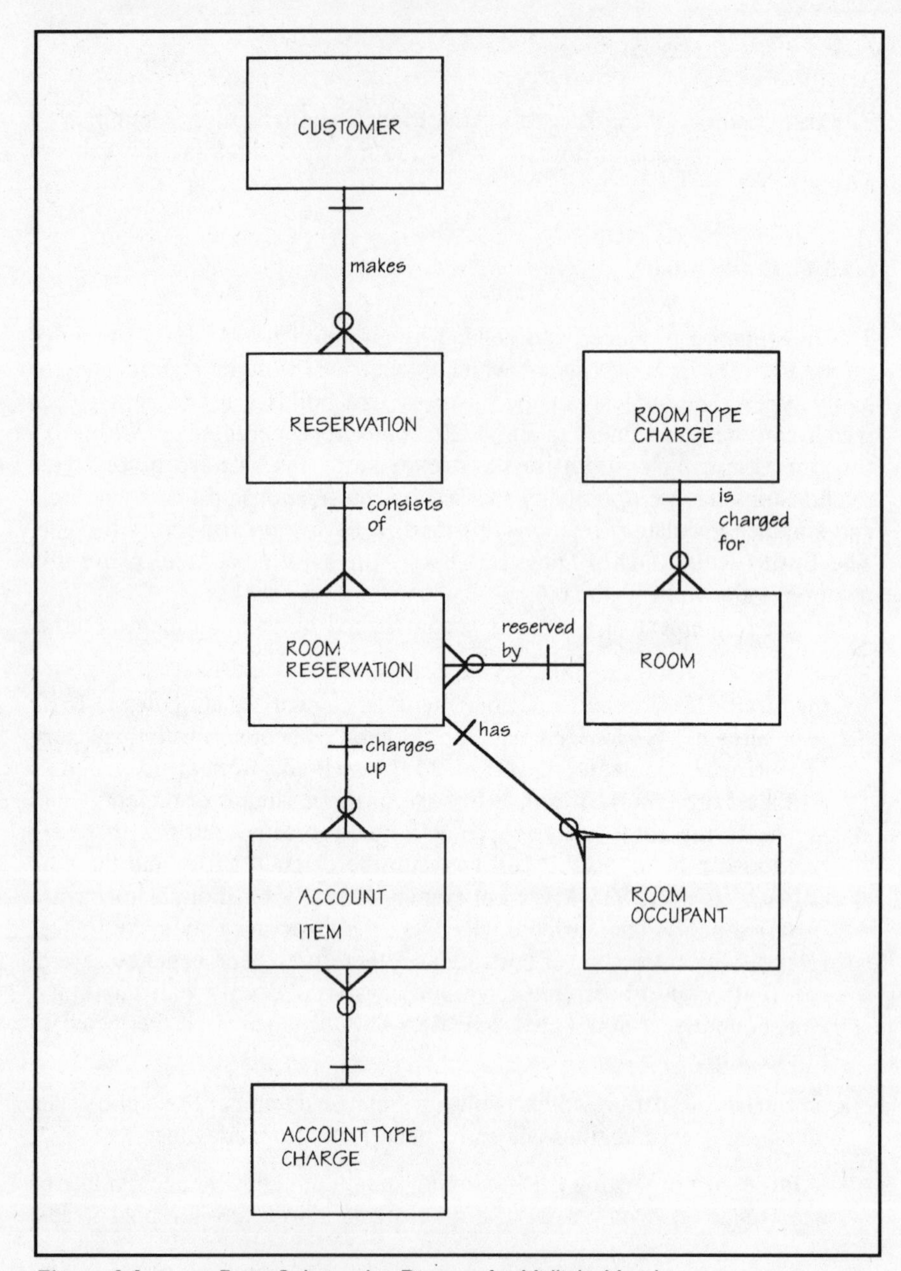

Figure 6.6 Entity Relationship Diagram for Midlinks Motel

Activities, Functions and Processes.
IE acknowledges two forms of **Business Activity: Business Functions** and **Business Processes**.

Activity is a generic term which can refer to either a function or a process. Characteristically all activities whether they be functions or processes have input to them, do work on the input (i.e. transform the input) and produce output. So what distinguishes a function from a process?

A **Function** is defined in IE as the highest level of activity, a group of business activities which together completely support one aspect of furthering the mission of the organisation. A Function is concerned with resources, products or support activities and can be distinguished from a process since how often it is done is not meaningful, nor can a person be directed to do a function. Take, for example: the function *Customer Service* for Midlinks Motel. How often is *Customer Service* done? Do *Customer Service is too general to be meaningful*. Customer Service is a Function which has to be described in terms of its component Processes (such as Register Guest, Accept Reservation, Collect Payment, etc.). Functions are named in a general way, e.g. Customer Service, Purchasing.

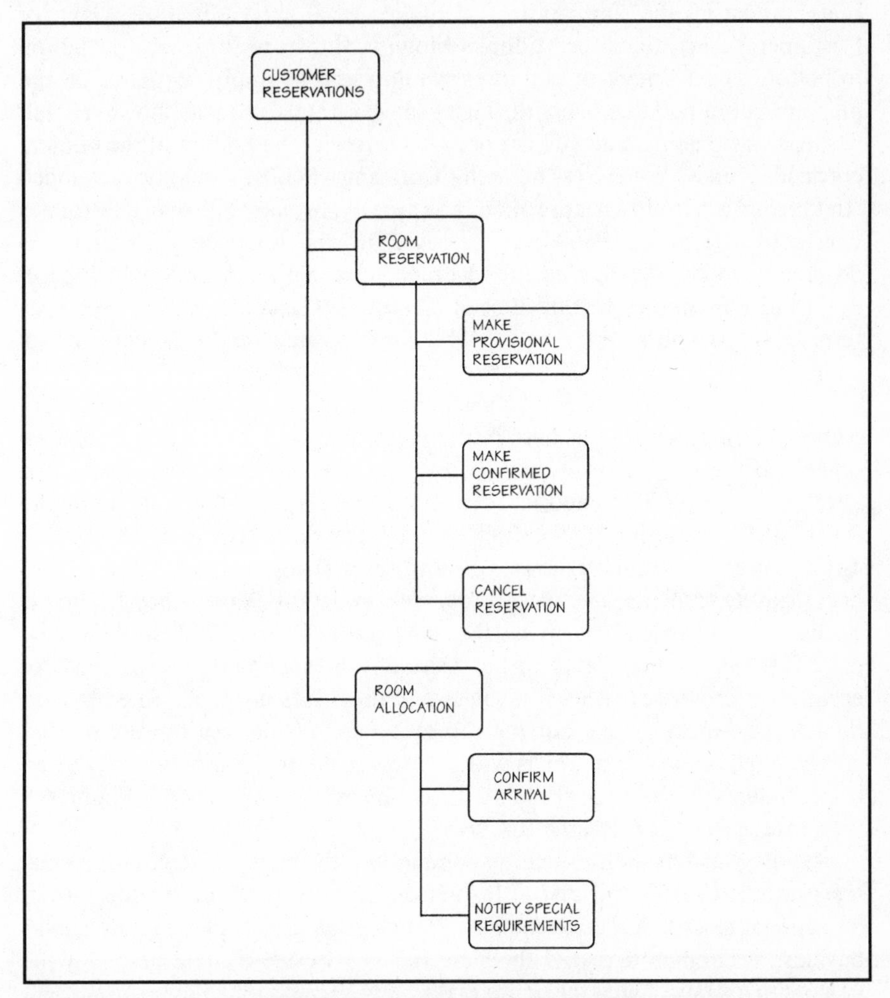

Figure 6.7 Activity Hierarchy Diagram for Customer Reservations

A **Process** is an activity which, by its execution, takes input as specific attributes of entity types, transforming the particular entity occurrences by amending their attribute values and outputting the entities in their new state.

By direct contrast with functions, Processes are named in a specific way using a verb/noun construct, e.g. Register Guest, Cancel Reservation.

Activity Analysis

Within BAA the activities are defined and dependencies between them are identified via **Activity Analysis**. Activity Analysis consists of **Activity Decomposition** and **Activity Dependency** analysis:

Activity Decomposition: Functions and processes are decomposed to increasing levels of detail using **Activity Hierarchy Diagrams** (AHDs) which use soft boxes to represent functions or processes, with connecting lines indicating the subactivities of higher level activities. (An AHD for Customer Reservations for Midlinks Motel is shown in Figure 6.7.) The top to bottom arrangement of activities would seem to imply sequence on the diagram but in fact this is not the case (some variants of IE do, however, add sequence by placing a downward pointing arrow at the bottom of the vertical connecting lines). Function Hierarchy Diagrams (FHDs) would be developed first, from a top-down approach, eventually breaking down to Process Hierarchy Diagrams (PHDs) at the lower levels. A series of diagrams is developed until no further useful decomposition can be made at which point Elementary Processes will have been identified. These elementary processes forms a single unit of work within which no one step can be executed on its own.

Activity Dependency: Business Activities do not exist in isolation from each other and the output from one activity may be required as the input for another activity. The second activity is therefore dependent upon the successful completion of the first activity. Dependency Analysis in IE is documented through **Process Dependency Diagrams** (PDDs). These document dependencies between Processes. Function Dependency Diagrams would not be created. By taking the independent branches of the PHD and concentrating on each level of the hierarchy a PDD can be developed for each set of processes on each level. Again processes are represented by soft boxes. External entities, outside the area of analysis, are represented by double-edged boxes. The PDDs will be used as the foundation for the design of procedures in the next phase of IE, Business Systems Design. Figure 6.8 presents the PDD for Reserve Rooms.

The diagram shows information coming into the business (e.g. provisional reservation details) and also information going out of the business (e.g. provisional reservation confirmation). Between processes the diagram shows business dependency, that is the state that the business must be in for the process to execute. What the diagram does **not** show is data flow.

It should also be noted that events are only identified on PDDs if they are essential for accurate process definition, a large arrow with the event written within its bounding shape being used to point to the process which it triggers.

Following the development of both Entity and Process Models, their interaction is analysed using techniques for interaction analysis.

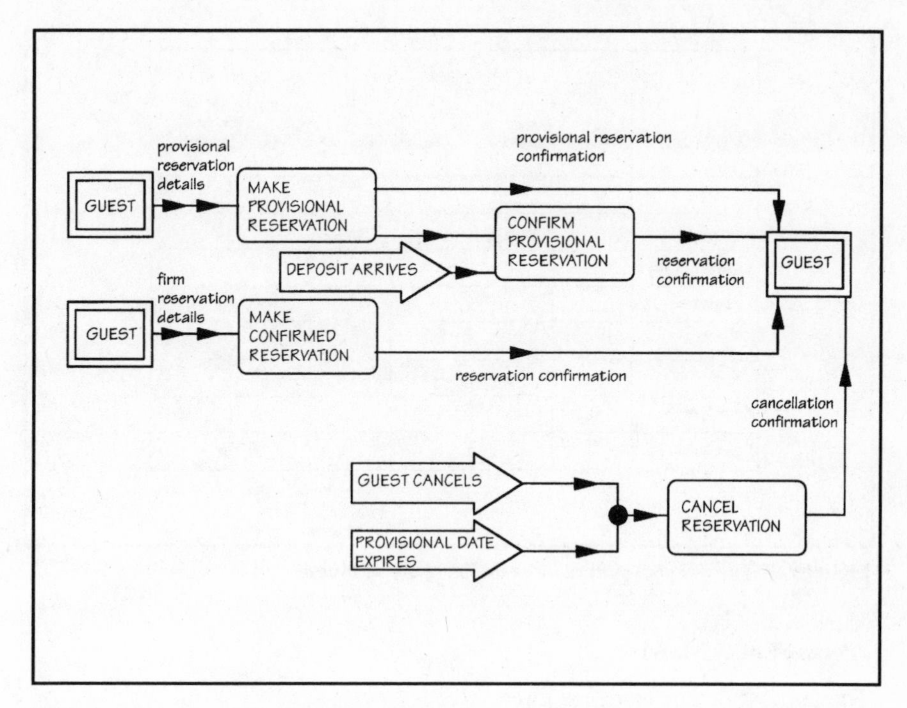

Figure 6.8 Process Dependency Diagram for Reserve Rooms – Midlinks Motel

6.4.3.3 Interaction Analysis

Having focused on data and activities separately our attention now turns to the way that data is affected by activities by using Interaction Analysis (IA) techniques.

This involves the use of:

- A **Process/Entity Matrix** to show which processes use which entity types; shows which processes **C**reate, **R**ead, **U**pdate and **D**elete entity types, (hence the memorable term CRUD matrix!). Just such a matrix for Midlinks Motel is shown in Figure 6.9, although here it is only part-completed.

- **Entity Life Cycle Analysis (ELC)** which details the possible changes of state which an entity occurrence can go through as a result of the action of processes;

- **Process Logic Analysis (PLA)** whereby processes are defined by the actions they perform on entities. By preparing the Process Action Diagram, within PLA, the analyst's understanding of the processing rules can be clarified. The level of detail of the Process Action Diagram forms the basis of code generation which can be automated via CASE technology.

ENTITY	PROCESS: Make Provisional Reservation	Make Confirmed Reservation	Cancel Reservation	Confirm Arrival	Add Charges	Receive payment
Customer	C/U					
Reservation	C					
Room Reservation	C	C/U	U	C/U		U
Account Item						
Account Type Charge						
Room	R					
Room Type	R					
Room Occupant						

C - Create
R - Read
U - Update
D - Delete

Figure 6.9 Process/Entity Matrix for Midlinks Motel

Process/Entity Matrix

The purpose of this matrix is three-fold:

- to verify the completeness of the Process and Entity Models. For example: does every process update or create at least one entity type? If an entity type is created by this business area, is there at least one process to create it?

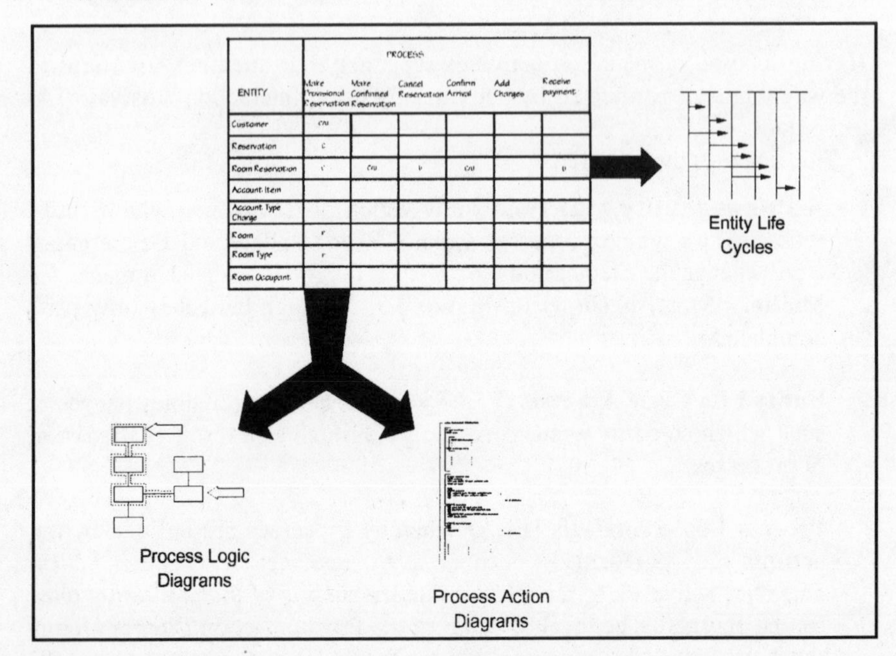

Figure 6.10 Relationship Between the Process/Entity Matrix, Entity Life Cycles, Process Logic Diagrams and Process Action Diagrams

- to provide a basis for more detailed techniques of interaction analysis;
- to define clustering of processes via Affinity Analysis thereby identifying business systems and facilitating understanding of data sharing/interface requirements. Affinity Analysis is used to identify groups of processes which are strongly inter-related by their usage of data (entities) and thus should be part of the same development.

The matrix is used in Entity Life Cycle Analysis and Process Logic Analysis the relationship between these being illustrated in Figure 6.10.

Entity Life Cycle Analysis

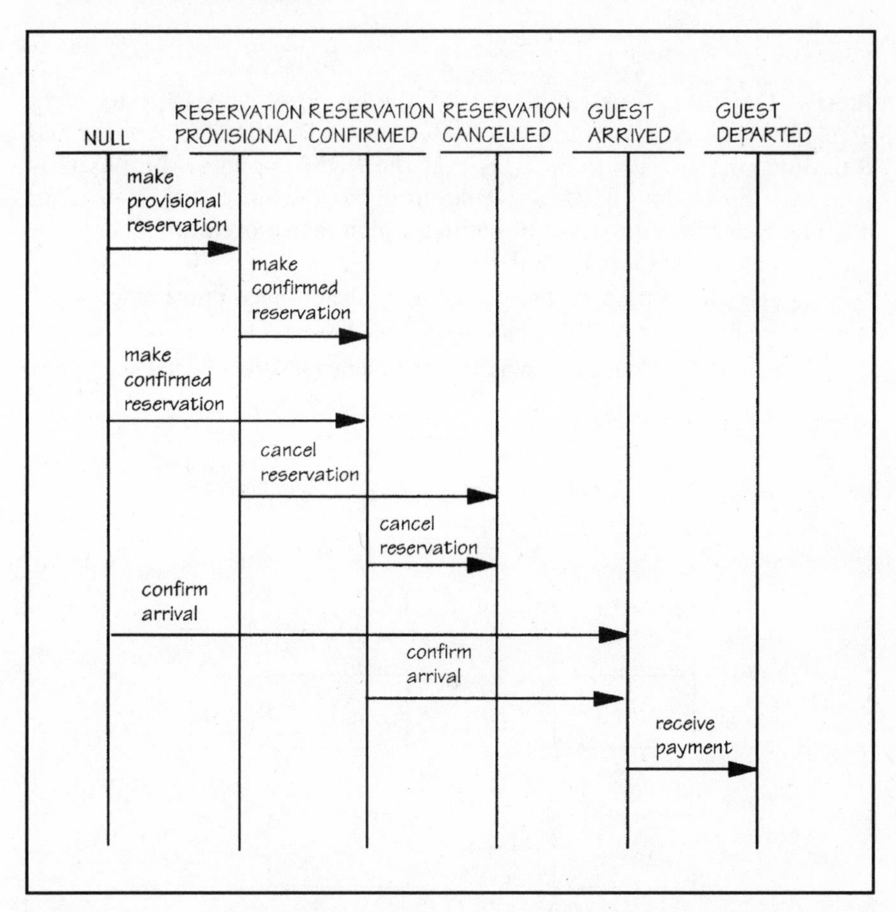

Figure 6.11 Entity Life Cycle State Transition Diagram for Room Reservation – Midlinks Motel

The underlying purpose of Entity Life Cycle Analysis is to identify the states that an entity may exist in, so that the state that an entity must be in for a particular process to begin execution can be determined. By reading along the row of any chosen entity on the Process/Entity matrix the Entity Life

Cycle is formed, linking the processes with their effect upon the entity. It also offers the opportunity to check that the processes required to place the entity into all of its possible states have been recognised. All entities will have a Creation, a number of Intermediate, and a Termination state, with processes which effect the transitions. IE uses **State Transition Diagrams,** one for each entity, to visualise the changes in state which an entity may undergo. Each transition is labelled with the name of the process which brings about the change. The State Transition Diagram may be augmented by the production of an **Entity State Change Matrix** which maps valid processes against entity states for a particular entity. The Entity Life Cycle State Transition Diagram for Room Reservation for Midlinks Motel is given in Figure 6.11.

Process Logic Analysis

Process Logic Analysis (PLA) defines the business processing of every elementary process including its effect upon Entities, Attributes and Relationships and also confirms that the ERM supports the business processes. By reading **down any column** of the Process Entity Matrix, for any one elementary process, all entities which that process affects can be identified for use in PLA. PLA, therefore:

- encourages rigorous understanding of the business processing;
- confirms that the ERM can support the business processing;
- provides the business logic for the system construction stage.

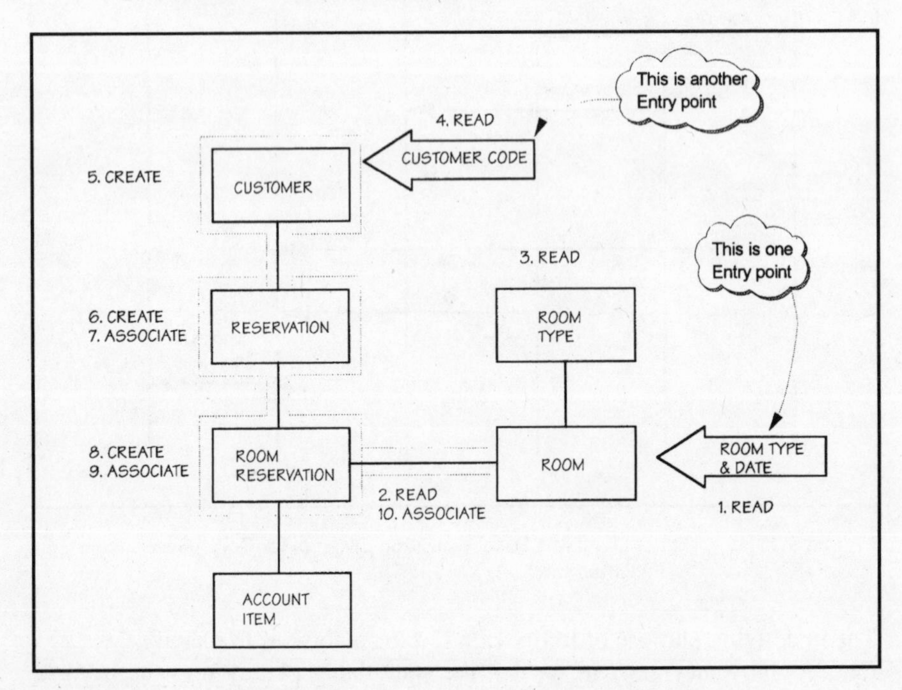

Figure 6.12 Process Logic Diagram for Make Provisional Reservation – Midlinks Motel

PLA uses a **Process Logic Diagram (PLD)** to show what happens to entities and relationships and the sequence of such actions for a single elementary process. Figure 6.12 gives the Process Logic Diagram for Make Provisional Reservation for Midlinks Motel.

The actions which can be taken against an entity type are Create, Read, Update and Delete with relationships between entity types being changes by Associate, Disassociate and Transfer. Preparation of the PLD enables the analyst to verify that the Entity Model can support the processing requirements and forms a useful start point for the more detailed Process Action Diagram.

A Process Action Diagram (PAD) is developed to define the business logic for *each* elementary process in terms of control conditions and simple conditions.

The inputs and outputs (Imports and Exports) of a process are described by **Information Views**. Information Views are shown diagrammatically in Figure 6.13 and the three types are described below:

- *Import Views* bring new information into the business, e.g. provisional reservation details;
- *Export Views* send information out of the business, e.g. provisional reservation confirmation;
- *Entity Action Views* provide or update information to or from the corporate data as represented by the Entity Model, i.e. defines the effects of the process upon the attributes of the entities which it affects.

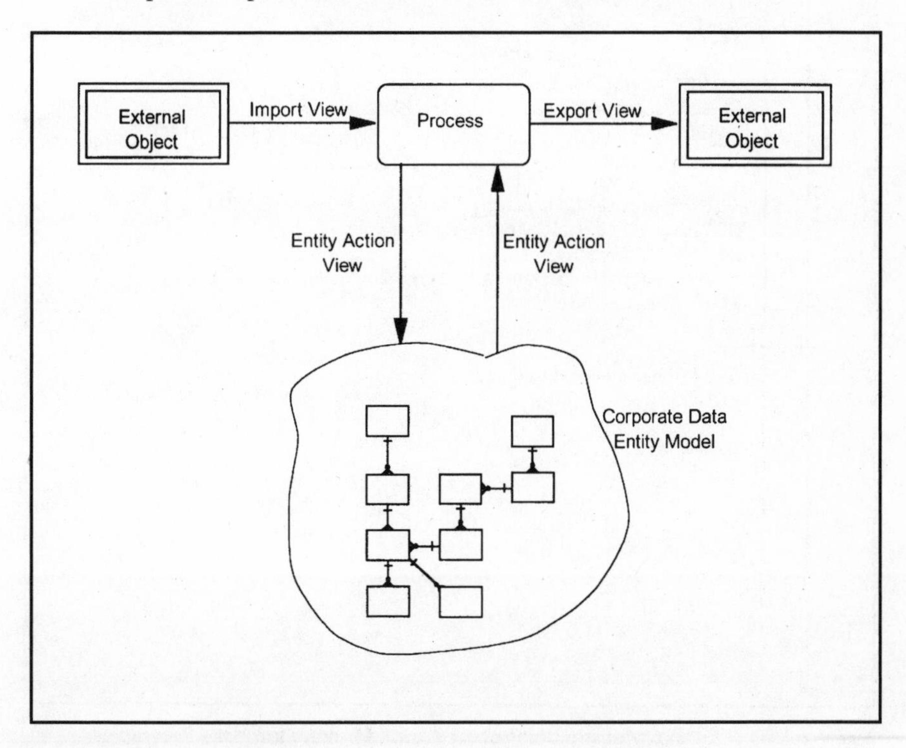

Figure 6.13: Information Views

The logic in the PAD may detail:

- selection conditions to determine which entity occurrences are to be acted upon;

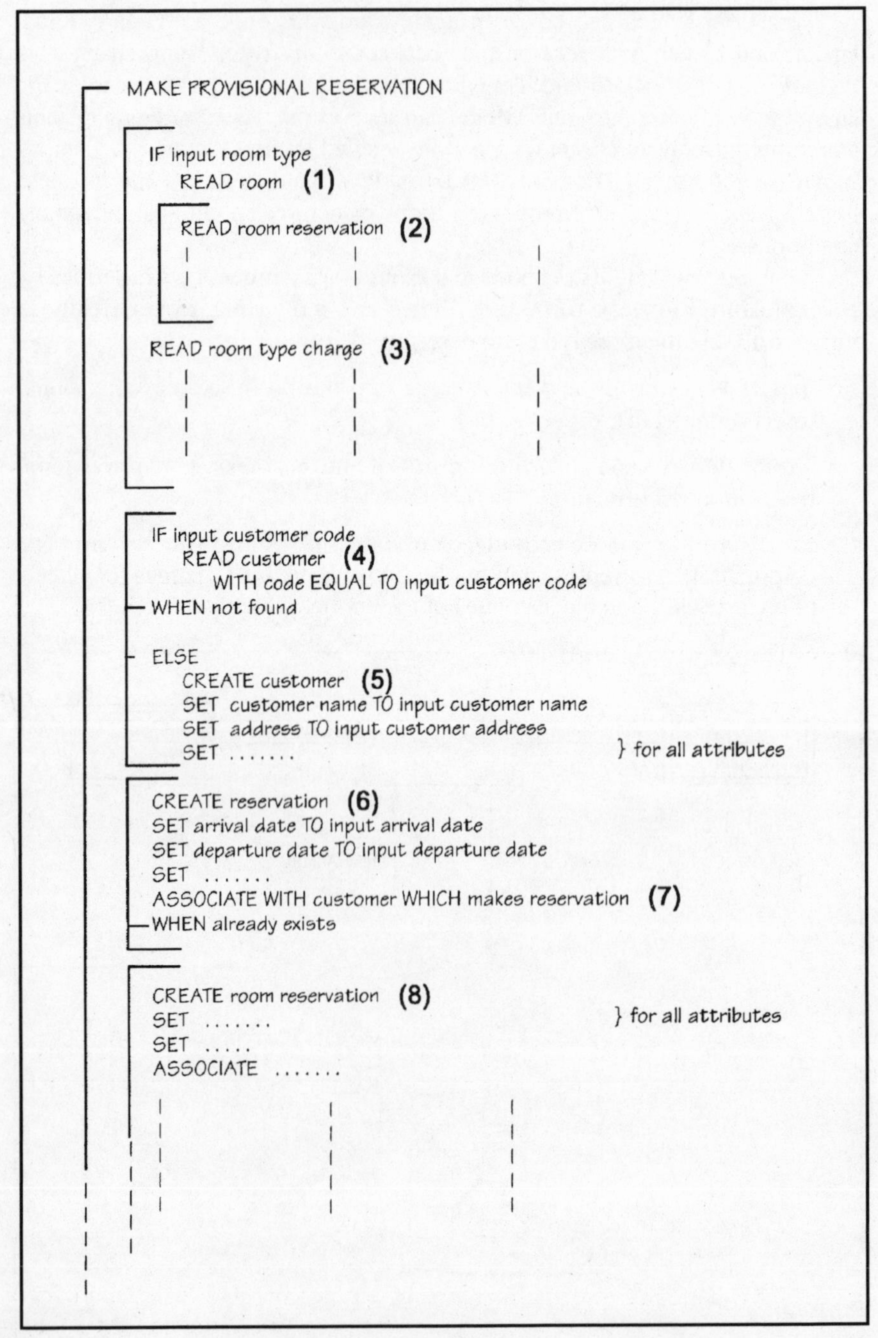

Figure 6.14 Extract from the Process Action Diagram for Make Provisional Reservation – Midlinks Motel

- execution conditions under which a particular action will take place;
- repetition of actions under conditional control;
- entity actions (Create, Read, Update, Delete);
- relationship actions (associate, disassociate, transfer);
- attribute actions (set, remove).

The Process Action Diagram of 'Make Provisional Reservation' for Midlinks Motel is part-created in Figure 6.14. It should be noted that the bracketed numbers in this diagram are not normally part of PAD. They are added here to aid clarity by correlating with the actions which bear the same number of the PLD of Figure 6.12.

6.4.4 Analyse Current Systems

The top-down approach of IE's is finally compared with the current systems via the bottom-up approach of **Current Systems Analysis** (CSA). The analysis of current systems is useful because it:

- highlights some problem areas to be solved in the new system;
- provides data and procedure models which can be used in completeness checking of the Business Model;
- supports transition planning from the current system to the new.

Current Systems Analysis uses the techniques of:

- Current System Procedure Analysis;
- User View Analysis;
- Canonical Synthesis.

6.4.4.1 Current System Procedure Analysis

Documentation of the current system consists of decomposition (as in function decomposition) of procedures and also Data flow Diagrams of the current system.

6.4.4.2 User View Analysis

User View Analysis draws together all the different views which users have of the data and current associations within the data. It collects the information from current data layouts on screens, files (manual and computer) and reports. Each user view consists of various groupings of attributes of an entity which a user needs to perform a function. A user view will contain one or more key fields which uniquely identify entity occurrences and thence values of associated fields. Each User View is represented on a Bubble Chart where the bubbles contain attributes and the arrowed lines show the associations between them. Figure 6.15 presents an example of a bubble chart for Midlinks Motel.

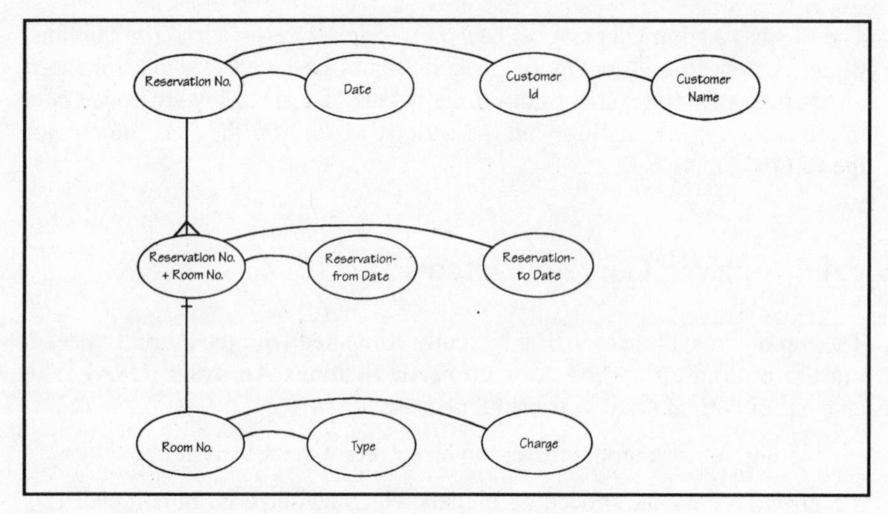

Figure 6.15 User View Analysis Bubble Chart – Midlinks Motel

6.4.4.3 Canonical Synthesis

From the collection of bubble charts the user views can be synthesised into a single composite view of the entire data by overlaying bubble charts and resolving alias names for the same attributes (fields). This technique is known as **Canonical Synthesis**. Key Groups occur in user views and they consist of a key field or a concatenated key field together with a collection of fields which are functionally dependent upon the key. These Key Groups are equivalent to entities, with the relationships between key groups being equivalent to entity relationships. From the canonically synthesised user views and the key groupings/relationships, an entity relationship diagram built from the bottom-up (cf top-down derivation from entity analysis) is created. This can be compared with the diagram produced from entity analysis thus allowing the potential for further entities to be added to our original ERD.

6.4.5 Confirm Business Area Module

Having developed the Business Area Model during BAA, **confirmation** is performed to check for correctness, completeness and stability within the model so that a firm foundation for Business System Design is assured. *Correctness* is addressed by:

- **Normalisation** of attributes to confirm that the correct attributes are assigned to the correct entities;
- **Process Dependency** checks to ensure consistency between PDDs and that every elementary process is on a PDD, is dependent on an event or another process and has entity actions defined with at least one input and one output;
- **Redundancy** checks to eliminate unnecessary attributes and relationships where for example an attribute can be derived;
- **Structured Walkthroughs** in which the deliverables at any phase of BAA are discussed with the users with the purpose of confirming understanding of the Business Area and their requirements of the new system. The walkthrough will lead to agreement upon the various aspects of the model and resolution of any problems within it.

Completeness is achieved by comparison with the current system which will potentially identify attributes and processes which have been overlooked. In addition the following matrices may be used:

- a **Process/Entity matrix** (CRUD matrix) to check for creation, updating and deletion of each entity;
- a **Process/Relationship matrix** to confirm the association and disassociation of entities;
- a **Process/Attribute matrix** to check that the values of attributes can be set, changed or removed.

Stability analysis involves assessing the model against any possible future business changes. The analyst will then discuss any necessary changes to the business model to make it more resilient to such future changes.

6.4.6 Plan for Design

BAA concludes with the definition of the Business Systems which are to be implemented, together with a plan for their implementation. One BAA Project may and almost certainly will lead to many BSD Projects. The **Process/Entity matrix** is established and then adjusted (sorted) such that the processes listed down the matrix are ordered using the Business Life Cycle sequence defined by the PDDs. The entities across the top of the matrix are now sequenced by the technique of **affinity analysis** (clustering) which leads to the production of a clustered Process/Entity matrix which aids the identification of business system areas.

A **Cost/benefit analysis** of the financial benefits of implementing the individual elementary processes is used to decide what Business Systems will be implemented and also the sequence of implementation for maximum business benefit. The natural sequence of implementation of Business Systems would be in the order prescribed by the Business Life Cycle since it would not make sense to implement a System which required information from another System which had not already been implemented! However, in many instances this logical sequence is countermanded for financial or political reasons.

In addition to deciding the sequence for implementation of Business Systems, the above techniques are used to determine the sequence in which the elementary processes will be implemented although the natural sequence may be overruled by the same factors discussed above.

6.4.7 Deliverables of Business Area Analysis

The deliverables from BAA are summarised in Figure 6.16.

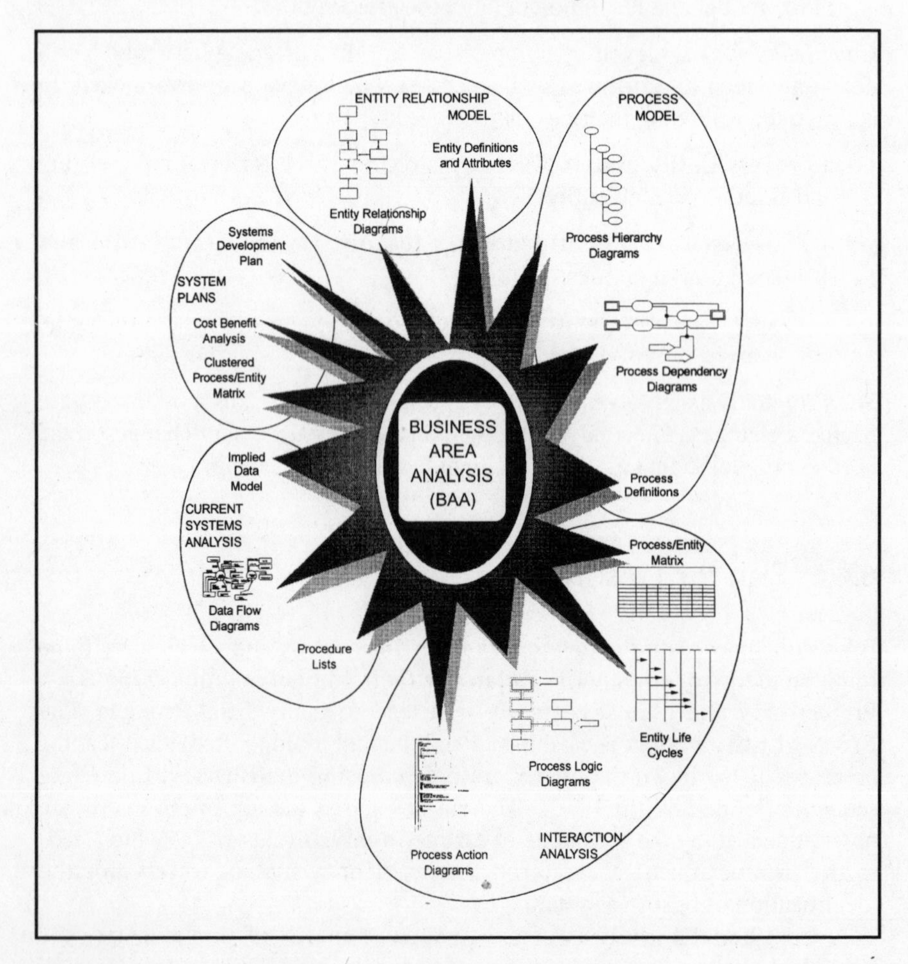

Figure 6.16 Business Area Analysis Deliverables

6.5 Business System Design (BSD)

Business Area Analysis subdivides the business area into small areas called design areas, each of which will form a new business system and each of which will be defined in terms of its processes, attributes and relationships which are to be supported. Business System Design takes the elementary processes from BAA and determines **how** they will be done from a user viewpoint with user dialogues, screens, reports and procedures. The Major tasks of BSD are identified in Figure 6.17 and are described more fully below:

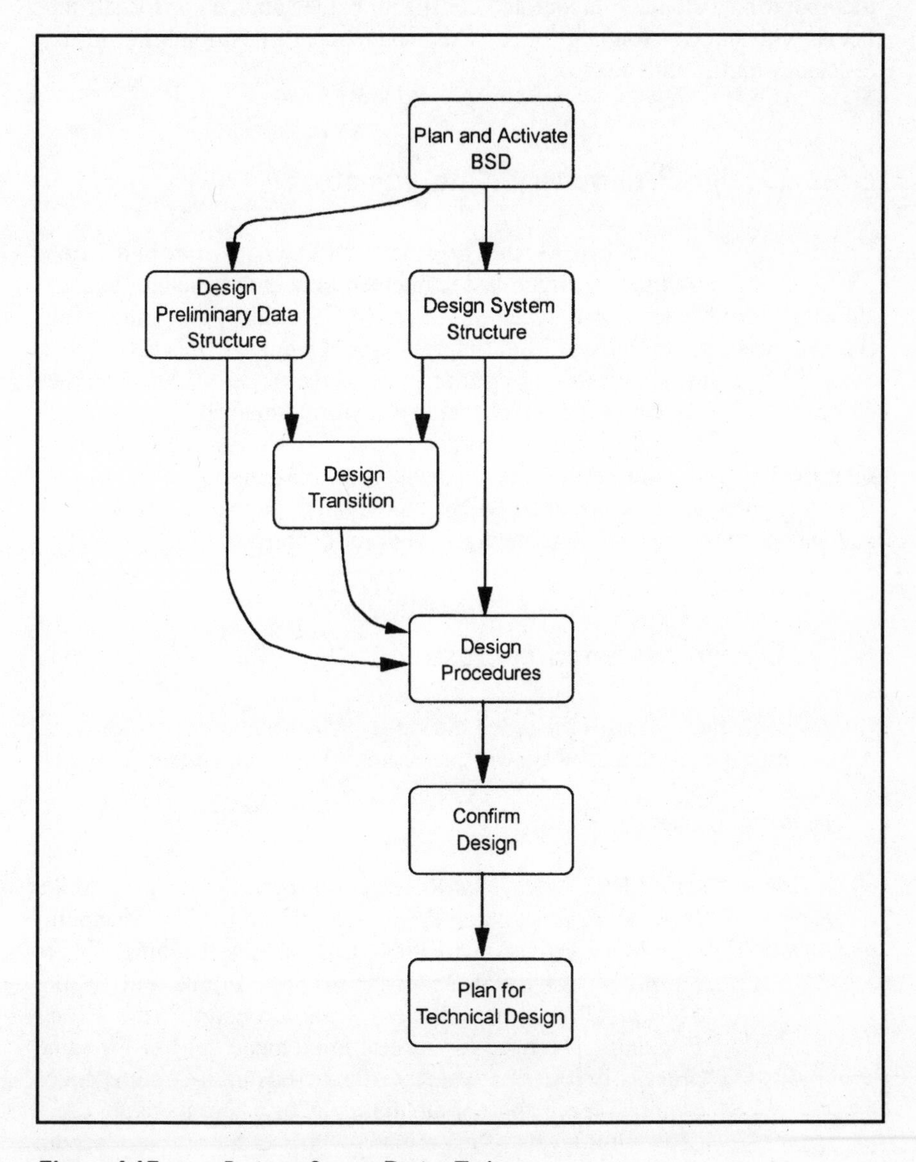

Figure 6.17 Business System Design Tasks

6.5.1 Plan and Activate BSD

As in ISP and BAA, the task here is to define the objectives, scope, resources, organisation and schedule for the project. In addition it is essential to define system standards which must be adhered to so that a consistent human/computer interface can be maintained both within the current project, and in relation to other systems already in place. The use of consistent command sets or function keys regardless of which procedure or system is being used will reduce user error. Additionally, screen designs (including window design where a Graphical User Interface is to be developed) incorporating a standard layout and consistent names and colour for attribute fields, with information/help lines in the same screen position will aid user efficiency and confidence.

6.5.2 Design Preliminary Data Structure

The Entity Relationship Model prepared in BAA is mapped into a preliminary Data Structure which is documented on a data structure diagram, showing record types (entity types), linkages (relationships) and entry points (key field access to entity). This diagram closely mirrors the ERD but is modified to comply with the structuring rules of the target DBMS. Detailed specification of the data structure elements is also documented.

Additional data items may be added during design, e.g. control totals, security data, last values for designed attributes such as employee number.

At this point, database designers will be brought in to provide the information and understanding needed to refine the design for the target DBMS.

6.5.3 Design System Structure

During BSD, the processes identified during BAA are mapped to procedures. A procedure is a meaningful unit of processing which supports a particular user task.

The mapping may be:

one to one: e.g. 'Make Provisional Reservation' process becomes 'Make Provisional Reservation Procedure'. This is the ideal mapping;

one to many: e.g. 'Make Provisional Reservation' process becomes 'Make *Simple* Provisional Reservation' procedure and 'Make *Multiple* Provisional Reservation' procedure (i.e. Block Booking). The need for several implementations of the same process is justified where different business circumstances require a more appropriate dialogue, or where there are users of varying levels of experience who require different screen information to guide them.

many to one: e.g. 'Make *Provisional* Reservation' process and 'Make *Confirmed* Reservation Process' becomes 'Make Reservation' procedure. This would be appropriate under circumstances where it may be organisationally easier to have a common dialogue to begin both types of reservation with entry of the type of reservation being made late in the dialogue.

zero to one: e.g. 'Reservations Menu'. These are designer-added procedures to provide for such requirements as menus, help facilities, sign on, data archive, database print facility, etc.

The overview picture of the design area is documented in a data flow diagram which shows the movement of data between procedures, stores and externals. This DFD is constructed by taking the PDDs from BAA and mapping them to **Procedure** Dependency Diagrams and then adding data stores identified from the preliminary data structure. By synthesising the procedure level DFDs, higher level DFDs are created. Hence these DFDs are created from the **bottom up**.

6.5.4 Design Transition

Here we are concerned with the transition from old to new system. The system design is refined to include procedures for:

- the interfaces to and from other systems;
- data conversion from existing files/databases and/or data collection from existing manual records.

6.5.5 Design Procedures

Each procedure in the system structure now becomes the focus of attention for detailed procedure design which will then go forward to provide the specification for program development either manually or via a code generator.

For on-line procedures, it may be necessary for the dialogue flow to be designed over more than one screen to support a particular procedure. This would certainly be the case where there was too much data to be displayed on a single screen. In this instance each procedure is defined as one or more **Procedure Steps** with a screen layout being designed for each procedure step.

Dialogue design therefore involves:

- defining the procedure steps which are required;
- defining the dialogue flows between procedure steps;
- designing screen layouts;
- defining the processing logic for each procedure step.

The design of dialogue is influenced by the work environment which is defined as either 'constant', where an established pattern of work exists (deviation from it being very unlikely), or 'dynamic' where a variable work pattern exists (very unlikely that a sequence of operations will be repeated frequently). Varying user roles and experience together with security access levels will further impact on the choice of dialogue style. A rigid pattern of interaction between user and computer will follow a fixed sequence of screens whereas if alternative courses of action are required in the system, branching actions with hierarchies of menus may be dictated.

The design of dialogues (which after all *is* the users' view of the system) should involve the use of **Prototyping** to demonstrate the dialogue to users, involving them in the refinement of the dialogue flows. **Dialogue Flow Diagrams** are created showing the legal flows which the dialogue may follow. Such a Dialogue Flow Diagram for Midlinks Motel is shown in Figure 6.18.

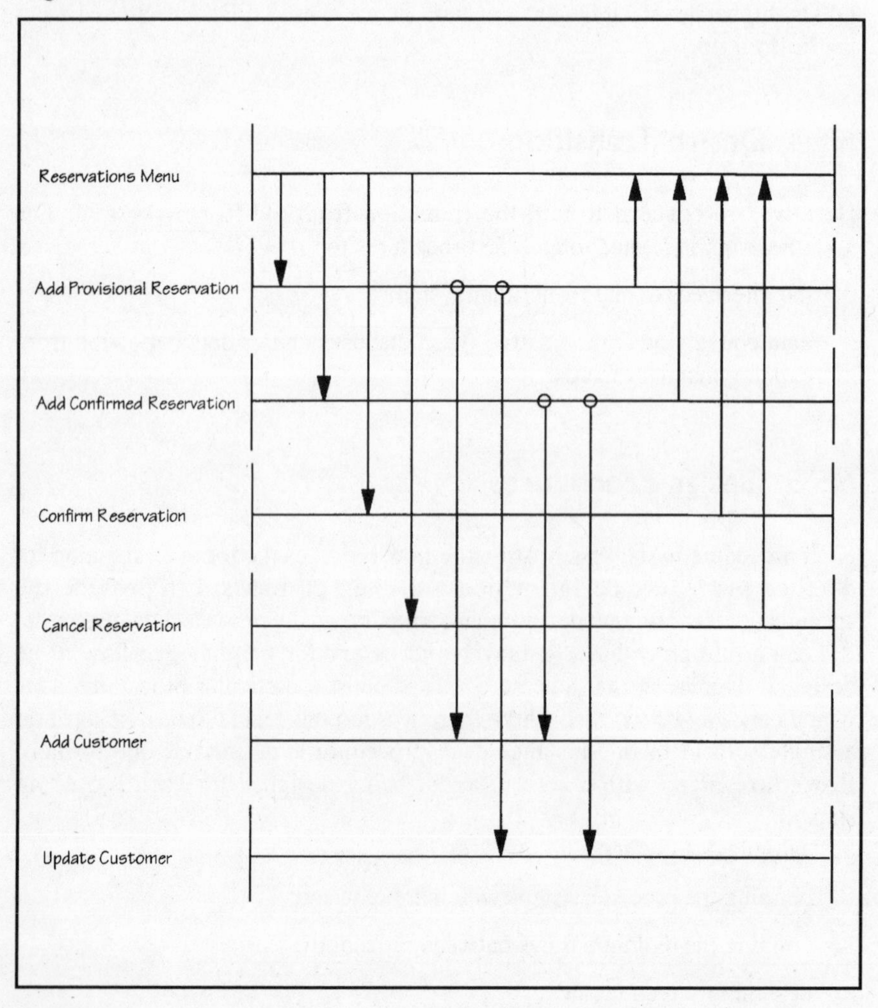

Figure 6.18 Dialogue Flow Diagram for Reservation System – Midlinks Motel

This diagram is the same in principle as the Entity Life Cycle State Transition Diagram except that it details the transitions between Procedures/Procedure Steps and therefore the flow of control.

The data access for a procedure step is documented using a **Data Access Diagram** which is similar to the Process Logic Diagram used in BAA. The data access diagram:

- confirms that the data structure diagram supports the procedure step's data access requirements;
- provides the database designer with information about the data access (what and how frequently) to enable the design to be tuned in accordance with system performance requirements;
- forms a stepping stone to the Procedure (Step) Action Diagrams.

The Procedure (Step) Action Diagrams are similar in nature to Process Action Diagrams defined in BAA but they define the system logic including the control of flows between screens, the execution of business processing and error handling. From these diagrams the program code can be generated either automatically or by using them as a specification for manual coding. Where the procedure is a report procedure the report layout is also specified and may be demonstrated using prototyping.

6.5.6 Confirm Design

The Business System Design is confirmed for completeness, correctness and usability with some initial consideration of performance to ensure that the system is still feasible and practical in the context of the proposed environment. The needs of users is also considered for transition to the new system, in particular the use of a skills matrix in a Training Needs Analysis to establish training requirements for the new system.

6.5.7 Plan for Technical Design

Preliminary plans are made for Technical Design, the next stage of IE. Implementation areas are defined by ranking groups of procedures on the basis of quantified benefits less the costs of their implementation whilst the sequence of implementation is decided by taking account of logical sequence, business priority and resource requirement.

6.5.8 Deliverables of Business System Design

The deliverables from BSD are shown in Figure 6.19.

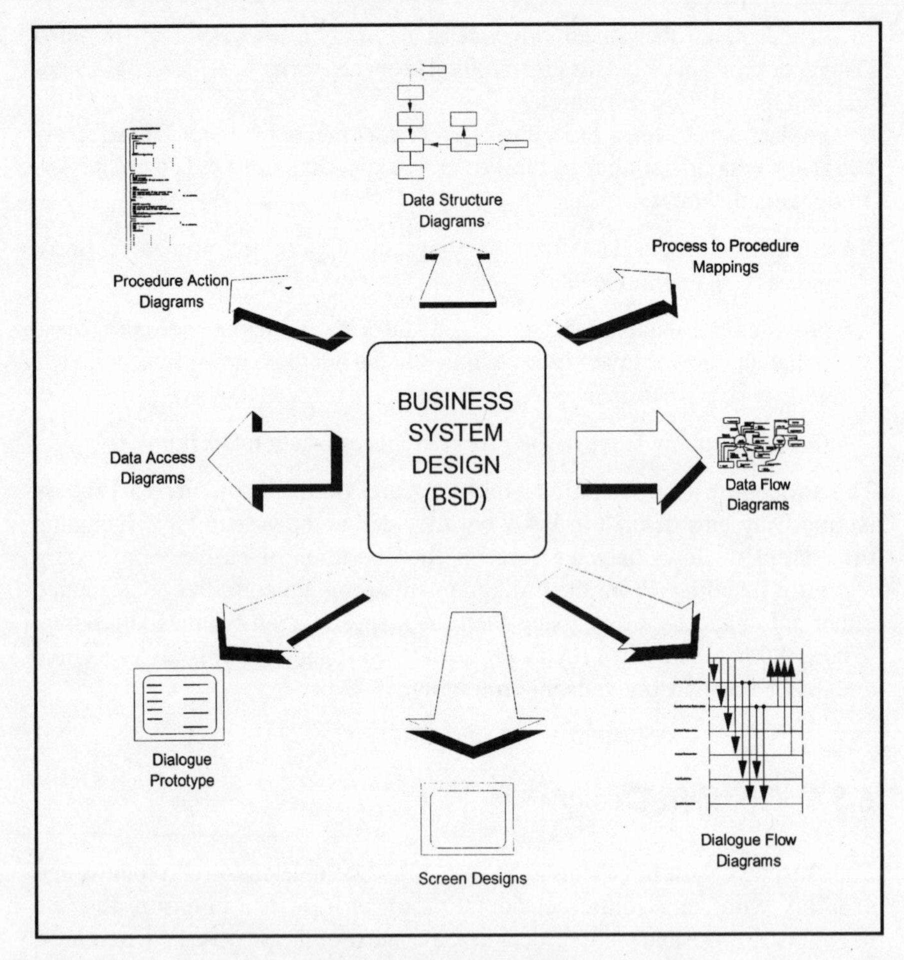

Figure 6.19 Business System Design Deliverables

6.6 Technical Design (TD)

The Technical Design stage of IE takes the computer-based aspects of the business system from BSD and advances the design of physical data structures, programs, operational procedures and interfaces in the context of the target environment. The detail required in this design will depend upon how the implementation is to be effected, more detail being required for high level language implementation than for 4GL implementation for example. TD provides a technical specification encompassing database and procedure design in line with the DBMS as well as the technical architecture for the system. Also indicated are the resource estimates for the construction and transition stages of the project.

The tasks which make up Technical Design are shown in Figure 6.20 and are described below. These are treated somewhat briefly as it would be usual with IE for much of this work to be automated by a CASE tool.

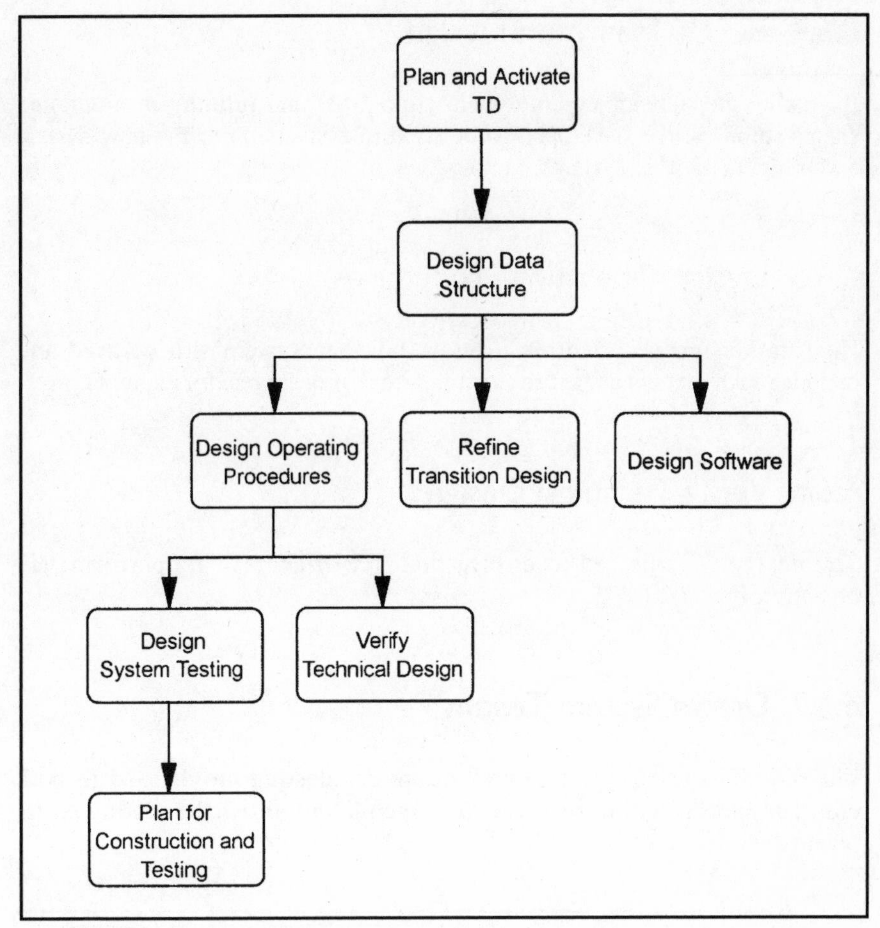

Figure 6.20 Technical Design Tasks

6.6.1 Plan and Activate Technical Design

The Technical Design is planned and the Technical Architecture, first defined in ISP, is reviewed so that the technology can be finalised for the system under development.

6.6.2 Design Data Structure

The data base design is now progressed from a technical perspective so that the physical design can be optimised for performance.

6.6.3 Design Software

At this point the Procedure Steps from BSD are packaged into software units (modules, programs, etc.) which are appropriate to the target environment.

6.6.4 Refine Transition Design

By taking the transition requirement from BSD and refining it, a detailed specification can be developed to be used in conversion to the new system and bridging to other systems.

6.6.5 Design Operating Procedures

Here the designer's attention turns to how the system will be used and includes such matters as security and fall-back procedures, for example.

6.6.6 Verify Technical Design

The design is reviewed to ensure that it will achieve its performance objectives.

6.6.7 Design System Testing

The requirements for system and acceptance testing are defined to both confirm understanding of the requirements and allow the testing to be planned for.

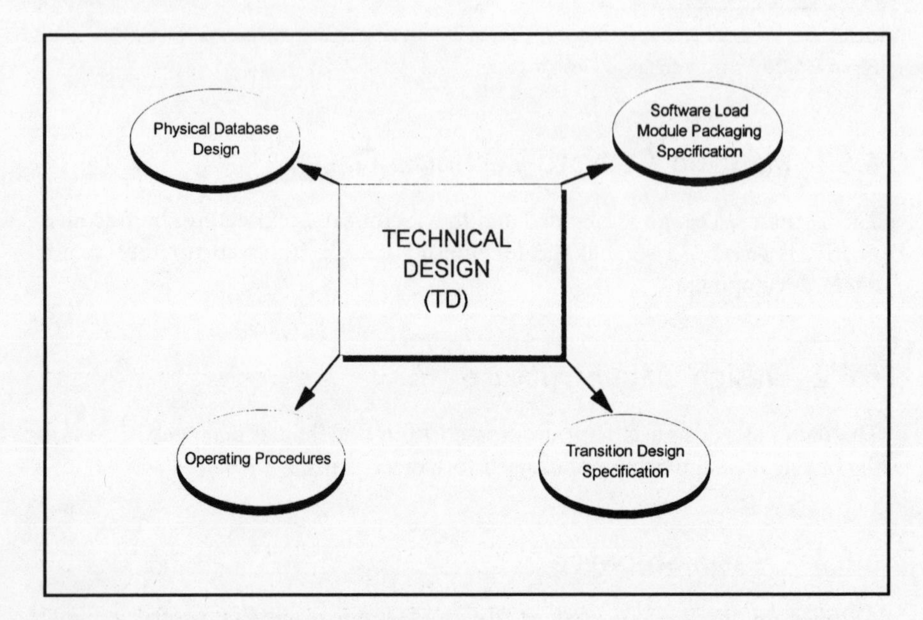

Figure 6.21 Technical Design Deliverables

6.6.8 Plan for Construction and Transition

Early consideration is now given to the environment for construction and testing at this point. The planning, undertaken here, will be further refined in the individual stages.

6.6.9 Deliverables of Technical Design

Technical Design culminates in the deliverables shown in Figure 6.21.

6.7 Construction

Construction is the stage where the designed system and the databases are built either by manual coding or via code generation from the design deliverables. Construction is complete when acceptance criteria have been satisfied and user acceptance obtained. The tasks within construction are as in Figure 6.22.

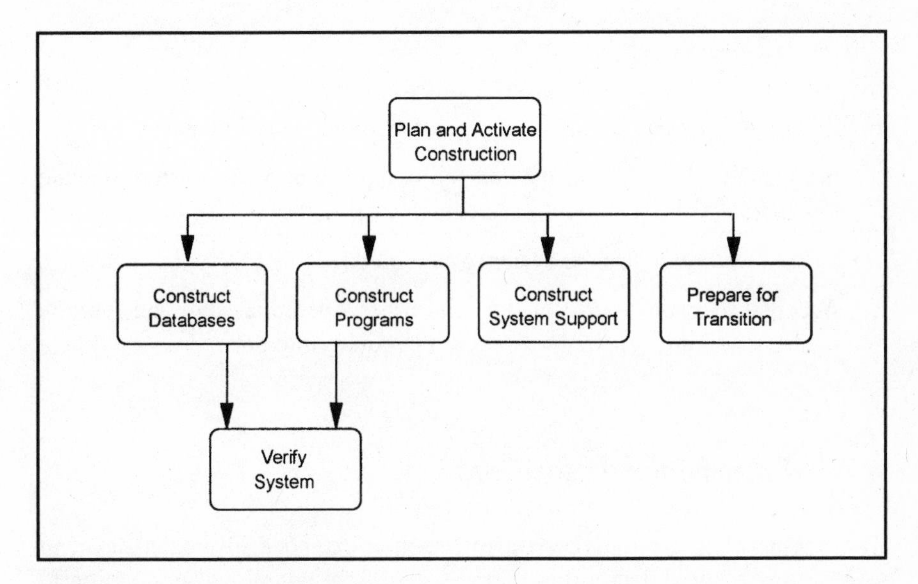

Figure 6.22 Construction Tasks

6.7.1 Plan and Activate Construction

The usual stage planning activities are undertaken and the environment for the construction effort is established. This environment includes the construction of the computing environment as well as the procedures which the construction team are to apply.

6.7.2 Construct Databases

The detailed database definitions are prepared, the databases set up and initial data entered if this is required.

6.7.3 Construct Programs

Programs and Screen Formats are constructed based upon the design specifications and these are then packaged according to the technical design. The programs are then subject to unit test.

6.7.4 Construct System Support

The purpose here is to develop system support materials. These materials include 'on-line help' facilities, user guides, training manuals, operations manuals, database administrators manual.

6.7.5 Verify System

In order to verify the system, three types of testing are carried out:

- Integration testing: to verify that the components of the system operate together;
- System testing: to test the system as a whole;
- Acceptance testing: performed by end-users, operations staff and auditors to check that the system does what they require it to do.

6.7.6 Prepare for Transition

The outline plans from earlier stages are now extended in greater detail to specify the required Transition effort. It should be noted that, in addition to IT staff resources, user resources will also need to be scheduled for Transition.

6.7.7 Deliverables of Construction

The deliverables of Construction are summarised in Figure 6.23.

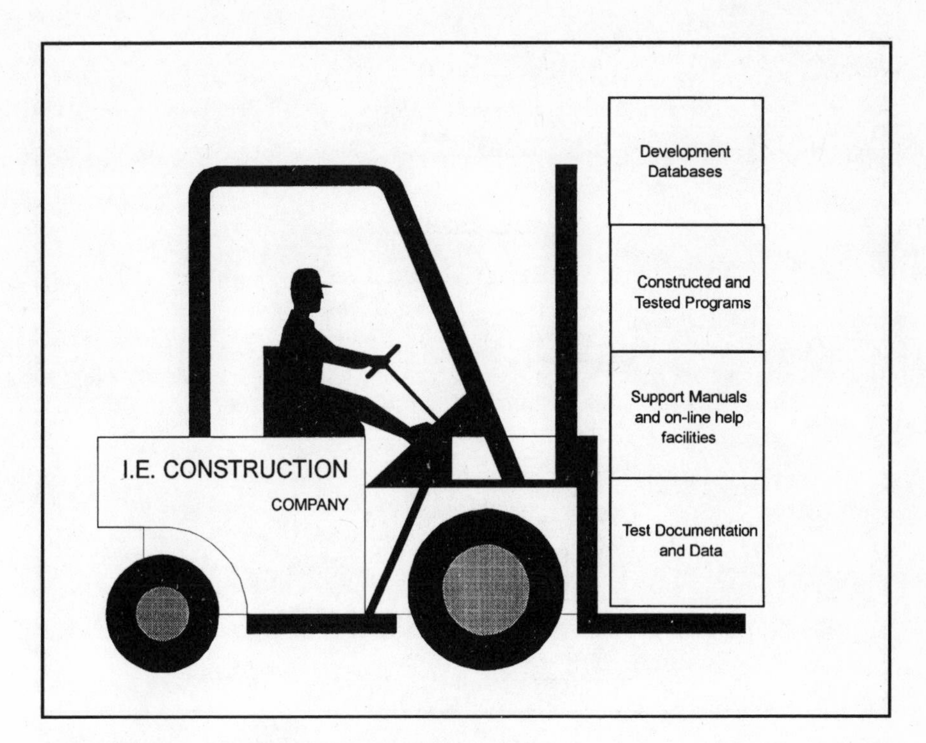

Figure 6.23 Construction Deliverables

6.8 Transition

Transition forms the controlled phasing out of existing procedures and files with their replacement by the new system. It is also accompanied by appropriate user training in the new system, the success of transition being measured over a defined period culminating in a post-implementation review. The tasks involved are indicated in Figure 6.24 and are further explained below.

6.8.1 Plan and Activate Transition

This task continues the activities from the last task within Construction, planning the resource requirement and transition schedule.

6.8.2 Conduct User Training

Users are trained on the system and with realistic training data loaded onto a training area of the system.

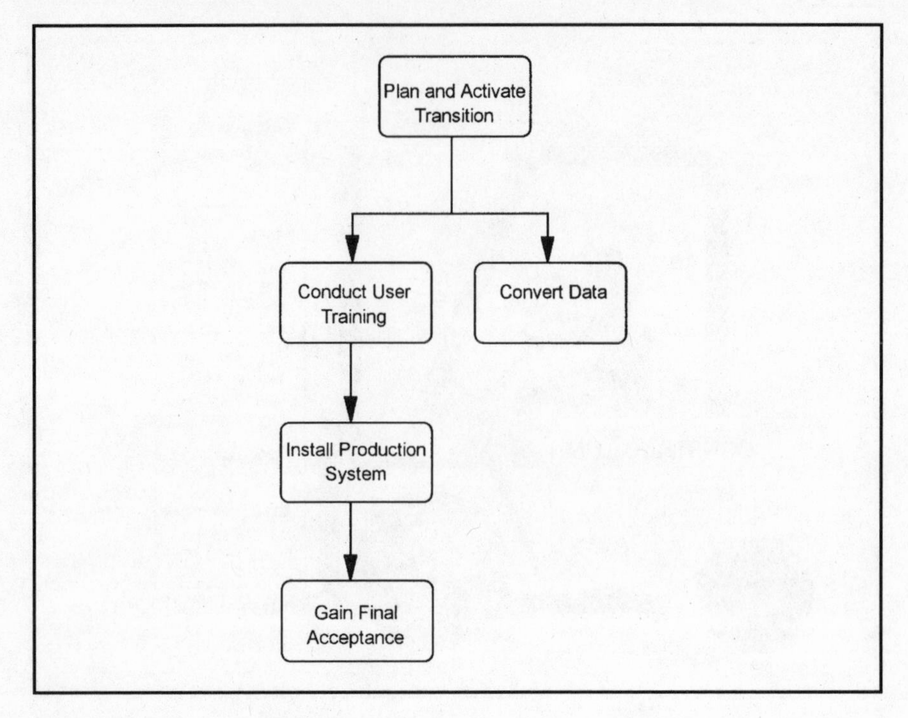

Figure 6.24 Transition Tasks

6.8.3 Convert Data

Data is loaded onto the production system databases either by data conversion from existing files or databases or by keying in data from manual records. The loaded data must be verified for quality and integrity to ensure that it meets the standard required by the new system. This task often requires a lot of user and IT staff effort.

6.8.4 Install Production System

Based on the planned migration strategy the system is installed. This may involve phasing the system in across user locations, across product ranges, running the old and new systems in parallel or adopting the 'Big Bang' approach whereby the old system is replaced by the new system 'overnight'.

6.8.5 Gain Final Acceptance

Final acceptance of the new system is sought from user management with a formal 'sign-off' for the delivered system.

6.8.6 Deliverables of Transition

The transition deliverables are shown in Figure 6.25.

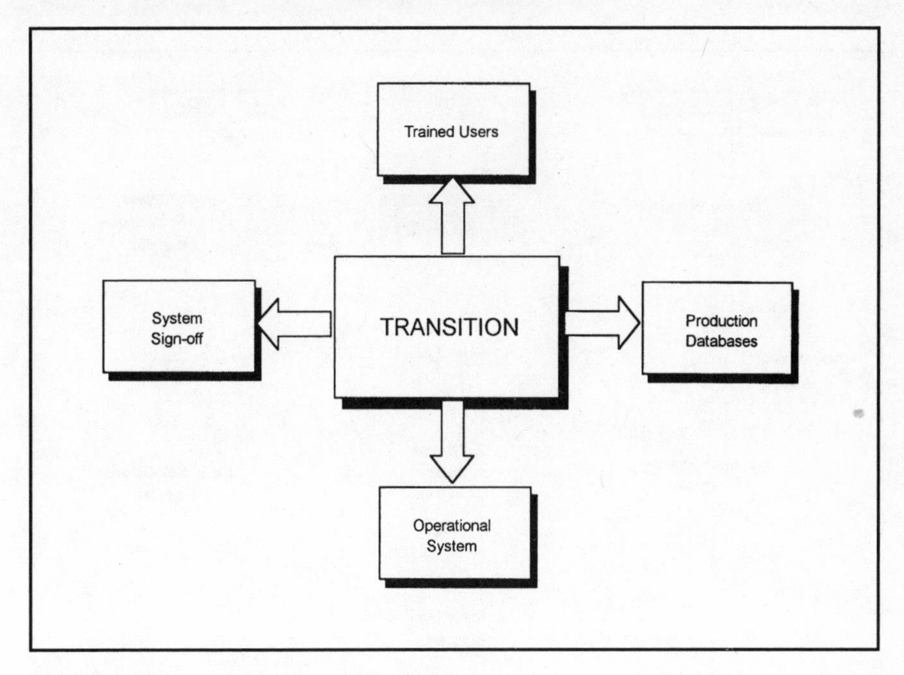

Figure 6.25 Deliverables of Transition

6.9 Production

The production stage of IE is the continued operation of the new system. In this period the system is reviewed and tuned to maintain performance and modified or enhanced to meet user requirements.

6.10 Conclusion

Information Engineering is an established method which exists in a number of slightly different variants brought about by divergent development from its original form. It exhibits a wider coverage of the Business Life Cycle than some other methods by encompassing Information Strategy Planning.

The method defines several development paths to suit an organisation's particular requirements for speed of development, level of investment and technical direction. It is well-documented and defined, with well-established, code-generating CASE tool support being available and strongly advised to realise the maximum benefit of rigour and productivity.

The Information Engineering complete development path has been described here, and the stages, tasks and deliverables have been considered in sequence. The major products of IE and their interactions are summarised in Figure 6.26.

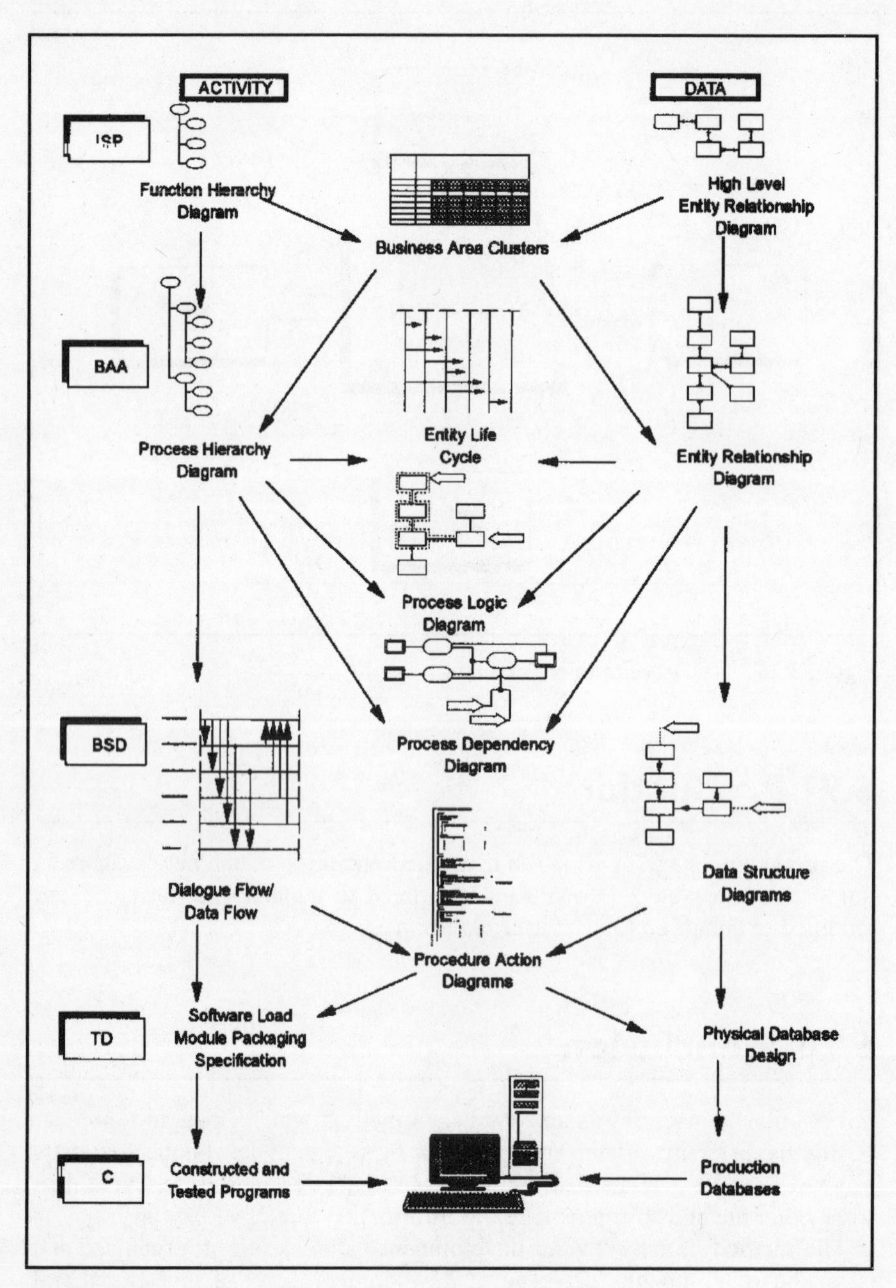

Figure 6.26 Summary of IE Products and Their Interactions

7

Yourdon Structured Method

7.1 Overview

Yourdon was amongst the pioneers in the field of structured methods, with books from Tom De Marco, Edward Yourdon and Larry Constantine, (Yourdon Press) setting the analysts of the late 1970s and early 1980s chattering with something approaching excitement. Here were a new set of techniques to tame the wild task of defining requirements! The present-day Yourdon Structured Method (YSM) has changed and matured considerably over the intervening years, moving from a stance which placed most of the emphasis on the data flow diagram and its supporting process descriptions and data dictionary (De Marco 1979) to one which now includes data and event analysis. YSM is not the most precisely defined nor fully documented of the methods considered in this book. The best definition of it is found in Edward Yourdon's very readable book (Yourdon 1989). The method contains many techniques in common with other methods considered here, but also has a few additional ones of its own up its sleeve. These, together with a different flavour of usage of the common techniques, lend it particularly well to use within the real-time environment, whilst not preventing it from being a method for use in a commercial information systems environment. It holds the analyst's hand further into the realms of design than many other methods, continuing to the point where processes are allocated to specific physical processors, and diagrams are created to show the hierarchical structure of program modules.

It is widely used on both sides of the Atlantic and, until the arrival of SSADM, was the most widely taught structured method in the academic world within the United Kingdom.

YSM has expanded and matured considerably over the years. The classical approach to the method involved a progression from the existing, physical system, through to a logical existing system view. New requirements were then added to create a logical required system view. Finally, physical implementation details would be included to give a physical required system model.

This approach was fundamental to all structured methods in their early days and is described in more fully in Chapter 1. It was the underlying philosophy of early Yourdon. However, the Yourdon method has, since the late 1980s, moved away from this structure. Edward Yourdon (Yourdon 1989) recommends that 'the systems analyst should avoid modelling the user's current system if at all possible' and move as quickly as possible to modelling the required logical system.

Thus, modern YSM adopts a view of the required system in terms of:

- required data: the entity relationship diagram provides a logical view of required data;
- required processes: data flow diagrams are the main tool for obtaining the view of required processes;
- events: both update and enquiry events affecting the required system are identified, and their effects analysed using state transition diagrams.

and accompanies the idea of modelling the existing system with a health warning!

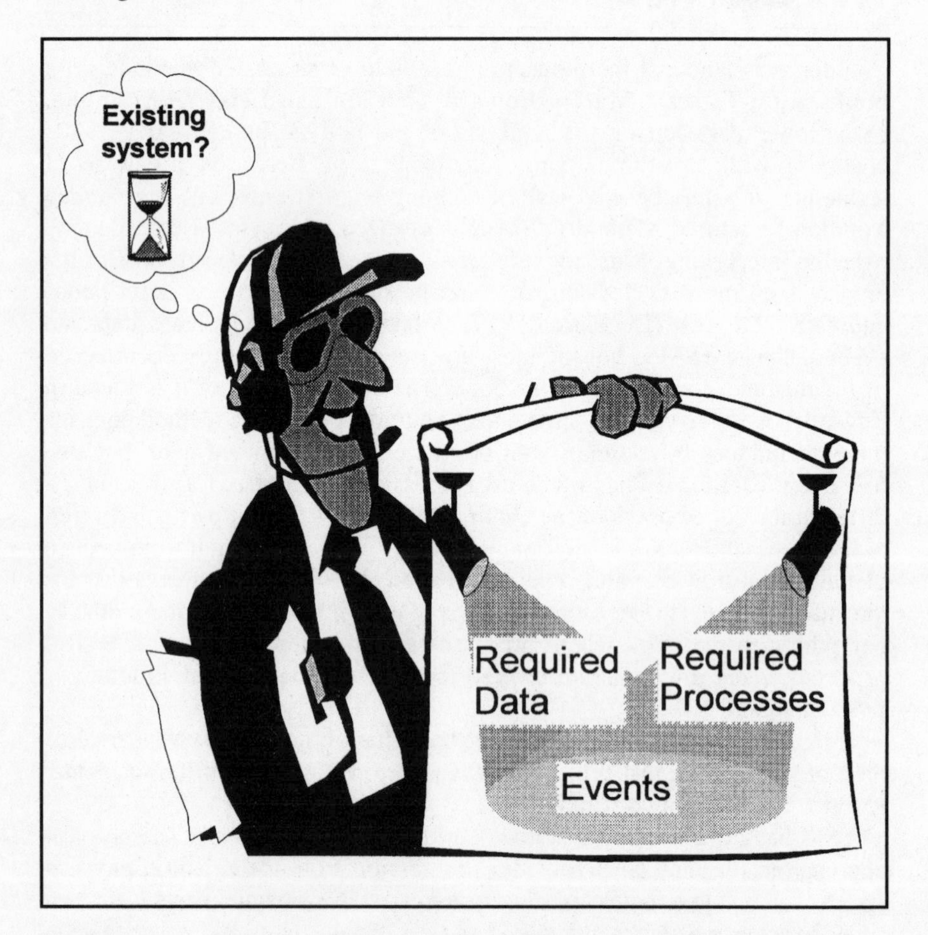

Figure 7.1 The Modern Emphasis of the Yourdon Structured Method

YSM is based on the development of 5 models:

- the current implementation model (optional);
- the essential model;
- the user implementation model;
- the system implementation model;

- the program implementation model.

The essential model is further subdivided into:

- the environmental model;
- the behavioural model.

and the systems implementation model encompasses:

- the processor model;
- the task model.

The relationships between these models are shown in Figure 7.2

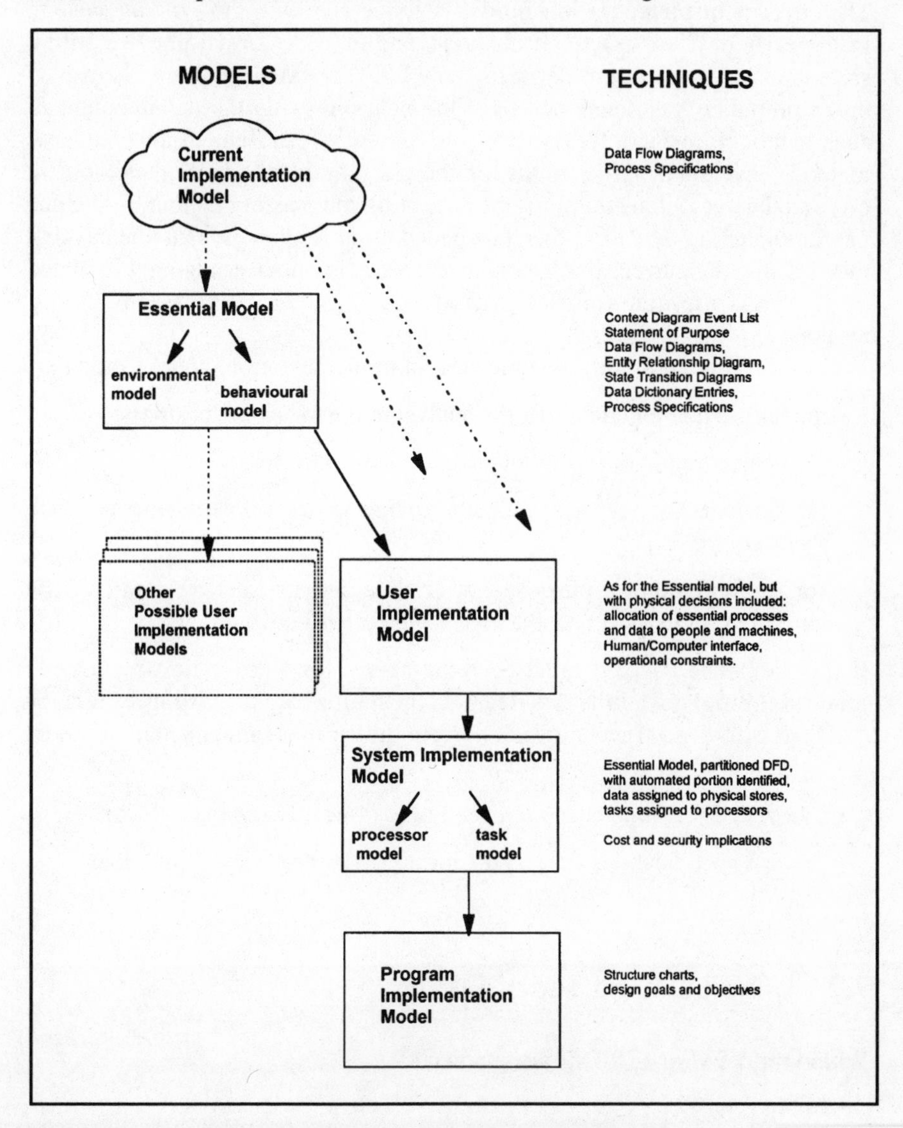

Figure 7.2 Models Within Yourdon Structured Method

You will see from Figure 7.2 that the current implementation model, if present, provides information to the implementation models the system, as well as to the essential model. There are potentially many user-implementation models proposed, of which one is selected, to go forward to physical implementation. Each type of model, with the techniques involved, is described below.

7.2 The Current Implementation Model

The **current implementation model** is concerned with the existing system. In the early days of structured methods, and indeed, normal practice before structured methods, was to begin by modelling the existing system. However, much prejudice has developed over the years towards the development of such a model because 'born-again' systems analysts, clutching their new method, have spent five months, of the six months available development time, analysing and meticulously documenting the existing system. In the one month remaining they have then proceeded to hurriedly automate the existing mess! Thus, the current implementation model is rather grudgingly included in the YSM method and designated not to be used unless absolutely necessary!

Good reasons for developing the current implementation model would be:

- the user is not confident that the analyst understands the business;
- the analyst really **doesn't** understand the business!
- the users do not have a sufficient overview of the whole business to state their requirements;
- for implementation purposes, it is necessary to see the detail of the current systems with which the new system will have to interface.

If this model is used, it must be remembered that the main objective is **understanding**, not a fully detailed documentation, of the existing system.

The recommended techniques to create the current implementation model are:

- a 'levelled set' (one or more levels) of data-flow diagrams (DFDs);
- process specifications for just the most critical or complex functions.

7.2.1 The Levelled Set of DFDs

Within the YSM, the DFD is also known as:

- bubble chart;
- bubble diagram;

- process model;
- work-flow diagram;
- function model.

The use of it is very much as described in Chapter 3. The YSM diagramming conventions are shown in Figure 7.3.

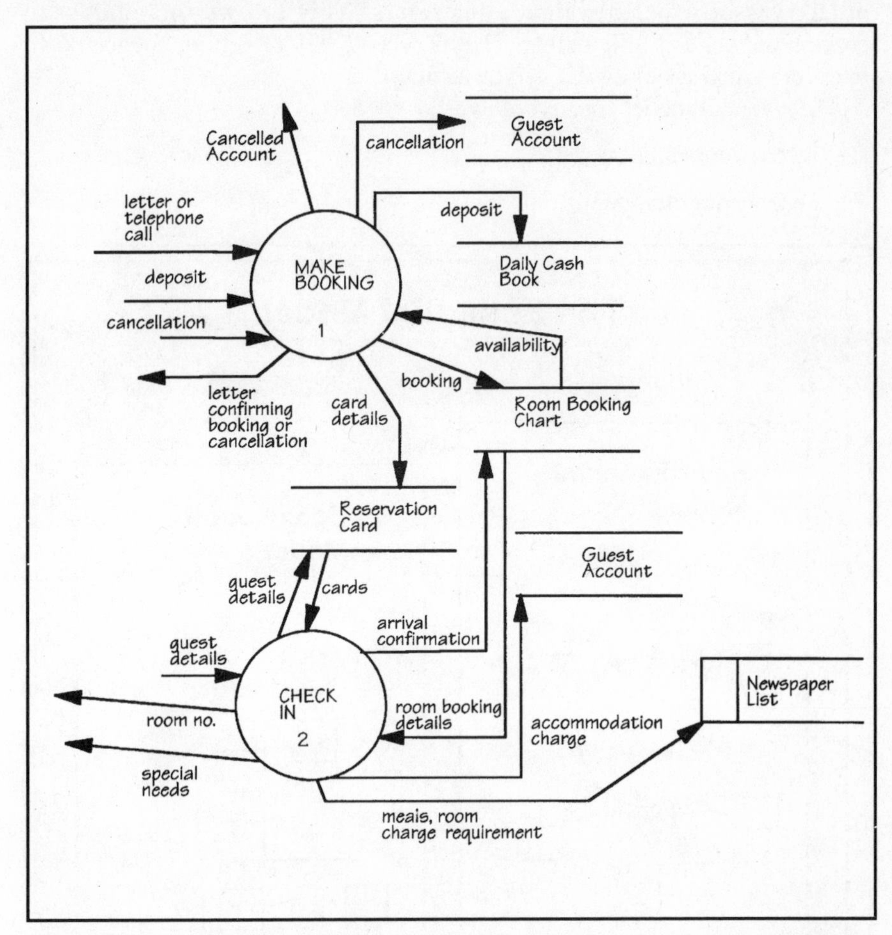

Figure 7.3 Midlinks Motel Current Implementation Model – An Existing System Data Flow Diagram

The external entities, which are called 'terminators' in YSM are not shown on lower levels of DFD, but would appear on the context diagram, an example of which appears later in this section in the discussion of the essential model.

Having drawn the existing system DFDs, it would be necessary to 'logicalise' these, removing all physical constraints on processing and data (as described in Chapter 3). This process is much less formalised in YSM than in other methods, with the implication that 'you shouldn't be doing this model anyway!'.

7.3 The Essential Model

The **essential model** is a 'required system logical' model. It is a diagrammatic representation of **what** the required system should do, and with what data. It excludes **how** it will be implemented, where, or by whom. Such concerns are the subject of later models. It takes no account of performance, storage objectives or other physical constraints. This allows the analyst to concentrate on, and fully define, what is needed, before getting bogged down in the complexities of how it will be achieved.

The essential model consists of two sub-models:

- the environmental model;
- the behavioural model.

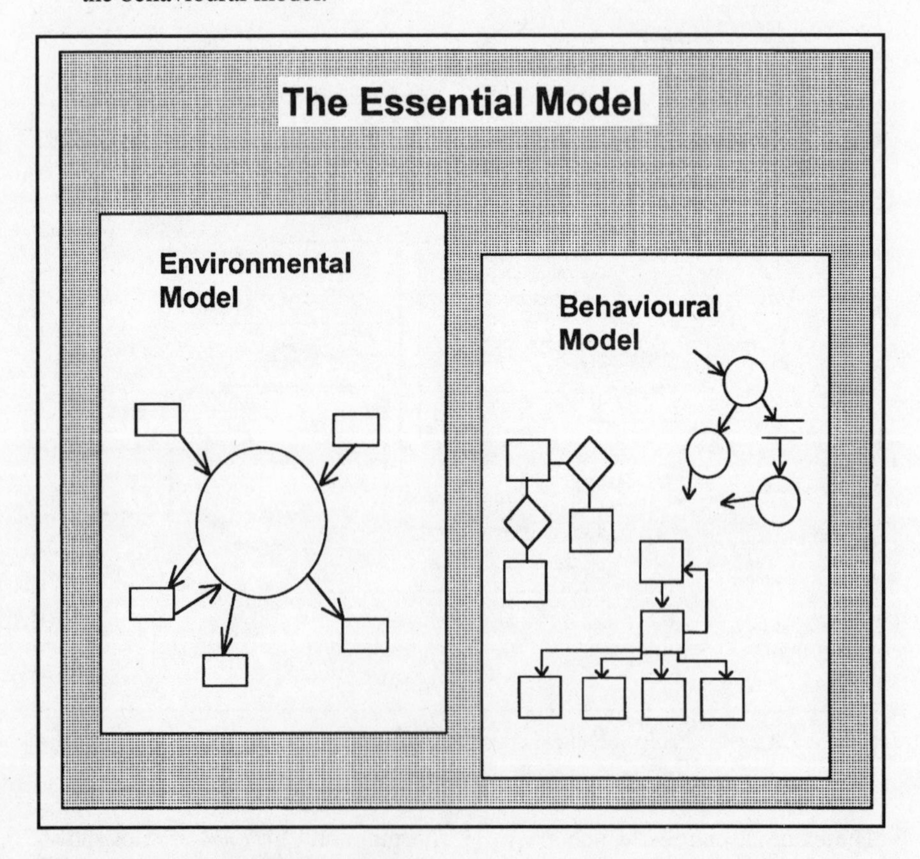

Figure 7.4 The Essential Model

7.3.1 The Environmental Model

The environmental model defines the boundary between the system (both automated and human aspects) and the rest of the world. It consists of:

- a context diagram,
- an event list;
- a statement of purpose.

7.3.1.1 Context Diagram

The Context Diagram shows the system boundary, set in terms of the objectives of the system and of the business. The context diagram is drawn to define a boundary within which lies everything which may be subject to change. This does not mean only the aspects which will be automated. The boundary should also, at this point, encompass areas which are likely to be affected clerically or subject to reorganisation and/or automation as a result of the project. The area inside the boundary is called the **Domain of Change.** The exact position of the boundary is usually negotiable at the start of the project. There will be a 'grey area' within which, at first, it will be uncertain which aspects will be subject to change and which will not. The process of producing the context diagram usually helps considerably in reducing the grey area, by presenting the business user with a clear picture of what is considered to be inside or outside the analyst's area of investigation. Everything within the boundary will be investigated; everything outside will not. The context diagram is frequently a document which is the focus of lively debate as users negotiate their territory around the boundary!

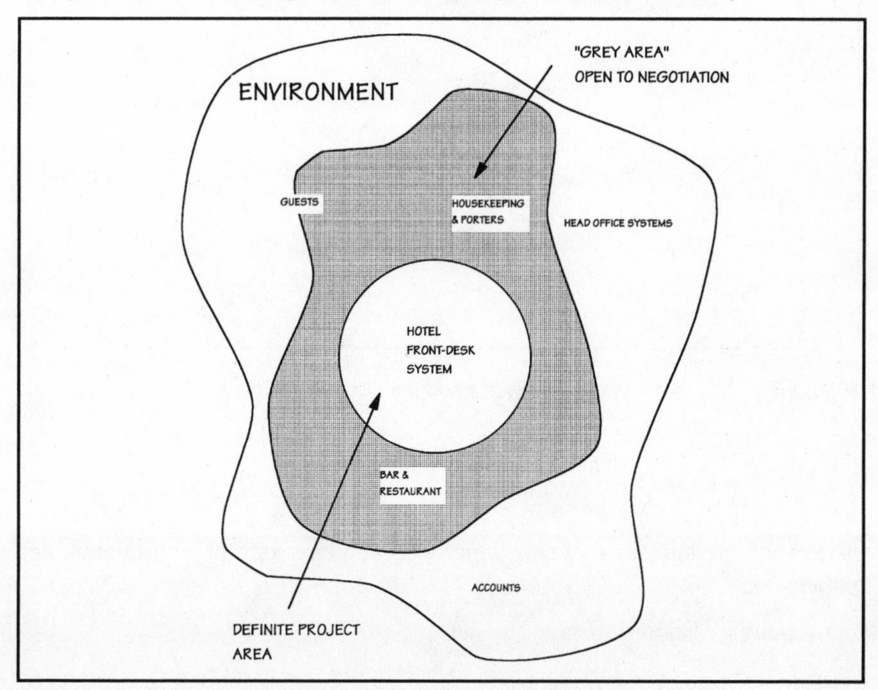

Figure 7.5 Initial Attempt at Defining the System Boundary

The context diagram is developed using the notation shown in Figure 7.7. It represents the required system, the interfaces to the domain of change, and the people and organisations outside of the domain of change with which the system will communicate (these are called 'terminators'). The inputs and outputs across the system boundary can then be used as a starting point to the identification of events. Unlike most other methods, YSM allows data stores to be shown outside of the system boundary, but disallows the use of flows connecting terminators.

The context diagram will be an iterative document, enhanced continuously as further detail and requirements are identified.

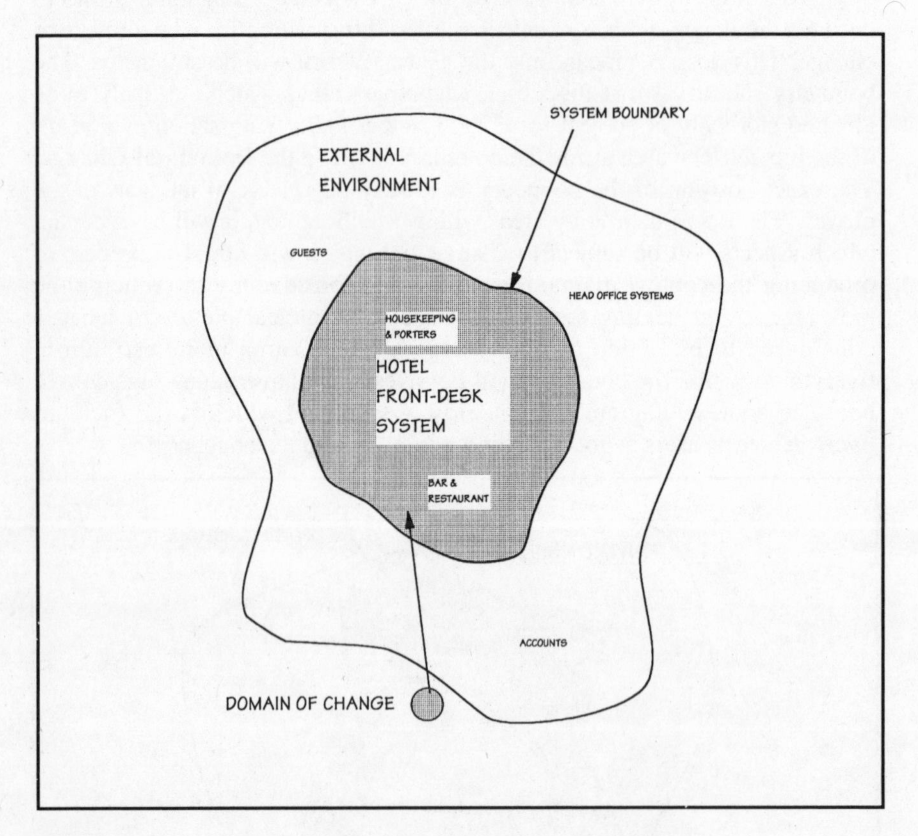

Figure 7.6 The System Boundary and the Domain of Change

7.3.1.2 The Event List

The event list defines all events which occur in the external environment and either:

- require a response from the system;

or

- make a change to the system.

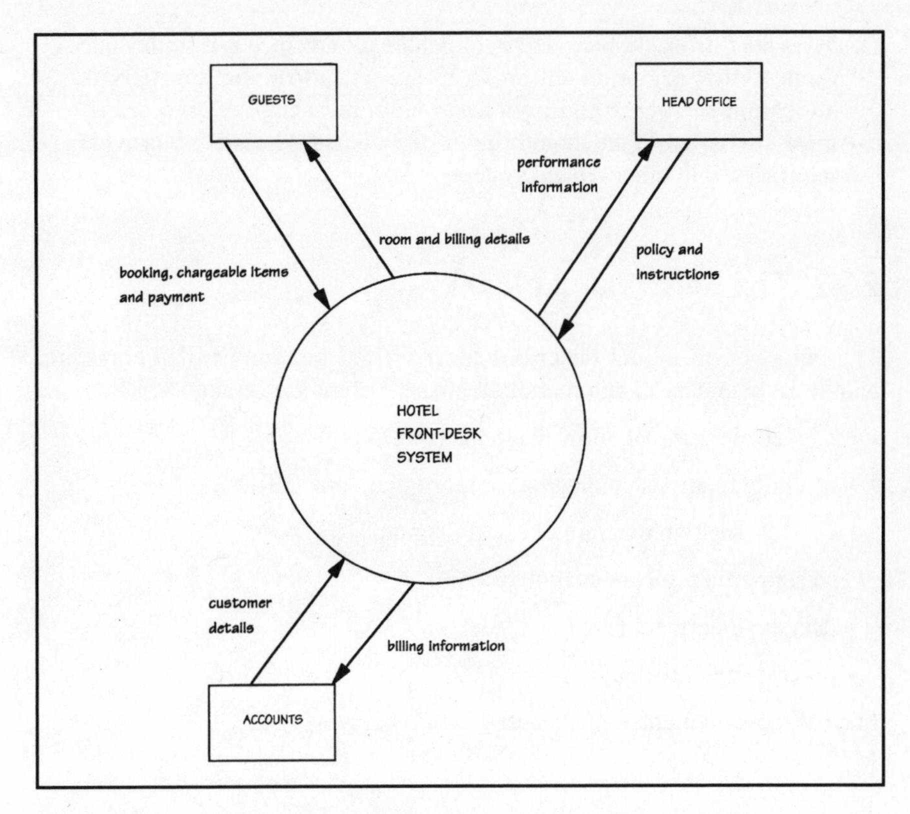

Figure 7.7 Midlinks Motel Context Diagram for the Essential Model

Typical events, taken from the hotel scenario would be:

- guest arrives (at which point changes to the system would include updating reservation, guest and guest account information);
- guest checks out (at which point, system changes would incorporate updating of guest account and reservation information).

YSM takes a wider view of events than certain other methods in that it looks at not just update events but enquiry events as well.

7.3.1.3 The Statement of Purpose

The statement of purpose is a concise definition of the purpose of the system. It is intended for top management and users, to act as a focus for more detailed analysis. The statement may be several sentences long but should not exceed a paragraph. It cannot fully define an entire system, unless the system is very small, but should at least state the aspects of the business which the system is to cover, together with all interested parties. The statement may also state the benefits to be achieved by the system. It is supported by, and in turn supports, the context diagram. An example of a statement of purpose is:

"The hotel front desk system is to handle all advance and immediate requests for accommodation from guests, allow the appropriate allocation of rooms and produce timely and accurate invoices. It must also assist in the monitoring of unsatisfied demand and provide interfaces with other related systems."

7.3.2 The Behavioural Model

The behavioural model describes the required behaviour of the system, within the boundary of automation. It consists of major techniques:

- a levelled set of data flow diagrams (DFDs);
- an entity relationship diagram or set of diagrams (ERDs);
- a state transition diagram or set of diagrams (STDs);

and supporting documentation of:

- data dictionary entries;
- process specifications.

Each of these elements is considered in turn below.

7.3.2.1 Data Flow Diagrams (DFDs)

DFDs are included in most structured methods, with only minor differences in their implementation, other than the degrees of emphasis given to them by the individual method. DFDs were the central tool of early Yourdon, but are now given an equal billing with the ERD and STDs. The notation was shown in Figure 7.3, in discussion of the current implementation model. Just the main differences in technique from other methods will be highlighted here.

One important difference in the use of DFDs between YSM and other methods is that flows of **control** (as well as data) and **control processes** can be shown. This is particularly useful in describing real-time (process control) systems. A flow of control is a signal from within the system (i.e. not triggered by an outside event) which kicks off a dormant process. A control process can be thought of as a 'supervisor' which co-ordinates the activities of other processes. Its only inputs and outputs are control flows. Control flows and control processes are not normally found in commercial information systems but are common within real-time process control systems.

In the development of DFDs, Yourdon recommends not the top-down approach, but a 'middle-out' sequence of development based on *event partitioning*. Top down development, it is suggested, can lead to arbitrary partitioning, too strongly based on current organisational divisions or the number of analysts on the project.

The event partitioning approach is as follows:

- draw a bubble for each event in the event list;
- name it;
- give it appropriate inputs, outputs and stores;
- check the result for consistency with the context diagram (note that the terminators shown on the context diagram are omitted from lower level DFDs).

Initially, the analyst should not try to re-level the DFD; but should develop a process bubble for every event (two bubbles if there can be more than one result of an event).

7.3.2.2 The Entity Relationship Diagram (ERD)

Early Yourdon placed less emphasis on the data analysis aspects of the system than on processes and states. The real-time process control systems, for which YSM was initially used, seldom have complex stored data structures to manage. However, its use in commercial data processing systems have

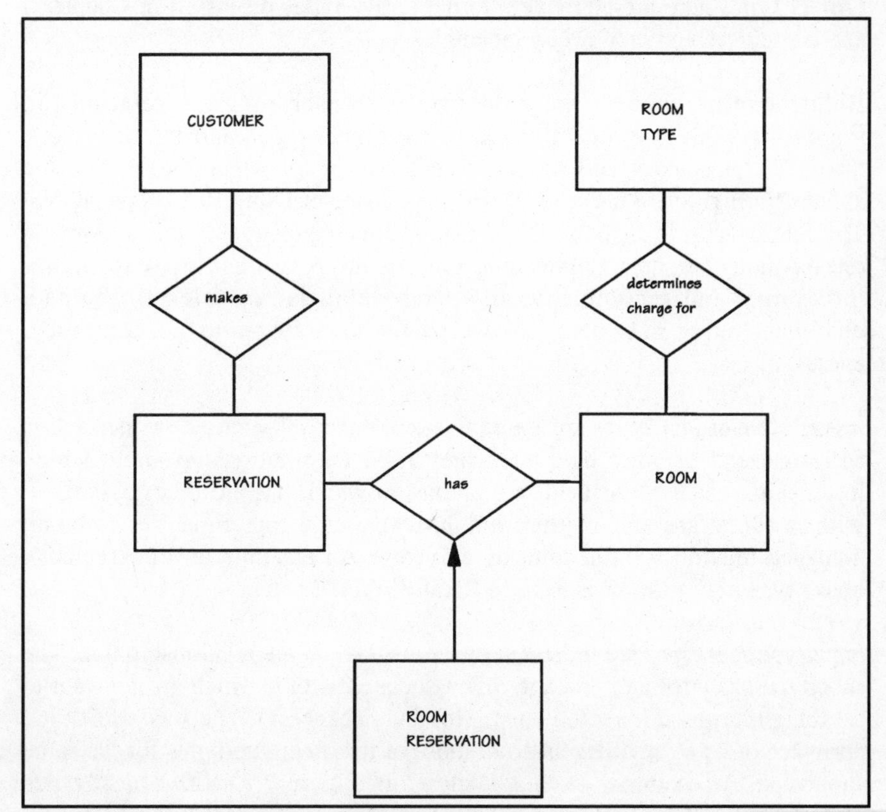

Figure 7.8 An Entity Relationship Diagram

encouraged the increased emphasis on data analysis which modern YSM recommends. The ERD should be developed in parallel with DFDs and state transition diagrams (STDs), none of these being the dominant model. The event list, developed as a part of the environmental model, can be useful in the development of the ERD. The logical stores, which are the object types within the ERD, become the data stores on the DFDs within the behavioural model.

The notation used by Yourdon for the ERD is derived from the works of Flavin (Flavin 1981) and is similar to that used elsewhere (Chen 1976), (Martin 1982) and (Date 1986). An example of the notation can be seen in Figure 7.7).

There are four major components of the ERD:

- object types (strangely not called 'entity types' in spite of the name of the diagram!);
- relationships;
- associative object type indicators;
- supertype/subtype indicators.

Object types here are equivalent to the **entity types** discussed in Chapter 3 and are represented in YSM by rectangles.

Relationships connect the object types to each other. A relationship represents a set of connections between object types and is shown as a diamond connected by a line to each object type for which the relationship applies. In Figure 7.8 a CUSTOMER makes one or many RESERVATIONS. The relationship is 'makes'. Yourdon acknowledges notations for showing one-to-many, many-to-many and one-to-one relationships (cardinality /ordinality) but recommends that such detail is relegated to the data dictionary rather than being shown on the diagram, to avoid clutter and confusion.

Associative object types are the equivalent of the link-entities or intersection entities found in other methods. They reflect a relationship about which information needs to be held: a relationship which has attributes associated with it. They are shown diagrammatically as a rectangle joined to an unnamed relationship diamond by an arrow. An example of an associative object type in Figure 7.8 is ROOM RESERVATION.

Supertypes/subtypes are indicated by a cross-bar on the relationship line: The supertype has attributes (data items) associated with it which are relevant to all the subtypes connected to it. However, each subtype has additional attributes of its own, different from those of the other sub-types for the same supertype. An example of this is shown in Figure 7.9. with object types CUSTOMER, GUEST and COMPANY. The object CUSTOMER is a supertype. It has two sub-types: GUEST and COMPANY. CUSTOMER has the attributes:

name, number, address. The sub-type GUEST will have these attributes, plus additional attributes, e.g. nationality, age, sex. The subtype COMPANY will have the common attributes name, number and address and has additional attributes: contact name, industry sector.

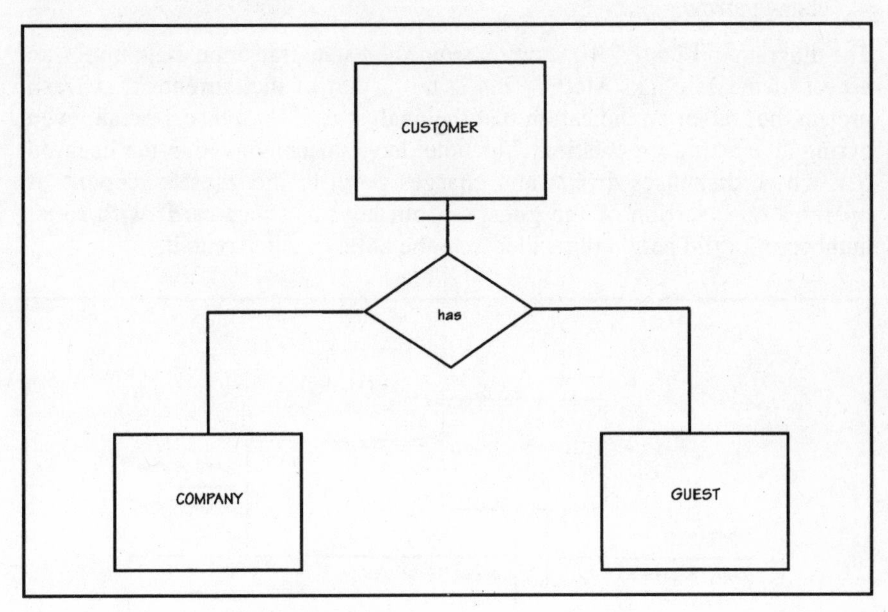

Figure 7.9 Supertypes/Subtypes

7.3.2.3 State Transition Diagrams (STDs)

State Transition Diagrams (STDs) are used to document the time-dependent behaviour of the system. They perform the same role as entity life histories (ELHs) and entity life cycles (ELCs) in other methods, but with a slightly different perspective. With ELHs, the emphasis is on the effects of events (which cause changes of state) rather than on the states themselves. Additionally, ELHs and ELCs address one entity at a time. It is possible to use STDs to analyse state changes of individual entities, but, for real-time systems, they would usually address the whole system and its state changes. This 'whole system' approach is feasible since real-time process control systems operate on transient streams of data and seldom have any significant logical data structure (ERD) associated with them. For small real-time systems, just one STD for the whole system may be developed. If this would be too complex, STDs can be levelled in a similar way to DFDs: any state on a higher level diagram becomes a lower level diagram's initial state. This 'whole system' approach does not preclude YSM being used on an entity by entity basis, where there is a complex data structure.

The STD documents the observable states in which the system resides subsequent to an event and whilst waiting for the next event. The notation of the STD comprises:

- states, represented by a rectangle;
- state changes, represented by an arrow;
- conditions and actions, shown as text along a line orthogonal to a state-change arrow.

The diagram in Figure 7.10 shows a simplified state transition diagram for an aspect of the Midlinks Motel. This is not a part of the current Front Desk project, but rather an indication that the analyst could not take a break, even during an evening's recreation! The hotel has a machine, within the licensed bar, which dispenses drinks and charges them to the guest's account. It operates on insertion of the guest's room key: a 'smart-card' with room number and valid period of residence in the hotel encoded onto it.

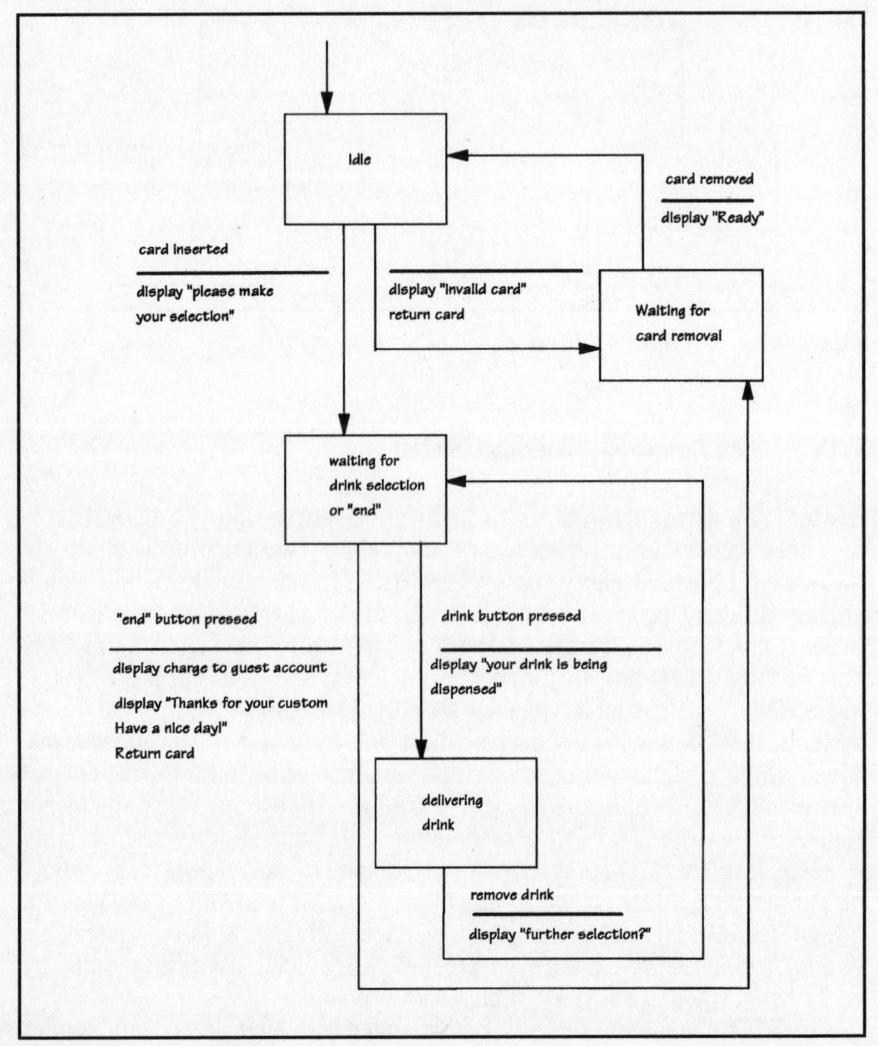

Figure 7.10 Midlinks Motel State Transition Diagram for the Drink Dispensing Machine

There are other states in which the machine may be, such as empty of a particular drink. The reader may wish to think about other valid states and try adding these to the diagram.

In real-time systems, STDs are used to model the inside of all control ('supervisor') bubbles. For the simple system, there may only be one such control bubble, hence the 'whole system' approach. They are also used in YSM for documenting the human interface of on-line systems. They form another view of the information presented in the DFD and as such should be cross-checked with the DFD for completeness.

7.3.2.4 Data Dictionary Entries

The **Data Dictionary** holds the supporting detail for each of the diagramming techniques described. Every dataflow and datastore, on a DFD, with its associated data items, must be defined in the dictionary. Every object type, attribute and relationship on the ERD must be defined there. Volumes of data, including trends and peaks, would also be recorded in the dictionary. Process specifications and descriptions of external entities are frequently also considered to be part of the data dictionary, making it a central repository of all detailed information needed to support the diagramming techniques.

7.3.2.5 Process Specifications

Process Specifications should be written for each lowest-level process on the DFDs. For control bubbles, the process is also described by the state transition diagram, with which the process specification should not conflict. The use of structured English is recommended to aid clarity and avoid ambiguity.

7.3.3 The Essential Model Completed

The Essential Model, containing the Environmental and Behavioural Models expressed in terms of the techniques discussed above, now encompasses:

- the essential policy and logic of the functions to be performed by the required system;
- the essential content of the data stored within, and moving through, the system;
- the essential time-dependent behaviour in order to handle events, signals and interrupts from the environment external to the system.

This will be used as a platform on which to build a series of user implementation models for the business to choose from. A copy of the essential model, prior to changes made to create the user implementation model, should be kept as a part of the system's documentation, since one essential model can form the basis of many different physical implementations.

7.4 The User Implementation Model

The **User Implementation Model** is a 'required system' model which is developed directly from the essential model, by adding physical decisions to it. It is created by augmenting, annotating and revising the essential model. As can be seen from Figure 7.2, several implementation models will normally be developed, giving the business user choices of alternative business and technical implementations. The essential model contained a complete description of **what** the system must do in order to satisfy the business objectives for the project. The user implementation model builds in some, although not all, of the **how**.

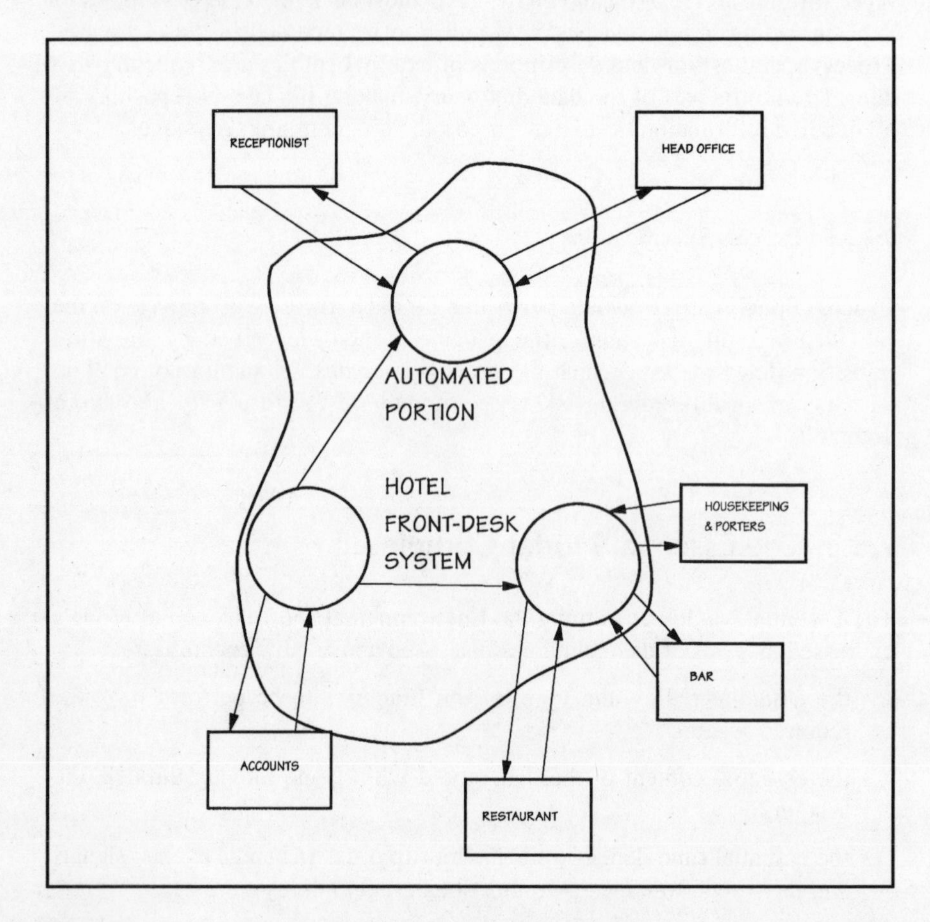

Figure 7.11 An Outline Essential Model, at the Start of its Conversion to a User Implementation Model, with the Automation Boundary Shown

The User Implementation Model addresses four issues:

- the automation boundary (or boundaries, since the automated portions of the system may be disjoint bits) and the allocation of the essential model's processes and data to people and machines;
- details of human machine interaction (with the use of state transition diagram notation for dialogue design);
- additional manual activities, to handle input and output, long-term storage, security, backup and recovery and other physical aspects;
- operational and environmental constraints, such as:
  - security and audit requirements;
  - volumes and throughput and response time requirements;
  - requirement to use particular software, languages or hardware;
  - environmental unfriendliness (e.g. excessive humidity or dirt);
  - required reliability metrics.

At this point, manual activities are taken outside of the automated portion and buried within the terminators.

The user implementation model falls into the 'twilight zone' between analysis and design. It goes beyond just describing **what**, but does not yet fully describe **how** the system is going to be implemented. Once the essential model is established, and a particular user implementation option has been selected, the design of the system can begin. The most important models to the designer are the systems implementation model and the program implementation model. Each of these is described below.

7.5 The Systems Implementation Model

The purpose of the **Systems Implementation Model** is to show how the user implementation model will be allocated to hardware and software. The designer must first decide how the user implementation model is to be physically split between processors and physical stores and then develop the systems implementation model to show:

- the allocation of processes and data stores to processors and storage hardware;
- the allocation of each process and data store to individual tasks (sometimes known as partitions or job steps) within each processor.

The Systems Implementation Model consists of:

- the processor model;
- the task model.

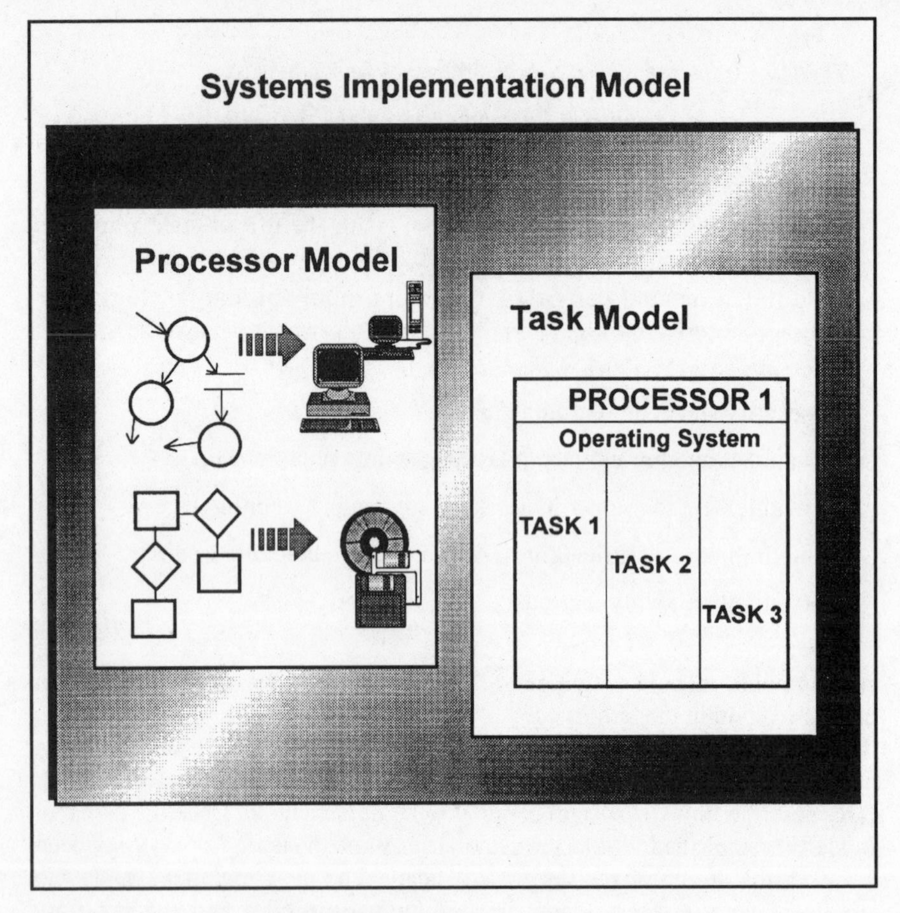

Figure 7.12 Components of the Systems Implementation Model

7.5.1 The Processor Model

The **Processor Model** illustrates the designer's decisions on the allocation of DFD processes to physical processors and logical data stores to particular physical data stores on specific hardware. The designer's choice of implementation depends heavily upon what is to be achieved by the system, but options would include:

- assigning all processes to one processor (mainframe approach);
- assigning one processor per DFD bubble (real-time, distributed approach);
- assigning data stores dedicated to each processor;
- sharing data stores between processors, which may require the duplication of certain data for practical reasons, whilst acknowledging the problems that data duplication brings.

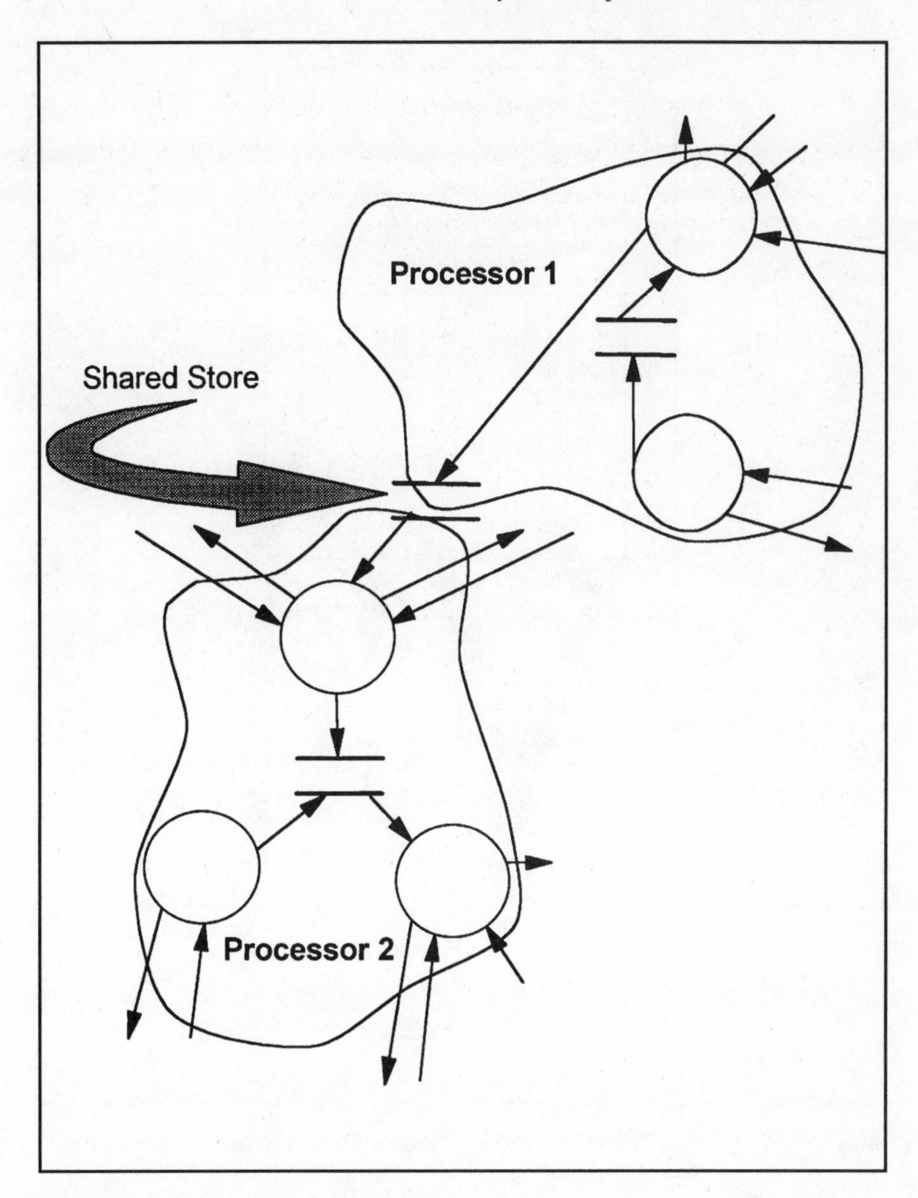

Figure 7.13 The Processor Model

Considerations of cost, performance, reliability, security, and many other constraints and requirements identified earlier in the project would dictate the appropriate implementation choices.

7.5.2 The Task Model

The **Task Model** shows the portions of the User Implementation Model to be allocated to each task (partition) within a processor.

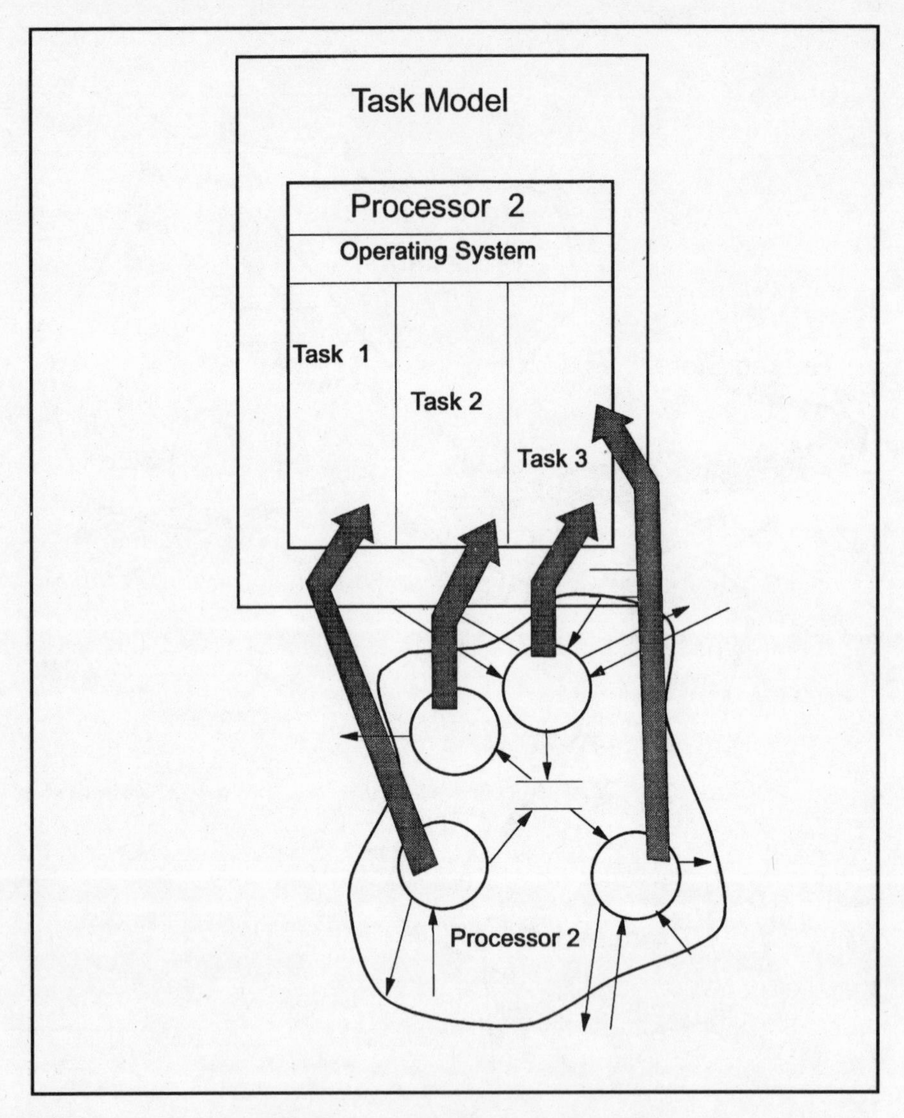

Figure 7.14 The Task Model, Mapping Processes from The User Implementation Model DFDs

7.6 The Program Implementation Model

Here we have reached the level of the individual physical task (partition) within a processor. Within the individual task, only one activity can take place at a time. The model for organising activity in a single synchronous unit is the **Structure Chart**, which shows the hierarchical organisation of modules within one task. Each structure chart will usually map the activity from one DFD bubble, but the mapping could be from more than one or part of one.

7.6.1 The Structure Chart

The Structure Chart has the following notation:

- a module, represented by a rectangle;
- a connection: a line joining two or more modules implies reference from one module to another (one module has called the other);
- a couple, represented by a short arrow with a circular tail, which implies data movement between modules. If the tail is open, this implies the passage of application data between modules. If closed (shaded), it implies the passage of control data.

It can also show:

- sequence, by the ordering of boxes from left to right;
- selection, by a diamond-shaped symbol on the branch to two boxes;
- iteration, by the use of an arrow circumscribing a connection between boxes. These constructs are illustrated in Figure 7.15. The diagram can be annotated with comments, if required. Other subtle embellishments to the chart which the designer may find useful are explained within (Yourdon and Constantine 1989).

There are several strategies for the conversion from DFD to structure chart. Classic and well-documented 'cook-book' strategies include transform-centred design and transaction-centred design, documented in Yourdon and Constantine's book (Yourdon and Constantine 1989). The designer has a considerable amount of flexibility in performing this task, and the 'best' solution is dependent on the particular hardware and software platform for the system. However, some general characteristics of a good transition from DFD to structure chart are possible:

Each module on the Structure Chart should be:

- a complete logical procedure (usually one bubble on the essential model DFD;
- able to be built, coded and tested in isolation;
- of limited size and complexity;
- have a single entry and exit point;
- have a simple interface to other modules;
- return to the calling module after execution;
- be predictable in outcome, always performing the same function.

Modules should not modify code in themselves or other modules. A module should not, in general, call more that half a dozen lower level modules. To exceed this means increased complexity of code, which can lead to maintenance difficulties.

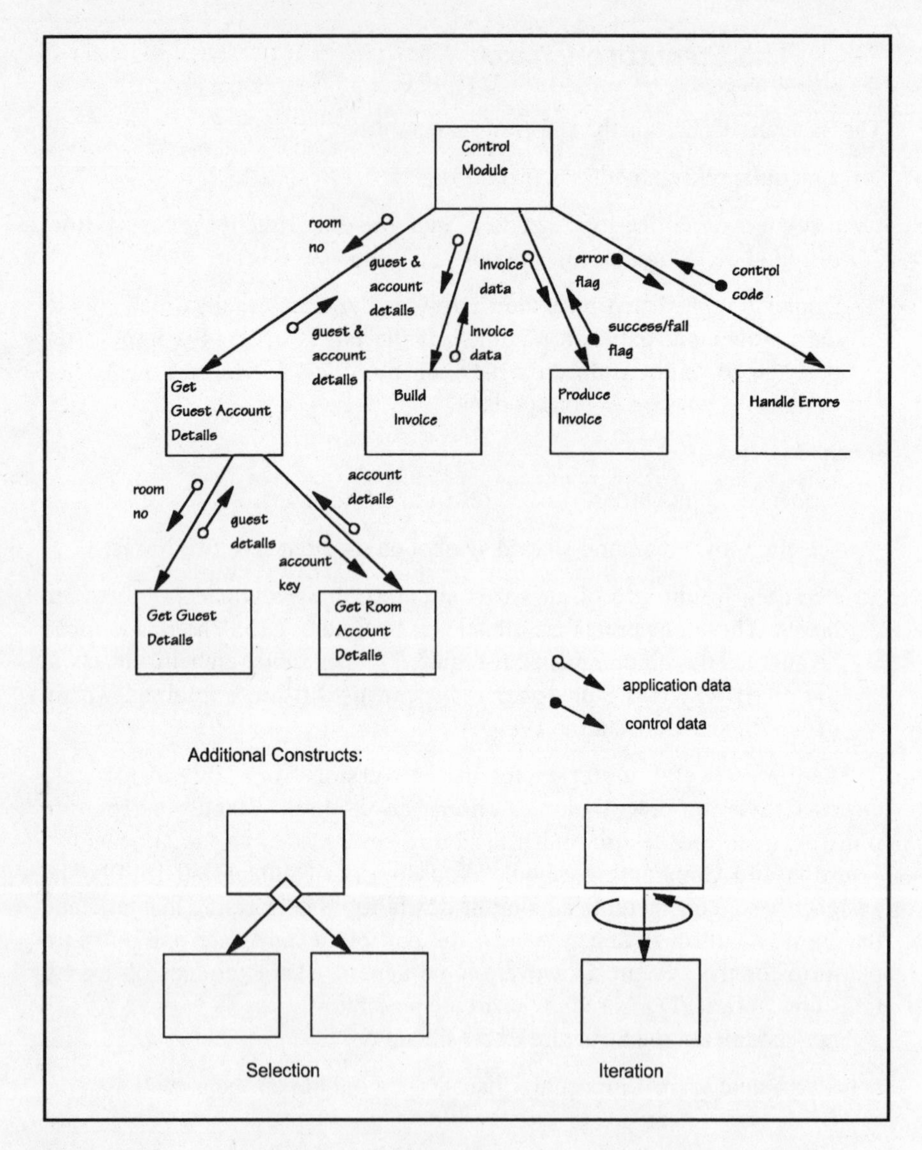

Figure 7.15 Midlinks Motel Structure Chart for the Production of an Invoice

Modules should be highly cohesive, and lowly coupled to other modules, which are the underlying principles of this and all structured methods.

7.7 Conclusion

The Yourdon structured approach to systems analysis and design was amongst the earliest players in the field. At that time it was concentrated on processes but was light on data analysis. Its use within the real-time community led to its refinement in the area of state transition diagrams and more recently, its use within commercial data processing systems encouraged the development of the data analysis aspects. In common with methodologists' current thinking, modern YSM has sought to diminish concentration on the existing system and to focus much more on the required system, taking the three-views approach of data, events and processes. The method ventures further into the definition of the physical system than do most other methods, giving useful pointers and diagramming techniques for structuring programs and partitioning processing and data. Its informality of documentation and lack of 'rules' makes it both flexible in use and conducive to individual customisation. However, it relies on the experience and expertise of the practitioner to select the correct tools for each aspect of the job, and to take account of the project tasks which are outside of the method, such as quality assurance and project management.

8
MERISE

8.1 Overview

MERISE is an information systems design and development method, most widely used in France. It was defined during the late 1970s, when the French Ministry of Industry successively funded several projects which brought together the expertise of a consortium of consulting firms, including CGI-Informatique, Bull, and Sema. The purpose was to develop an information systems design method which would be used by both private firms and civil service to produce data processing applications. The objective was not to devise a completely new method, but rather to assemble the accumulated wisdom and best practice from existing methods of the time and to bring these together in a method which would address the whole of the system life cycle. In the early 1990s, MERISE was estimated as holding 45 per cent of the structured methods marketplace in France (Source: Euromethod Phase 3a Information Pack 1993).

MERISE is in the public domain and adaptable enough to be used for a wide variety of applications. Its thrust is towards Management Information Systems. It provides a simple set of techniques within a framework, and shares with other methods a view of the developing system in terms of:

- **events:** Petri nets are used to model the synchronisation of events. This captures the dynamic effects of processing on data;
- **data:** a static view of the data and relationships is modelled using a traditional entity modelling technique;
- **operations (processes):** These are modelled using Petri nets.

It has no one official standards manual and has several different 'dialects' which manifest themselves in the different naming of some aspects and the manner in which stages are defined. Here, we have tried to adopt one consistent approach to the terminology, but have stated the common alternatives in brackets.

The MERISE life cycle approach is strongly oriented towards the perspectives of *actors*, i.e. the human users and at some stages the computer processors themselves. Processing is defined in relation to particular actors. Data is considered in terms of its meaningfulness as information to a human actor.

The human actors identified are also allocated responsibilities as *partners* in the decision-making and development process. The partners, (illustrated in Figure 8.1) each have their own world view, and set of concerns and responsibilities in relation to the project.

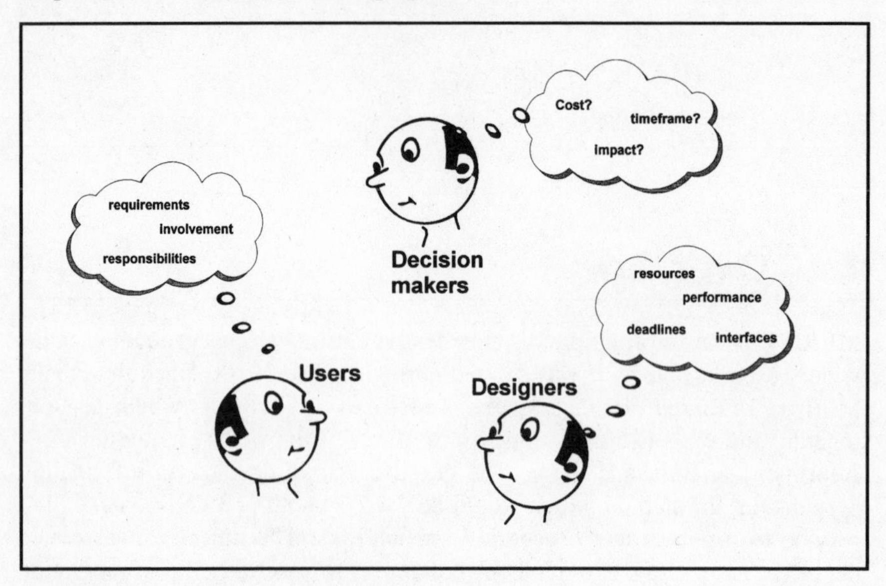

Figure 8.1 Partners in MERISE

8.2 Features of MERISE

MERISE is described as a *life-cycle method* because it aims to address the whole of the system life-cycle, from initial conception through to installation and maintenance, and to address this on a broad basis. It takes a view from a company-wide strategic basis in the first instance, focusing down to individual applications and full detail as the life cycle progresses. The method contains tasks related to modelling, project management, decision making and quality assurance.

The structure of MERISE is based on a definition of the organisation's interacting systems and of the role of the information system within these (see Figure 8.2).

The structure of the typical organisation is defined as a system with three levels of sub-system:

- a guidance (management) subsystem, which provides human guidance, business rules and policies;
- an operating subsystem, which performs the everyday business of the organisation (taking orders, supplying products, collecting payment, etc.);
- an information subsystem, which provides control information and records the business rules set by the management system.

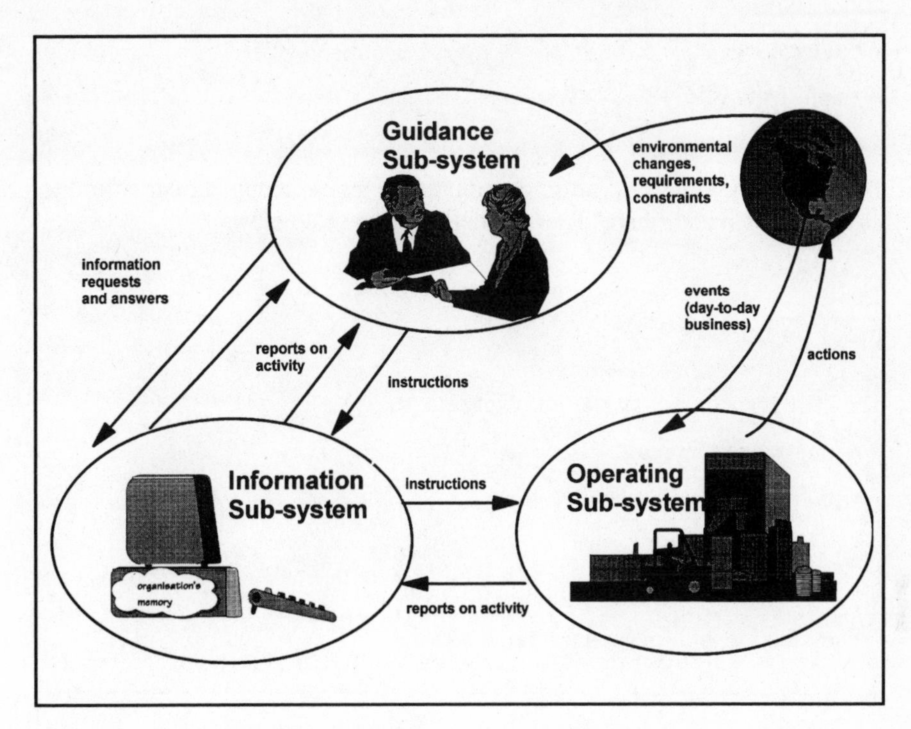

Figure 8.2 The Interacting Sub-systems Within an Organisation

The information system must accurately model:

- the environment of the organisation;
- interactions between the operating (product- or service-producing) systems and the customers and suppliers;
- rules of behaviour of the organisation: its budgets, sales conditions, policies, etc.;
- control information received from the operating sub-system;
- the rules of the guidance (management) sub-system.

The information system essentially plays the part of the organisation's memory, in terms of the activity of the operating sub-system and the rules and directives of the sub-guidance system. These ideas are not revolutionary but are not often immediately obvious with other methods. MERISE explicitly states them.

8.3 The MERISE Framework

The development framework provided by MERISE has three concurrent cycles:

- life cycle;
- approval cycle;
- abstraction cycle.

These cycles interact and no information systems development can afford to disregard any one of them. They are described in turn below.

8.3.1 The Life Cycle

The life cycle consists of the following stages:

- long-range planning;
- preliminary (initial) study;
- detailed study;
- implementation and installation (launching);
- production and follow-up (maintenance).

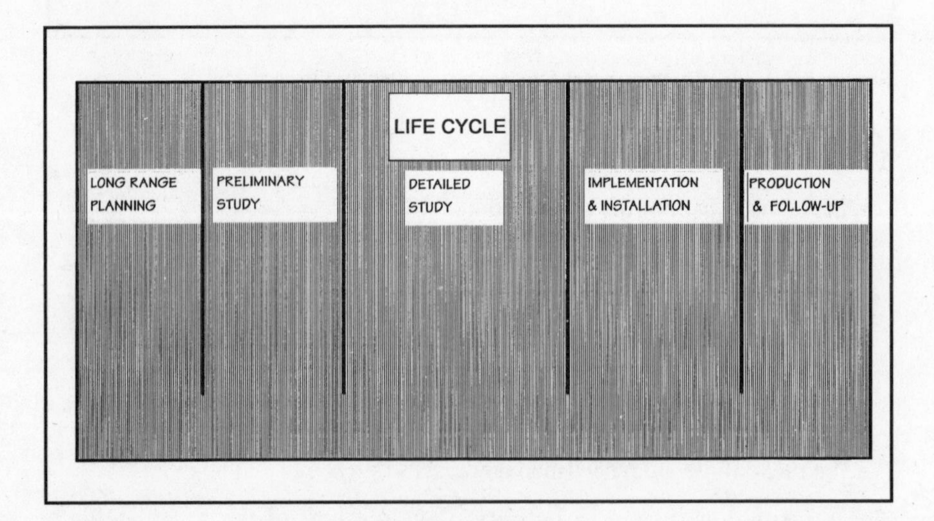

Figure 8.3 The MERISE Life Cycle

MERISE acknowledges that the preliminary study may be limited or non-existent with projects which are small or represent a low risk to the business.

MERISE breaks each stage down into phases, as shown in Figure 8.4. For each phase, there is a further sub-division into sections, each containing:

- a task plan with the definition of the role of each partner (actor);
- the structure and contents of the deliverables for the phase.

The list of tasks and deliverables is intended as a reference document to be adapted for the specifics of each project.

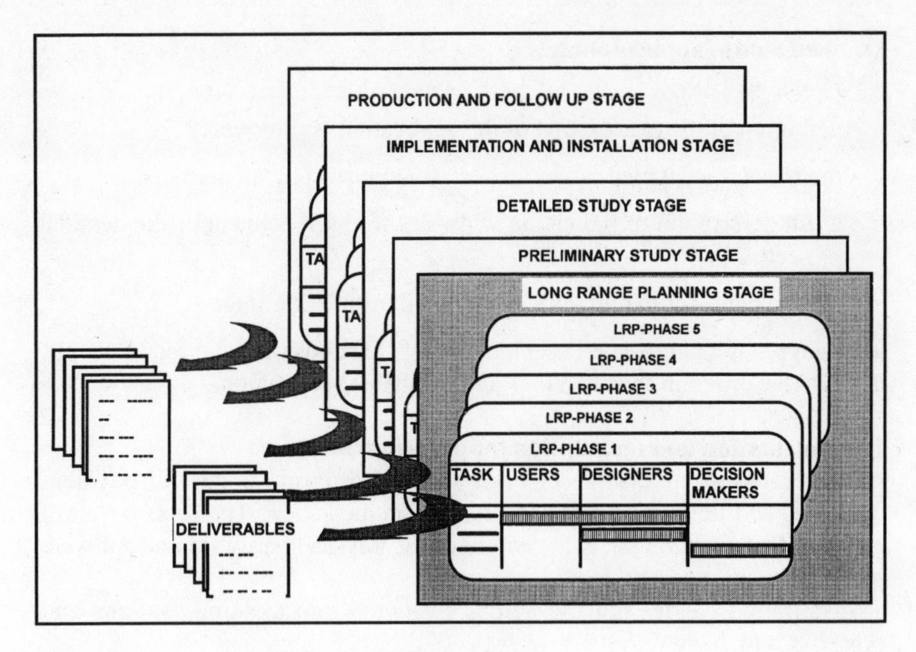

Figure 8.4 MERISE Stages, Phases and Tasks

Long-range planning study (company level)

This involves

- the partitioning of the company (or parts of it) into Business Areas;
- redefinition of the scope of the company's information systems;
- definition of objectives and priorities;
- definition of a strategy for evolution;
- formulation of outline annual plans (from the corporate strategy plan, usually covering the next three to five years).

Preliminary study (business area level)

This is concerned with:

- understanding the real needs, by gaining sufficient knowledge of the existing environment to evaluate the relative importance of each need;
- proposing solutions that integrate with the business objectives set by the decision makers;
- presenting the cost benefit analysis and choosing one solution from a number of options;
- defining a development scenario, according to this stage's evaluation of the work needed, and breaking down of the task into sub-projects.

The stage must begin with a clear definition of the business and information system objectives, aligned to the company's long range strategy plan. One Preliminary study may result in many separate projects at the detailed study level.

Detailed study (application level)
This stage involves:

- definition of the exact scope of the application (sub-project);
- the issue of a detailed project report, for approval by all partners;
- a full description of the nature and rules for each element of the detailed project;
- outline plans for the implementation and installation stage.

This stage results in detailed specification of functions data and requirements, and also the program specifications and database specifications.

Implementation and installation (application level)
Implementation covers the development of program code and physical databases and the provision of technical documentation. The tasks within it are very dependent on the exact nature of the physical hardware and software platform being used.

Installation carries out the actual cut-over from existing systems and procedures to the new system.

Production and follow-up (business area level)
This stage, which may consider more than one application within a Business Area, is concerned with the post installation assessment of the achievement of objectives, and with the metrics associated with systems maintenance.

The *life-cycle*, as defined above, will be subject to decisions made at intervals during and at the end of stages. These are considered next, as part of the *approval cycle*.

8.3.2 The Approval Cycle

The *approval cycle* recognises the necessity for identifying decision points during the information system development project (Figure 8.5). It is concerned with the formal acceptance of the products of the MERISE stages and the eventual end-product, both technically and from the perspective of usability. The approval cycle identifies a hierarchy of decisions which have to be made during the life cycle and identifies the 'actors' (partners) who must make the decisions.

The decisions identified within the approval cycle may be of the following types:

- *identification decisions,* related to the major objectives and major 'actors' of the project;
- *management decisions,* involving the choice of major events, functions, and data to assist the guidance system;
- *organisational decisions,* determining, for example, whether the system

will be centralised or distributed, the degree of distribution and the degree of automation;

- *financial decisions,* which address the cost/benefit analysis of the project;
- *technical decisions,* concerning hardware, software and network choices;
- *processing decisions,* deciding which aspects of the system should be on-line or batch, and the ownership of data.

The hierarchy and relative importance of decisions related to a particular organisation's guidance system will have a significant effect on the *life cycle.* In MERISE, the hierarchy of decisions is based on initially deciding upon the things least likely to change. These are represented by the *identification* decisions, followed by the *management* decisions. The *organisational* and *financial* decisions are usually considered next, in parallel. The *technical* and *processing* decisions will usually emerge as a result of the preceding decisions. It follows that the effect of changes to management and identification decisions after the commencement of the project will be fundamental and far reaching.

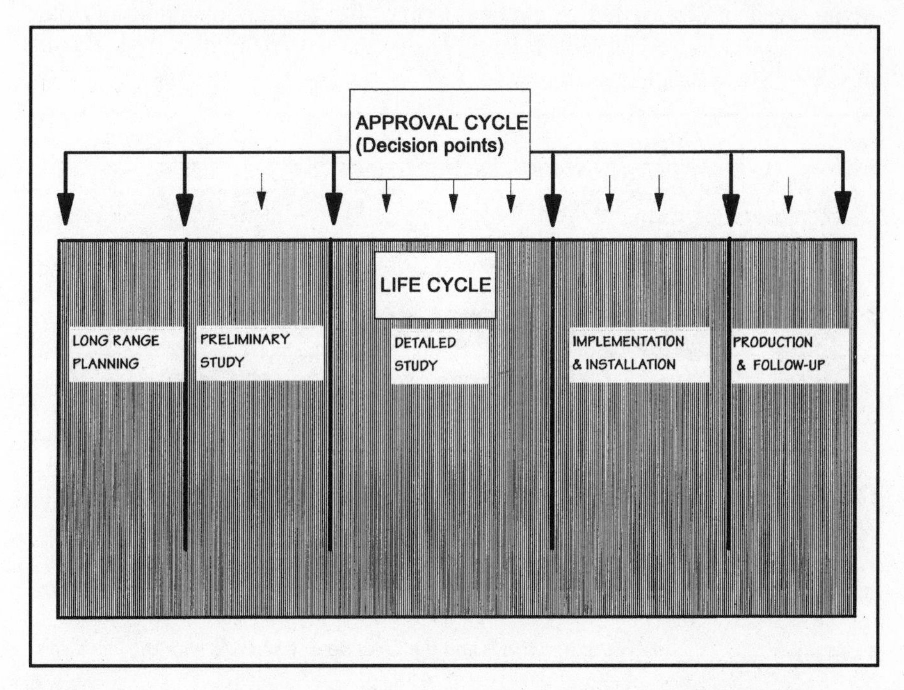

Figure 8.5 The Interaction Between the Life Cycle and the Approval Cycle.

Quality Assurance of the project is effected through the approval cycle. This is based on Quality Assurance standards as defined in IEEE 730 (20). It starts during long-range planning when a quality plan is produced. The quality plan identifies actors and their responsibilities as well as the deliverables for each task, with objectives and quality criteria.

The *life cycle* and *approval cycle* provide the framework to support the third cycle of MERISE, the *abstraction cycle.*

8.3.3 The Abstraction Cycle

The *abstraction cycle* is concerned with the diagrammatic modelling of the developing system. The concept of abstraction is commonly used in engineering disciplines to isolate certain essential elements and ensure the description of a consistent system in terms of these elements before introducing other elements. Thus, with the *abstraction cycle* we are able to identify, and consider separately, the rules of processing, and the classification and behaviour of data required by the business, before introducing further requirements and constraints imposed by any particular physical platform. The use of levels of abstraction will:

- allow verification of the consistency of the IS at pre-defined levels;
- allow simulation of the behaviour of the system at each level;
- allow the designer to take into account only one class of problem at a time.

Many levels of abstraction would be possible. The abstraction cycles of other methods considered in this text typically have two levels: logical and physical. MERISE chooses three.

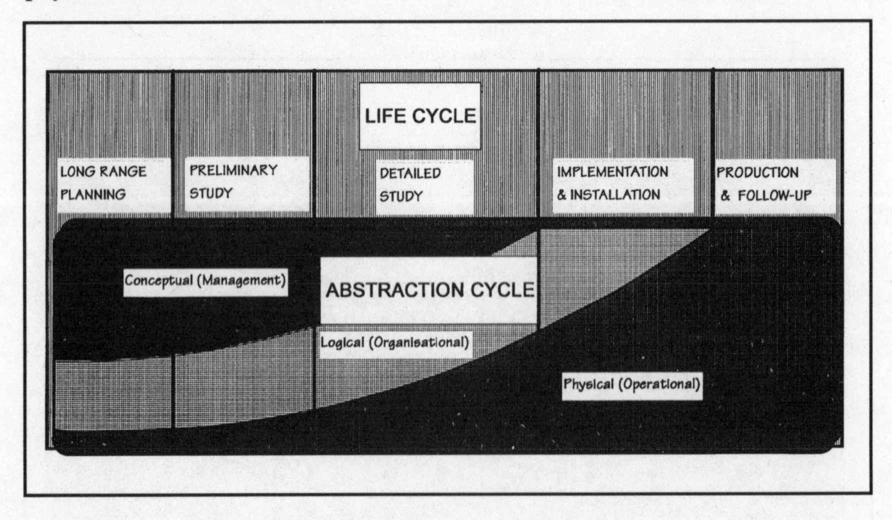

Figure 8.6 The Abstraction Cycle and the Overlap of Its Levels with the Life Cycle

The MERISE abstraction cycle has three levels of abstraction, on the basis that the organisational level is too rich in detail to be combined with either of the other levels. The levels are listed below, together with some alternative descriptions found in the various interpretations of MERISE:

- conceptual (management);
- logical (organisational);
- physical (operational or technical).

These levels relate to both processes and data. They represent, in simple terms:

- what (conceptual);
- who does what, where and when (logical);
- how (physical).

The **Conceptual (Management) Level of Abstraction** is management-decisions oriented. It embodies the object classes and behaviour rules, defined in accordance with the guidance sub-system objectives. It describes both static and dynamic data and has techniques for building diagrammatic models for both data and processing.

The outputs from this level are:

- a conceptual data model;
- a conceptual (management) process model.

These are described further later in this chapter.

The **Logical (Organisational) Level of Abstraction** represents organisational choices and decisions. It describes the resources and data used by the actors (either human users or computers) through specific procedures. The procedures described here can be manual, automated or a mixture of the two. The *logical* level in MERISE is a different concept from *logical* within other methods where logical is more akin to MERISE's *conceptual.*

At this level of abstraction, the data organisation is described in physical terms (records, sets, pointers, tables).

The outputs from this level are:

- a logical data model;
- a logical (organisational) process model.

The **Physical (Operational, Technical) Level of Abstraction** embodies technical choices and decisions made, according to technical targets and constraints (performance, response times, storage constraints, etc.)

At this level, data structures are described in the specific language of the chosen Database Management System (DBMS). Procedures become transactions and programs.

The outputs from this level are:

- a physical data model;
- a physical (operational) process model.

8.4 The MERISE Models

MERISE provides a set of simple graphical modelling techniques in order to facilitate communication between the partners and provide a means of representation and description of the requirements. The major modelling techniques are the Entity Model and Petri Nets, which are illustrated within the explanations of the models below.

MERISE proposes the use of six models to describe the developing system. These are the:

- Conceptual Data Model;
- Logical Data Model;
- Physical Data Model;
- Conceptual (Management) Process Model;
- Logical (Organisational) Process Model;
- Physical (Operational) Process Model.

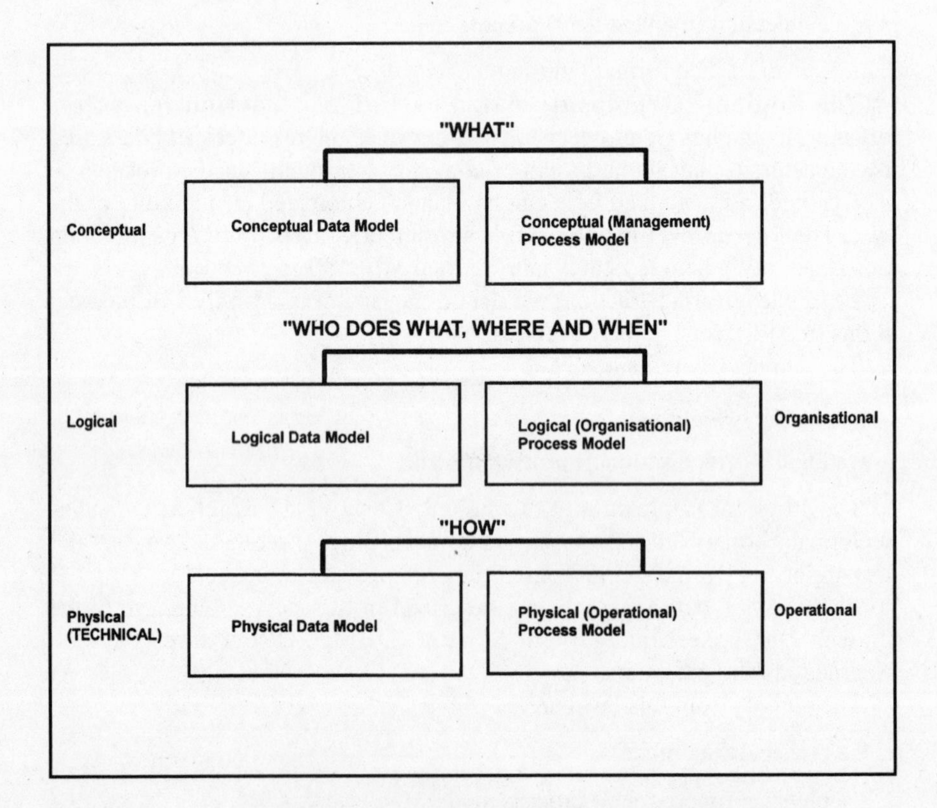

Figure 8.7 The MERISE Levels of Abstraction and Their Models

At each level of abstraction both internal models of the system and external models of the user interface can be developed. Their relationships to the abstraction cycle are shown in Figure 8.7.

Each type of model, with the techniques involved, is described below. The Data models are considered first, followed by the process models.

8.4.1 Data Models

MERISE defines data as being a logical or physical description of an object (entity) in terms of characters, bits, etc. Data is objective. Information, on the other hand, is a description of the same data, by or for a given actor. The same data may constitute information to one actor but meaningless 'noise' to another. Information embodies objects, classification, behavioural rules and facts related to these.

8.4.2 The Conceptual Data Model (CDM)

The CDM models objects (entities) and their attributes and relationships. It represents the same concepts as the entity model, described in Chapter 3, including the concepts of entity (object) type and occurrence, unique identifier and recursive (self) relationships. Many to many relationships must be resolved into their constituent one-to-many relationships. A set of formal rules for normalisation is defined in MERISE to ensure that the entities are in at least fourth normal form. These rules are consistent with the standard Codd and Boyce-Codd rules well-documented by Chris Date (Date 1985). The diagrammatic conventions for a conceptual data model are as shown in Figure 8.8.

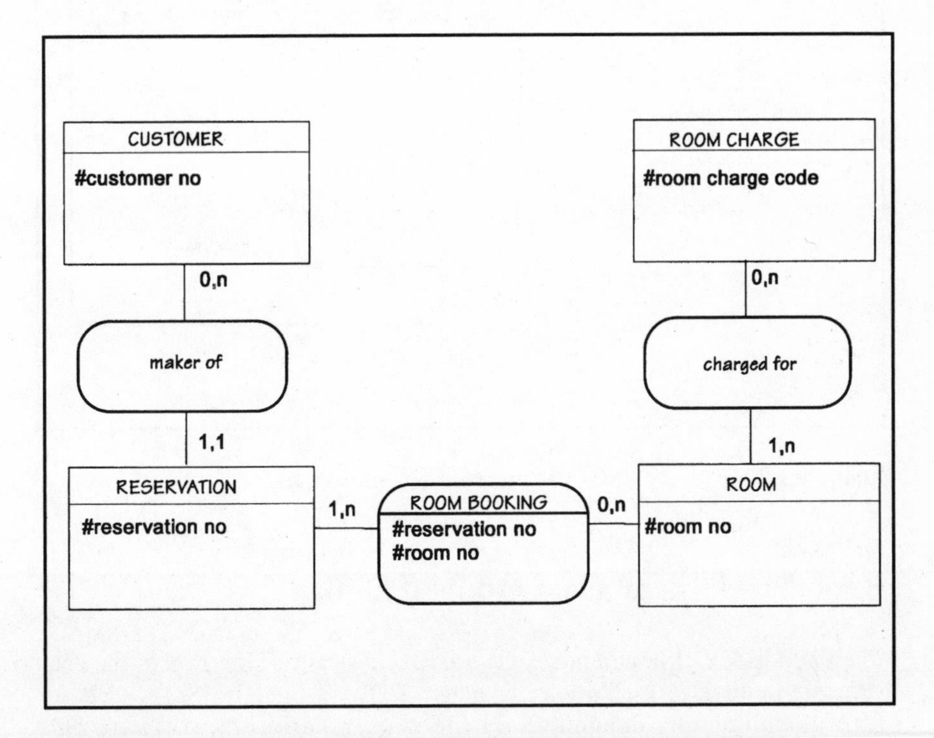

Figure 8.8 Diagramming Conventions for the Conceptual Data Model

8.4.3 The Logical Data Model (LDM)

The Logical Data Model is based on the Conceptual Data Model. Its purpose is to show constraints imposed by the chosen DBMS, and organisational constraints such as the degree of distribution of data. The data model is re-drawn showing the physical record-types and the sets, segments, tables, etc. into which these will be organised. Figure 8.9 illustrates the type of division into sets which would be relevant to a network-type DBMS. The entities RESERVATION, ROOM BOOKING, and ROOM CHARGE are each the owner of a set. ROOM is logically a member of two sets. It will need to be decided later which set ROOM should physically belong to. ROOM will then be linked to the other owner logically, probably by pointers which indicate the address of the other owner. ROOM BOOKING is a member of one set and the owner of another, which is allowable. Only binary sets are permitted, with one owner and one member.

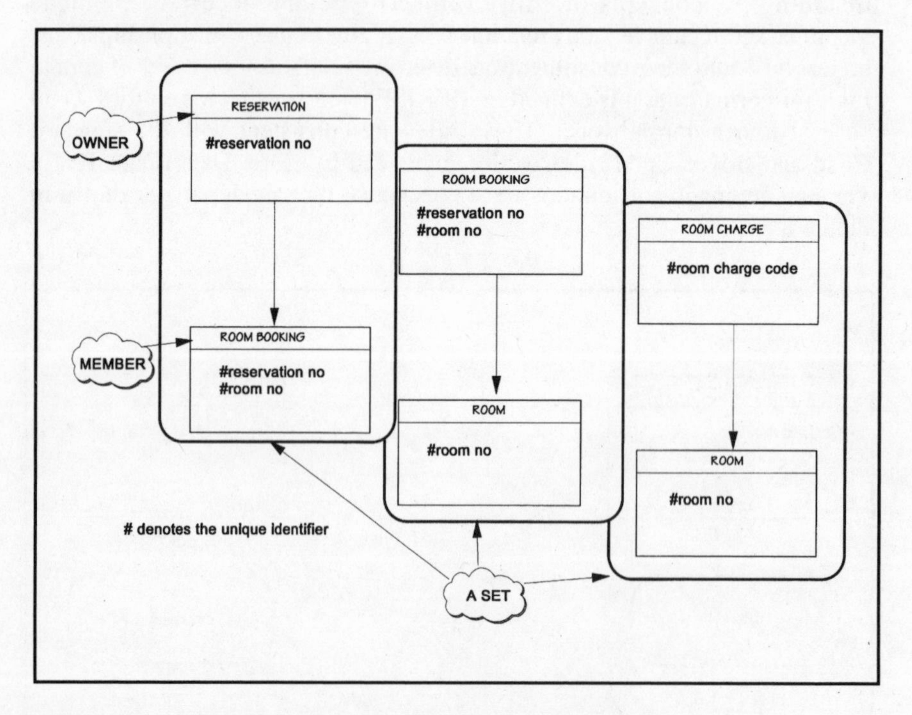

Figure 8.9 Logical Data Model (Network Class)

8.4.4 The Physical Data Model (PDM)

The PDM in MERISE comprises the actual, physical specification of the data structure in the Data Definition Language (DDL) of the DBMS. As such, its form is completely dependent on the chosen hardware, software and networking configuration of the physical system. MERISE gives a set of 'first-cut' rules for conversion of the LDM to a physical data model appropriate to a network or a relational DBMS.

8.4.5 Process Models

In order to define processing, MERISE combines two approaches:

- a **state-oriented** description, where changes to entity occurrences are expressed through the definition of a 'before' state and an 'after' state of the entity and rules for making the transformation from one to another, e.g. the occurrence of the event *guest arrival* changes the entity GUEST from a state of RESERVED GUEST to a state of ARRIVED GUEST.
- an **action-oriented** approach, where permissible changes are given by allowable action sequences, e.g. a guest must make a reservation and be associated with a room reservation before an account item can be raised. Thus:
  - insert GUEST;
  - insert RESERVATION;
  - insert ROOM RESERVATION;
  - insert ACCOUNT ITEM;

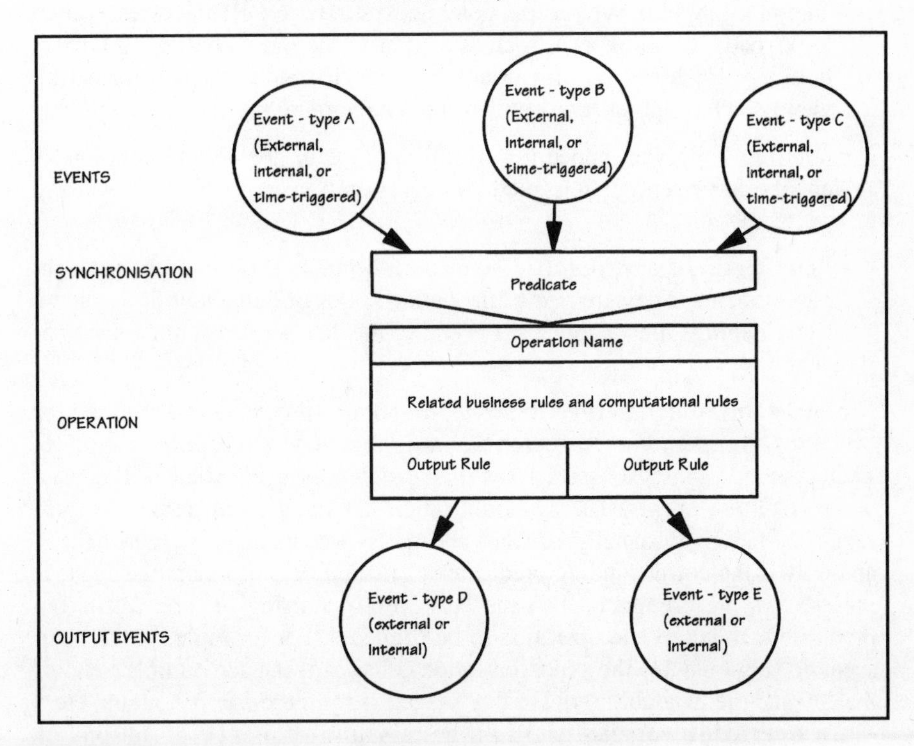

Figure 8.10 The General Form of a Predicate Petri Net

would be a valid sequence of actions, whereas:

- insert ACCOUNT ITEM;
- insert GUEST;

would not, if the ACCOUNT ITEM was for that particular GUEST, since the guest must have made a reservation and be registered to a room before any account items can be accepted.

In MERISE both approaches are combined to give rise to the definition of:

- event;
- synchronisation;
- operation (process).

These features are modelled using Petri Nets, as shown in Figure 8.10.

The definition of an **event** within MERISE is wider than that of certain other methods. An event is something which has happened, either in the environment external to the information system, or within the information system itself, which causes or requires some reaction from the information system or the environment. As such, enquiry, update and output events are included.

In MERISE, events may be:

- *external:* any event coming from the environment external to the automated system, such as the hotel guest's letter requesting a reservation to be made, or going to it, such as a letter to the guest confirming a room booking. Each event corresponds to a specific document (information view). Each event must be of interest to one or more actors;
- *internal:* any event appearing within the system, inside an operation, is an internal event. It is issued by the operation, e.g. invoice awaiting payment;
- *time triggered (artificial):* These define managerial choices or legal requirements. They are part of the defined rules of behaviour of a system. Certain things must happen at the end of the day, week, month, quarter or year.

Synchronisation describes the *condition*, or set of conditions, required for activating an operation. Synchronisation is the same concept as condition. Each operation must be associated with a synchronisation, which will consist of one or more events. The synchronisation is expressed in terms of a pre-requisite state (a Boolean predicate) and the synchronisation is fired once if and only if the corresponding predicate is satisfied. If more than one event is involved in the predicate, then the last event occurring in time fulfils the predicate and causes the operation to be triggered. For example, in the hotel scenario (Figure 8.11) the synchronisation consists of the arrival of the end of the day and the availability of the day's receipts from the bar, restaurant, etc.

An **operation** (or process) is a permissible sequence of actions or structured set of management rules activated either by one event, or by the synchronisation of several events. The operation constitutes a success unit: once started it is non-interruptible; all resources needed for its execution must be available before its commencement. An operation is a black box, producing a predictable output or defined reaction from the system in response to a particular synchronisation. The operation may use information not present in the input data of the event, by reference to the information base (database). For example, (Figure 8.11) once receipts are available at the end of the day, these will be input to the operation *Add charges to guest account*. The GUEST ACCOUNT details on the database will be accessed to ensure that the guest's stay is valid. If not an error message will be output. If valid, an occurrence of the entity GUEST ACCOUNT ITEM will be created.

Processes and the dynamics of data (state changes) are modelled using Predicate Petri nets. Synchronisations can be expressed using a predicate, and if required the number of event occurrences expected and the frequency of the event. The predicate can be combinations of logical and/or conditions to define the trigger or triggers for the process. The duration of a synchronisation and the life of an event can also be defined, for simulation purposes.

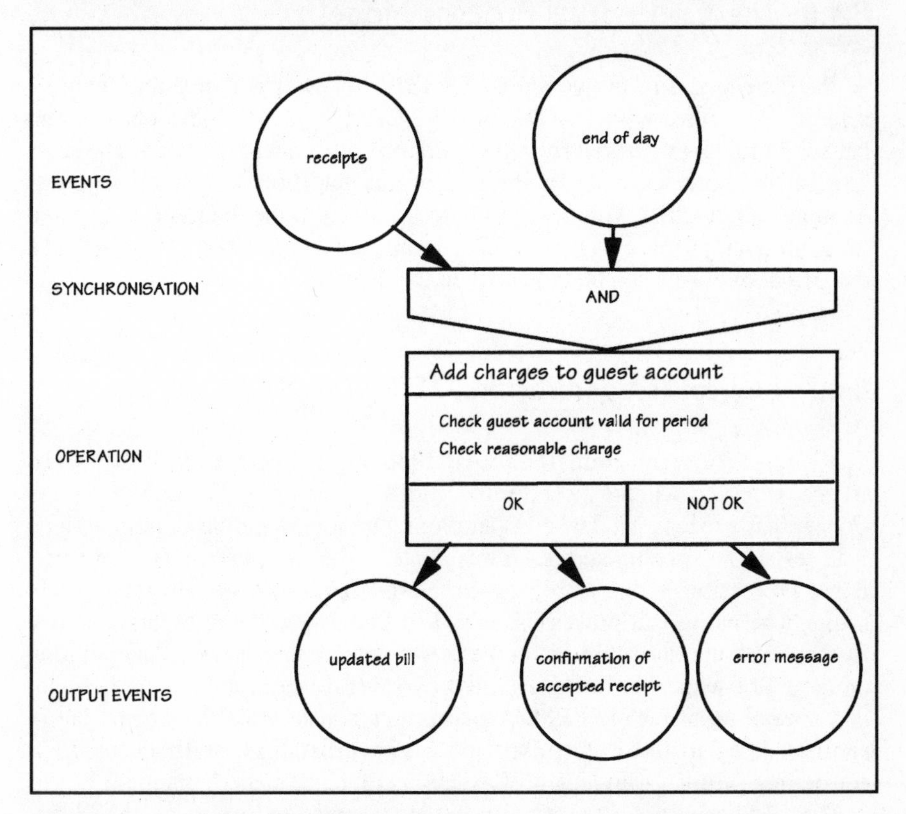

Figure 8.11 Petri Net for the Conceptual Process Model for One Operation Within the Hotel System

The following rules for drawing Petri Nets are specified by Rochfeld (Rochfeld 1983):

- an operation must not require the intervention of other events once started;
- any permanent information is not an event and must not appear in the synchronisation predicate;
- any event must be an internal, external or artificial event, as defined earlier;
- any internal or external event must concern objects defined in the conceptual data model;
- all properties used by the predicate and local conditions must be part of the message data within the event(s). Consultation of the information base is **not** done at the synchronisation level, but such access is allowable during the operation.

8.4.6 The Conceptual Process Model

At the **conceptual** level, we are concerned with **what** is happening, without reference to who, where or when. The CPM models object classes and behaviour rules, as derived from the guidance sub-system. Figure 8.11 shows how, in the hotel case study, the end of day addition of charges to guest accounts is modelled. Volumetrics could be added to the diagram, if required for simulation purposes. Other information, such as necessary delay between the arrival of events can be indicated on the diagram.

8.4.7 Logical (Organisational) Process Model (LPM)

The Logical (Organisational) Process Model is concerned with the **what**, **where, who** and **when** aspects of the system. To develop the LPM from the Conceptual Process Model, all operations previously defined are examined with respect to specific actors. In this context, the computer is also an actor. Batch processing is one actor, real-time processing may be several separate actors. The individual processors in a distributed system are separate actors. At this level, the operation on the Petri Net now models a **task.** An operation on the CPM which will be performed by several actors in sequence is split into several tasks for the LPM. A **procedure** is a set of tasks carried out in sequence, by a given actor, inside a given process, with given time constraints, with a given means of execution (e.g. automated, manual).

The LPM comprises information views corresponding to an individual actor's procedures.

The Logical Process Model will not show all physical detail, but will indicate types of program, types of resource, distribution of processing and the split between manual and automated tasks. Figure 8.12 illustrates this and is derived from the CPM presented in Figure 8.11.

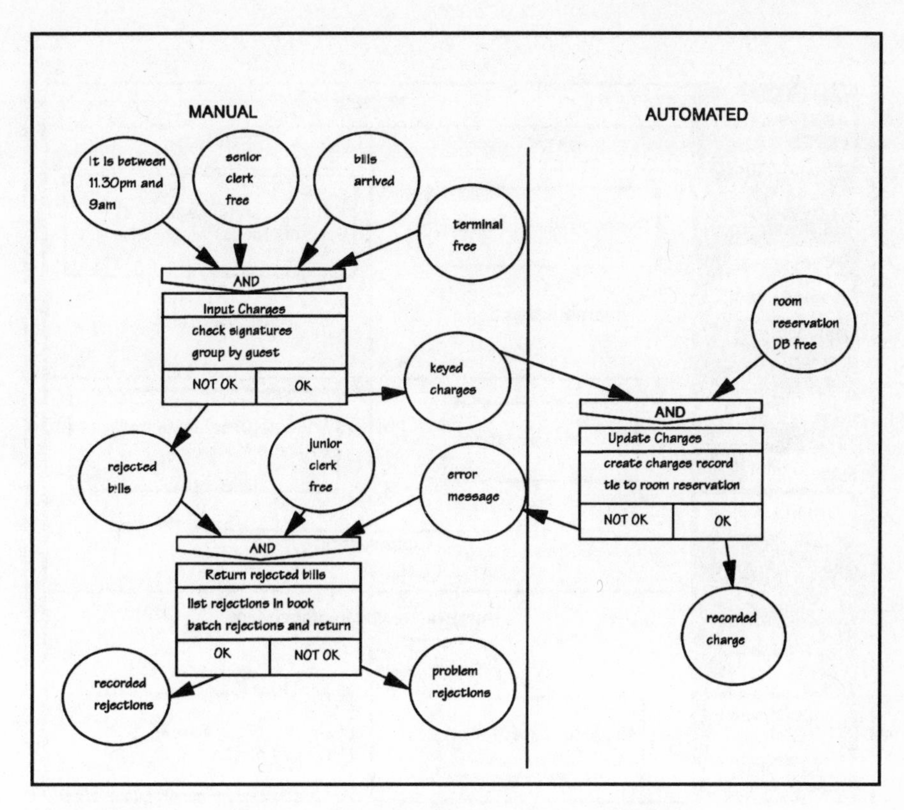

Figure 8.12 Logical Process Model for Part of the Hotel System

All events previously defined in the CPM must be accounted for in the LPM. New organisational events, concerned with organisational resources, are introduced. Example of additional, specifically organisational, events are:

- executive/supervisor's authorisation received;
- free computer terminal becomes available to the operative.

8.4.8 The Physical (Operational) Process Model (PPM)

The Physical (Operational) Process Model is concerned with **how** the system will be implemented. The Physical Process Model diagrammatically represents data and programs. It represents files (permanent, temporary), input and output formats and mechanisms; programs and their order of execution. It includes transaction processing codes, program specification and communications network design. For the PPM, a further restriction is introduced. Only the computer is taken into account and only automated procedures are considered. Each procedure now becomes a **transaction** and the structure of the programs is defined. Each logical screen becomes a physical screen or screens.

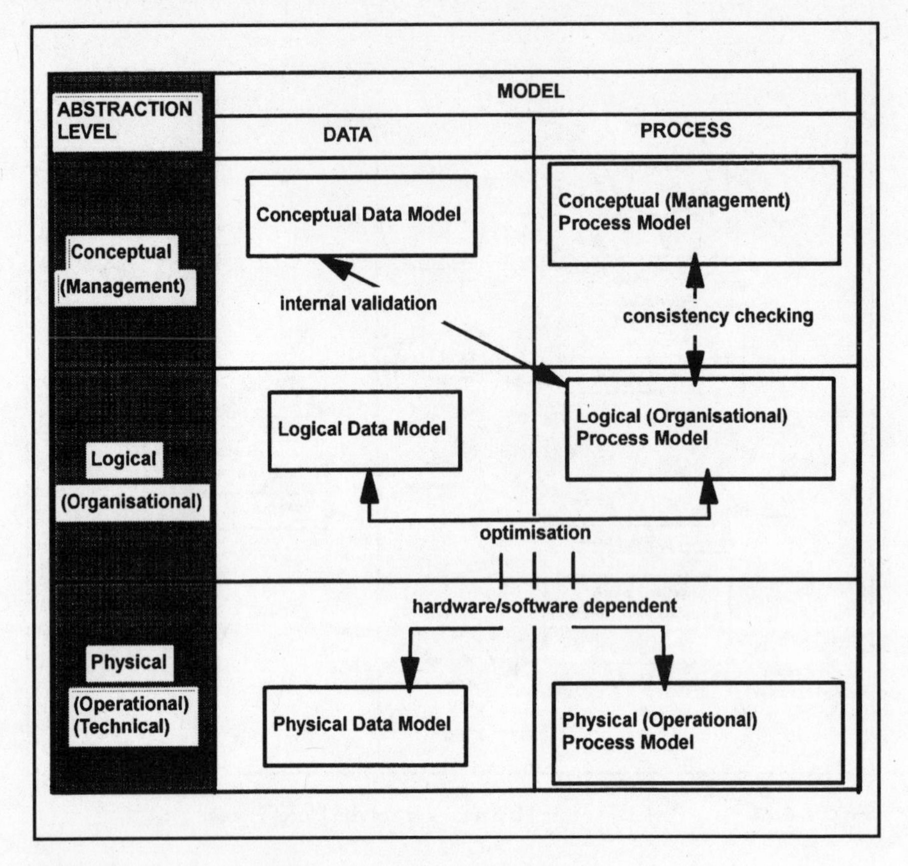

Figure 8.13 The MERISE Models and their Interactions

At this point, MERISE suggests that any structured programming tool can be used including Nassi Schneiderman diagrams, HOS Structure charts, or whatever is useful for the level of programming language to be used. These are well documented elsewhere. The reader may wish to refer to Martin and McClure (1985) for further information.

8.4.9 Cross-checking Between the MERISE Models

There is considerable scope for cross-checking between the models, as illustrated in Figure 8.13.

8.4.10 Where Do We Start Modelling? (The MERISE Sun Curve)

Inside any given *life-cycle stage*, MERISE recommends a traditional approach to the analysis and design sequence, starting with outline models of the existing system at the physical (operational) level, based on documents and screens used and progressing through the levels of abstraction from physical to logical to conceptual and back again for the future system. These models are each validated and approved by the *partners' (approval cycle).*

The MERISE 'sun curve' (see Figure 8.14) is a pleasant graphical representation of this, indicating that 'at the end of the day' the future system emerges. MERISE emphasises the iterative nature of the development process, which, in relation to the sun curve, indicates a certain amount of backwards time-travelling!

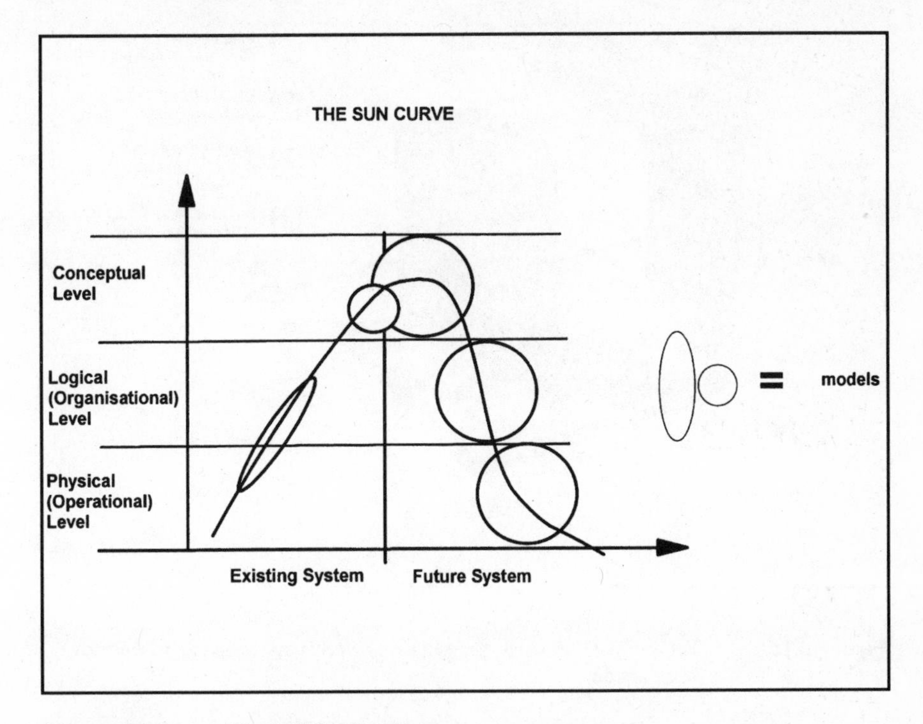

Figure 8.14 The MERISE Sun Curve

In Figure 8.14, the compression of the ellipse representing existing system models indicates that an outline model only would be developed. Its position, spanning the physical and logical levels indicates that separate models for the physical and logical levels may not necessarily be produced. The conceptual model for the future system would be based on that for the existing system, with the requirements of the future system added. From this point, the models for the future system would be fully-developed. The logical and physical models would be built, derived from the conceptual model. At each level of abstraction, both data and process models are indicated.

Figure 8.15 shows the multi-faceted nature of MERISE, with levels of abstraction, scope across the business and precision modelled with the MERISE life cycle stages. This illustrates the different emphasis and level of detail of the stages. It can be seen that models drawn during the Long-Range

Planning stage will be at a high level (strategy and overall solution) with more emphasis on the conceptual and logical levels of abstraction than the physical. Such models will be company-wide in scope.

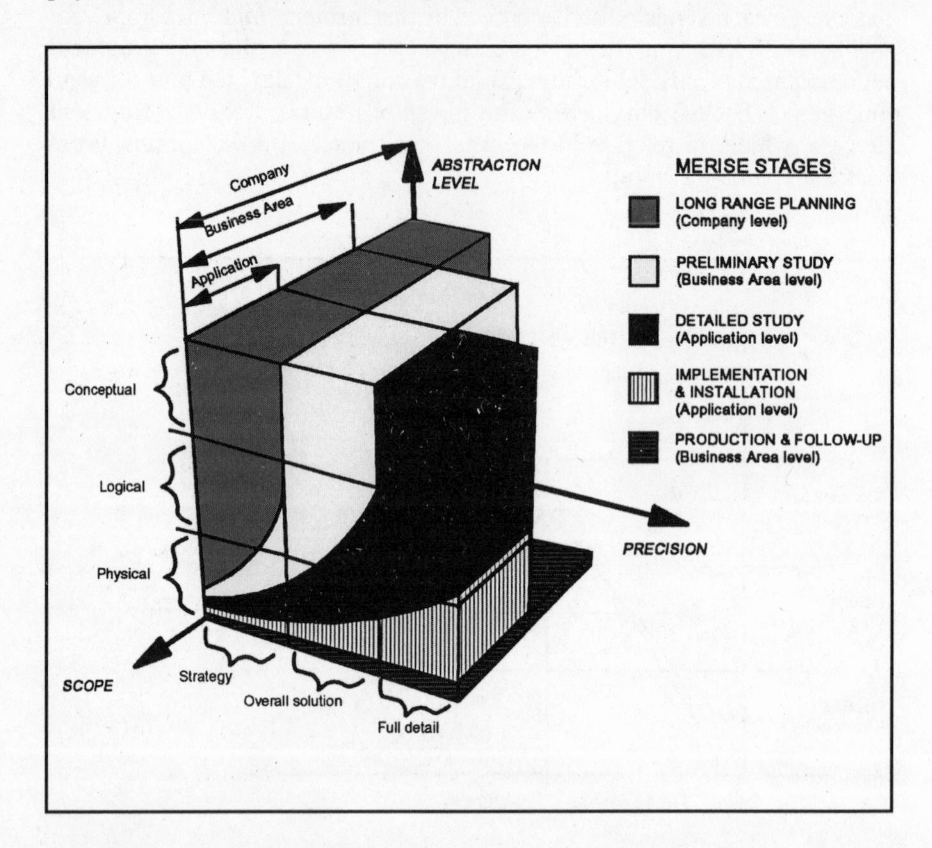

Figure 8.15 The MERISE Life-cycle Stages, Their Precision, Scope and Levels of Abstraction

8.5 Conclusion

The development of a project with MERISE involves the interaction of three cycles, life, abstraction and approval. MERISE covers the system life cycle from initial conception, in line with organisational strategy, through to implementation and maintenance. Within the life cycle, there are stages, phases and tasks. The tasks reflect the responsibilities of users, designers and decision makers.

The abstraction cycle supports three levels of abstraction. The conceptual level models what is required. The logical level brings in organisational considerations and embodies who does what, where and when. The physical level completes the physical picture of how the project will be developed.

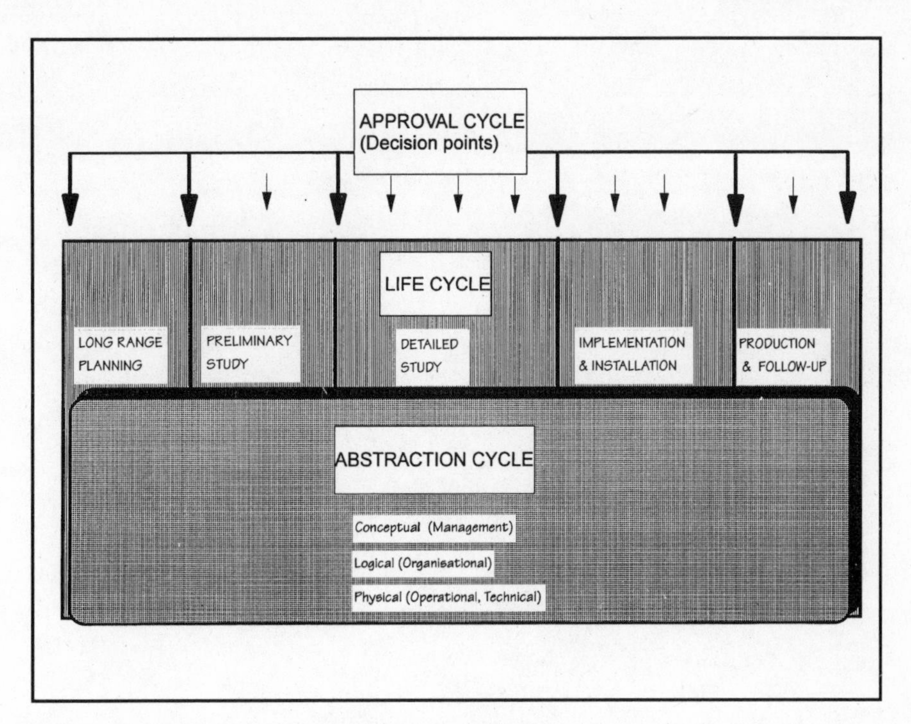

Figure 8.16 The Interacting Cycles of MERISE

The approval cycle controls a hierarchy of decision types and decision points. It also covers aspects of the project concerned with quality assurance.

Figure 8.16 shows the interaction of the three concurrent cycles of MERISE. MERISE is strongly oriented to the actors' viewpoints and their definition and usage of information. It provides for the modelling of events, synchronisations and operations at many levels.

MERISE is a well-established and mature method. However, because there is no one official standard, MERISE has developed in line with particular supplier and consulting organisations' interpretations of it and thus it has many dialects. It is a medley of these dialects which has been presented here.

9

The Comparison

9.1 Overview

In the preceding chapters, we have looked at five structured methods individually. Now we come to the point where we must set them side by side and see how they compare with one another. In making a comparison of these methods, we are aiming to:

- provide a consideration of the contents of chosen methods;
- highlight their differences in techniques, approach and framework.

We must emphasise at the outset that we are making a *comparison* rather than a full evaluation. An evaluation is often extremely context-dependent. For an organisation to make an evaluation to determine the **best method for its purpose**, many additional features would need to be considered.. These would include such factors as availability of support, acceptability, current skill levels of staff, as listed in Chapter 2. Thus, this chapter will **not** conclude with any one method emerging as the triumphant winner. Rather, it will provide the reader with an assessment of the coverage by each method of our chosen list of features and a checklist of the techniques present in each method.

9.1.1 The Choice of Comparison Techniques

The options which were open to us for making our comparison fell broadly into the categories of:

- experiment under tightly controlled conditions;
- experiment under field conditions;
- survey of information already available;
- comparison of selected features.

These options are included in the DESMET evaluation methodology, described in Chapter 2.

Any comparison of structured methods must acknowledge the difficulty of making a practical comparison under each of these options.

Experiment

The scientific experimental approach would be to change one of the variables within the experiment and observe the result, repeating the experiment to confirm the conclusions. Ideally we should use each of the methods under comparison to develop a system to provide a solution for the same problem. However since one component of any human problem area is itself a biological system (Human) there is at least one component which can perform somewhat inconsistently under experimental conditions. The process of using one of the methods will affect the problem area itself, including the analyst(s) undertaking the comparison. Additionally, the problem area is a moving target since with time the environment of the problem area will have changed. Immediately, several other parameters have been altered. Such problems may be addressed by developing tightly controlled experimental conditions for conducting the comparison or by taking account of such problems within a field experiment comparison. Comparison by experiment is one of the options defined in DESMET (see Chapter 2), with experimental design being the subject of one of its modules, which provides advice on the identification and, where possible, the management of these variables. The use of a series of experiments, for our purpose of providing a comparison of framework, techniques and approach, would not have been a cost-effective method of achieving the level of comparison we required.

Survey

Survey of organisations' experiences with structured methods forms another potential comparative method but clearly would be a major undertaking with its own set of problems. The variables (project size, analyst skill levels, measures of success, etc.) would in some way have to be factored out statistically to achieve a fair comparison. Further the sample size would have to be significant and the results subjected to statistical analysis. Within the context of this book, survey was not a viable option.

Feature Analysis

Feature Analysis, as stated in DESMET (Chapter 2) is likely to be the most frequently used means of evaluating methods. We also consider it to be the most practical approach for a book which compares methods and hence that is in essence what is presented here. However, the comparison is kept to technical features of the methods. The organisation doing a full evaluation would consider many more features, and assign scores in relation to the impact of the various features in its particular environment.

9.1.2 The Framework for the Comparison

The comparison presented in this chapter illustrates the features which we have chosen.

The purpose of the comparison is to highlight differences in techniques, approach and framework, between the five methods, in order to give the reader a better understanding of the methods. Such understanding may then form a foundation for customising a development method to your own circumstance or assist in the evaluation of the individual methods and promote an informed choice of method. The methods in their respective chapters have been described with reference to a consistent scenario, the Midlinks Motel case study. This has been done to make comparison easier.

The features against which our **Feature Analysis Comparison** will be made are as follows:

- the methods and the Business Life Cycle;
- the underlying philosophy of the methods;
- the 'structuredness' of the methods;
- the user role in the method;
- what size of system is the method aimed at?;
- the techniques within the methods;
- CASE tools and the methods.

These were introduced in Chapter 2. We have made our assessment of performance against these features in a general way, on the basis of a valuation of (H)igh, (M)edium or (L)ow **or** on the basis of (Y)es or (N)o etc., as appropriate.

9.2 The Methods and the Business Life Cycle

Business Strategy Planning
Information Strategy Planning
Feasibility Study
System Analysis
System Design
Program Coding
Module Testing
System Integration & Test
Implementation
SOFT SYSTEMS/ MULTIVIEW
STRUCTURED SYSTEMS ANALYSIS & DESIGN VERSION 4
INFORMATION ENGINEERING
YOURDON STRUCTURED METHOD
MERISE

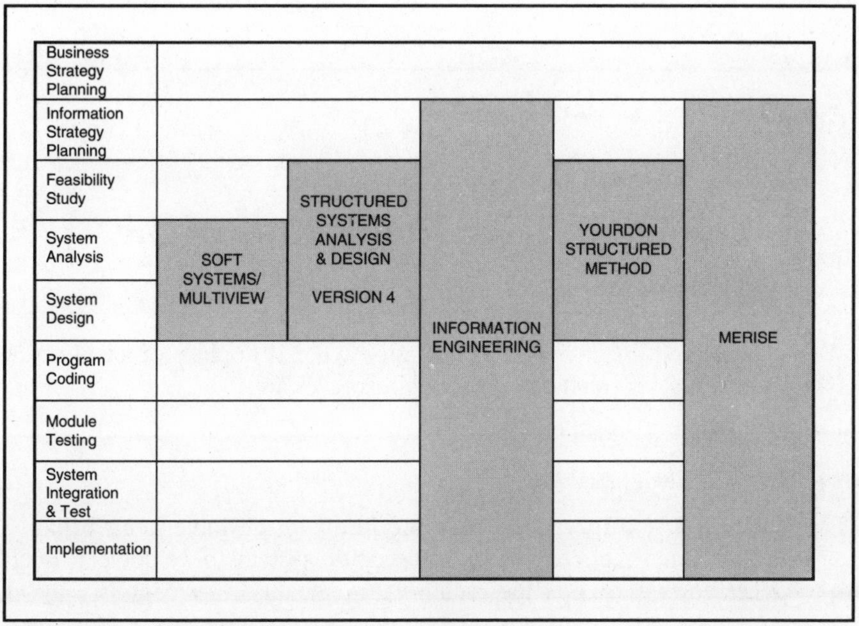

Figure 9.1 Coverage of the Business Life Cycle

The Business Life Cycle forms one aspect of the framework for comparison, within which the features of the methods have to operate, and which was laid down in Chapter 1. The first aspect of the methods to be established is the extent to which each method covers the Business Life Cycle. Figure 9.1 shows the stages of the life cycle over which the five methods operate.

Information Engineering and MERISE display the greatest coverage of the stages of the Business Life Cycle largely because one of their aims is to define the whole organisation's data, rather than just the project's data. IE explicitly states that this information is a resource like any other held by the organisation. IE was developed with the intention that it would be supported by a CASE tool through to code generation and implementation. SSADM (V4) and Soft Systems/Multiview both cover the Systems Analysis and Systems Design stages. Feasibility Study is formally embraced by SSADM whilst Soft Systems/Multiview, via its contingency approach, could also be used to carry out a feasibility study although it is not a formally defined stage (Wood-Harper *et al.* 1985, Avison and Wood-Harper 1990). Yourdon Structured Method embraces the early part of program coding via its use of structure charts.

Soft Systems /Multiview	SSADM (V4)	Information Engineering	Yourdon Structured Method	MERISE
M	M	H	M	H

Figure 9.2 Life-cycle Coverage

9.3 The Underlying Philosophy of the Methods

Wood-Harper and Fitzgerald (1982) identify two philosophies which underlie systems development methods. These philosophies are:

- the SCIENCE paradigm;
- the SYSTEMS paradigm.

The **Science Paradigm** underpins methods which have been broadly classified as **Hard Approaches** to information systems development. Such approaches are based upon the premise that 'Reality' exists, consisting of concrete entities and immutable facts, which can be objectively identified.

The Science paradigm embodies the 'divide and conquer' principle of breaking a problem down into smaller and smaller, more manageable parts. The problem inherent in the hard approach is that it does not tackle situations where 'Reality' is not easily identified and structured, that is to say where the problem area is 'fuzzy' and therefore not easy to identify and state in a formal way.

By contrast, the **Systems Paradigm** forms the basis for those methods which fall into the so-called **Soft Approaches** to Information Systems Development. The argument here is that 'Reality' is subjective and exists as an individual's perception of the environment in which the individual operates. 'Reality' varies with an individual's perception of his or her environment and to achieve a valid 'Organisational Reality' requires the coalescence of perceived realities. This merging of views of 'Reality' necessarily requires participation of the individuals within the system. The Systems Paradigm is concerned with the whole system, holding the view that when small systems coalesce to form larger systems, the large system reveals new properties, not present in the constituent systems, formed from the union, i.e. the whole is greater than the sum of its parts. The consequence of the Science approach, by implication, is to lose these 'emergent properties' when systems are broken down. The next parameter of comparison is therefore to question whether the methods fit into the science or systems corner of the ring.

SSADM, IE, YSM and MERISE break the problem down using progressively lower-level diagrams to model the situation. IE does, however, step back to take input from the Business Strategy Plan for the survival of the *whole* organisation, using this to help develop an Information Strategy Plan. MERISE also gives consideration to the wider 'corporate policy and goals' (Rochfeld *et al.* 1985) which will impact upon the requirements which are identified. These features do not, however, lead directly to a solution for problem areas, the solutions coming from the 'divide and conquer' principle which is still the main approach adopted by both IE and MERISE. SSADM, IE, YSM and MERISE must therefore be placed in the Science corner. However, MERISE also acknowledges that data has different meaning when viewed from different users' perspectives and thus in the assessment we have allowed a definition of 'Science, but with Systems tendencies' (Science/(Systems)).

Multiview is a hybrid method using Soft Systems at its front-end which means that it can not neatly be placed in the science or systems corner. In analysing Human Activity Systems (Soft Systems Method) there is most certainly a systems approach. Multiview is heavily influenced by the work of Checkland (1981, 1990) in Soft Systems Method which sits at the spectral end of the Systems approach. Checkland (1988) argues that the historical engineering background of computer-based information systems, tied to the huge capital investment, inevitably promoted the systematic hard approach to systems development. However although technology has moved on apace the thinking which underlies system development has not. He argues away from these early goal-seeking approaches toward systems which take account of the political and social aspects of organisations. In its second stage where information is analysed Multiview moves into the science camp breaking down data and functions through hierarchical diagrams. Subsequently the allegiance reverts to the systems approach as Socio-technical aspects of the system are examined. Based on ETHICS (Mumford 1983a, 1983b, 1985), Multiview recognises the people as part of the system and due weighting is given to the effect that they have upon the problem area by encouraging their participation and considering the social implications (their role and tasks) in relation to technical options. The science approach is adopted in the last two stages, which are concerned with the design of the new system, albeit in the light of the impact of the earlier systems approach.

SSADM, IE and YSM consistently adopt a science approach with the intention of implementing a computer system as the solution to the problem area. MERISE has the science paradigm as its overriding approach but does have some systems leanings. Multiview may be perceived as a systems or science approach based method depending upon the outcome of a particular project. The authors of Multiview state that a computer system is expected to form part of the solution to the problem area but acknowledge that it is possible for Multiview, having used Soft Systems Method, to stop after the first stage producing an improved Human Activity solution. This latter case would leave the perception of Multiview as a Systems approach.

Soft Systems /Multiview	SSADM (V4)	Information Engineering	Yourdon Structured Method	MERISE
System/ (Science)	Science	Science	Science	Science/ (System)

Figure 9.3 System or Science Paradigm Underlies the Method

9.4 The 'Structuredness' of the Methods

This section deals with how 'structured' the methods are and thereby how tightly they prescribe the actions of the analyst.

A method is inevitably judged by the quality of system which it produces and this in turn is dependent upon a number of factors. One of those factors is the practical way in which the method is implemented, which is also tied to the skill of the analyst in using it. Here the assumption is made that the practice of each method follows the documented method rather than any variant of it.

The authors of Multiview expressly state that it is a *contingency* approach which permits flexible use of aspects of the method in relation to the requirements of a particular problem situation. However, such flexibility does carry a cost. One of the stated reasons for using a method is to tackle the problems associated with lack of control and non-standard methods of system development. A contingency approach to the definition of a method may compromise these benefits of a method.

YSM advocates certain specific techniques, and gives a development framework but does not rigidly specify the techniques which are to be used at each point in the framework. There is a method framework with 'mainstream' techniques which are recommended and other techniques which may be useful. However, the choice of what is done is left in the hands of the systems analyst and again control and standardisation are potentially compromised.

MERISE is very well structured providing a set of techniques within a method framework and specific rules for the performance of these techniques.

Both IE and SSADM are highly structured with explicit guidance for the analyst. IE needs to be well structured to deliver the required syntactically correct outputs for the **automatic** generation of code. SSADM is at the extreme end of the structured state of methods and via its Structural Model it specifies who does what, when, where and how! This is the corollary to the Multiview approach. Such high structuring with defined activities, paradoxically, requires skilled analysts to implement the method, together with a not insignificant amount of determination.

A consequence of a method being highly 'structured' is that rigidly defined outputs can be used as inputs to automate database definition, application definition and code generation.

Soft Systems /Multiview	SSADM (V4)	Information Engineering	Yourdon Structured Method	MERISE
L	H	H	M	H

Figure 9.4 Level of 'Structuredness'

9.5 The User Role in the Method

All of the methods acknowledge the importance of users to the development of new systems but to what extent do their views influence the type of system which is produced? This is important because users are experts in their particular area of the Business System and therefore hold information about what functions are needed and ideas about how they should be performed.

Multiview, in Stage 3, adopts the view that a high level of user participation is essential justifying this with sound reasoning. The overall contingency view of Multiview would however permit minimisation or even omission of the user influence depending upon circumstances. In the context of the small system environment in which Multiview operates this can be seen as a function of the overhead which user participation inevitably imposes. Judgement of the level of user participation needed is the skill demanded of the analyst here.

User participation is an expressly stated requirement in SSADM, the user role being as part of the team responsible for reviewing the products of the method. Users also have to 'sign off' the modules and stages as defined in the Reference Manual to signify their acceptance. User involvement is sought to establish requirements, define functions and in prototyping of critical areas of the system. User management is also responsible for major decisions, setting terms of reference to establish a development project through to Stage 6. However, such concerns as Project and Quality Management are deliberately excluded from core SSADM.

IE signifies the importance of users by promoting Joint Application Development and by relying heavily on diagramming techniques to encourage user involvement. It also uses Structured Walkthroughs with users and User Workshops to confirm understanding and ensure quality.

YSM acknowledges users, classifying them according to type. Whilst recommending the importance of developing a good relationship with users, YSM does not formally assign roles in the decision-making and development process. This is consistent with its flexible approach and therefore such roles may be assigned if thought appropriate.

MERISE's life cycle approach embodies the perspective of the human 'actors'. Processing definition is carried out in relation to particular actors and their views of information rather than just a standard definition of data. This forms a major consideration within the MERISE approach as users are formally identified to be partners in the decision making and systems development processes. The role of users is therefore significant.

None of the methods formally addresses *user training in the use of the method* (although the SSADM manual acknowledges the need for user knowledge) seeming to only consider training in relation to the operation of the new system.

Soft Systems /Multiview	SSADM (V4)	Information Engineering	Yourdon Structured Method	MERISE
H/M	M	M	L/M	M

Figure 9.5 User Role; Level of User Involvement

9.6 What Size of System Is the Method Aimed at?

Assuming that the system to result from application of the method is a computer system, then it is worth considering the size of system to which the methods will lead. At this point it is worth defining what we have taken to be the meaning of small, medium and large in relation to the size of a computer system. For our purposes a small system is developed by one or perhaps two analysts within six months, a medium system by up to five analysts in a year and a large system by more than five analysts in more than a year.

Multiview is designed for use in smaller sized projects where perhaps a stand-alone microcomputer system is sufficient. With this in mind Multiview considers the selection of application packages and their analysis in relation to the statement of requirements. Applications generators embodying such features as prototyping, screen design and database design are also discussed.

SSADM (V4) Core Method produces vast quantities of documentation, an overhead which is quite plainly inappropriate to small systems. The development of subject guides, such as SSADM for Microcomputers, may address this and increase the breadth of systems for which SSADM is appropriate.

IE is also documentation-heavy but is obviously intended for large systems since it is very much geared toward CASE Tool support and Code Generation, on a target specific DBMS/large mainframe platform. The cost of such tools is well outside the resources of a small organisation, to say nothing of the cost of equipment to support the tool and the system it produces. However with formalisation of multiple development paths, IE has also addressed the different scales of system which can be developed.

YSM was initially developed in relation to real-time systems which tend to be smaller, having a smaller data structure underlying them. By incorporating techniques for data analysis YSM has been brought into the area of commercial business systems. It is flexible in use and could be used for large- or small-scale projects.

MERISE, with similar origins to SSADM, is directed toward large-scale computer systems. Although the overhead of the method will depend upon the variant of the method which is used, this overhead will be large.

Soft Systems /Multiview	SSADM (V4)	Information Engineering	Yourdon Structured Method	MERISE
Small	Large	Large	Large Medium & Small	Large

Figure 9.6 Size of System at Which the Method Is Aimed

9.7 The Techniques Within the Methods

Here the various techniques will be discussed in relation to the methods which are encompassed by this text. Where there is no entry for a particular method against a technique, it does not use that technique but it may use another technique which accomplishes the same function. The figures indicate whether a technique is used [(Y)es] in the method or not used [(N)o].

9.7.1 Canonical Synthesis (Bubble Charting)

Canonical Synthesis is a technique specific to IE representing the combined user view of data attributes and the dependencies between these attributes. The Bubble Charts produced yield a comparable picture to the data analysis achieved by Normalisation of Entities. (Note: In YSM DFDs are also known as Bubble Charts but should not be confused with this!)

Soft Systems /Multiview	SSADM (V4)	Information Engineering	Yourdon Structured Method	MERISE
N	N	Y	N	N

Figure 9.7 Canonical Synthesis Used

9.7.2 Data Flow Diagramming

Soft Systems /Multiview	SSADM (V4)	Information Engineering	Yourdon Structured Method	MERISE
Y	Y	Y (minimal)	Y	N

Figure 9.8 Data Flow Diagrams Used

Data Flow Diagrams are part of all methods studied here except for MERISE although their significance is greatest to Multiview, SSADM and Yourdon and minimal to IE. MERISE uses Predicate Petri Nets rather than DFDs to model processes and events.

The principle behind DFDs is consistent (see Chapter 3), although their diagramming conventions vary between the methods. This is summarised in Figure 9.9.

	SSM/ MULTIVIEW	SSADM (V4)	INFORMATION ENGINEERING	YOURDON STRUCTURED METHOD	MERISE
PROCESS	1.1 Sophisticated Simplified			1.1	NOT APPLICABLE
DATA STORE	L2	M4			PETRI NETS USED
SOURCE/SINK/ EXTERNAL ENTITY					
DATA FLOW			"Data View"		

Figure 9.9 Comparative DFD Diagramming

9.7.3 Dialogue Design, Dialogue Flow

Dialogue Design and Dialogue Flow are present in all methods, with documentation of the types of user and the flow of data items within the working patterns. The means of documentation varies considerably, from the state diagrams used in IE and YSM to the largely form-based approach of SSADM. Thus, this feature is assessed on the presence of a dialogue design capability within the methods rather than a specific technique.

Soft Systems /Multiview	SSADM (V4)	Information Engineering	Yourdon Structured Method	MERISE
Y	Y	Y	Y	Y

Figure 9.10 Dialogue Design Used

9.7.4 Entity Modelling

Entity Modelling is common to all of the methods under the aliases Entity Model (Multiview, MERISE) Logical Data Structure (SSADM) and Entity Relationship Diagram (IE, YSM).

Soft Systems /Multiview	SSADM (V4)	Information Engineering	Yourdon Structured Method	MERISE
Y	Y	Y	Y	Y

Figure 9.11 Entity Modelling Used

The principle behind them is the same (see Chapter 3) whatever the method although as with DFDs the diagramming conventions are different. This is shown within Figure 9.12.

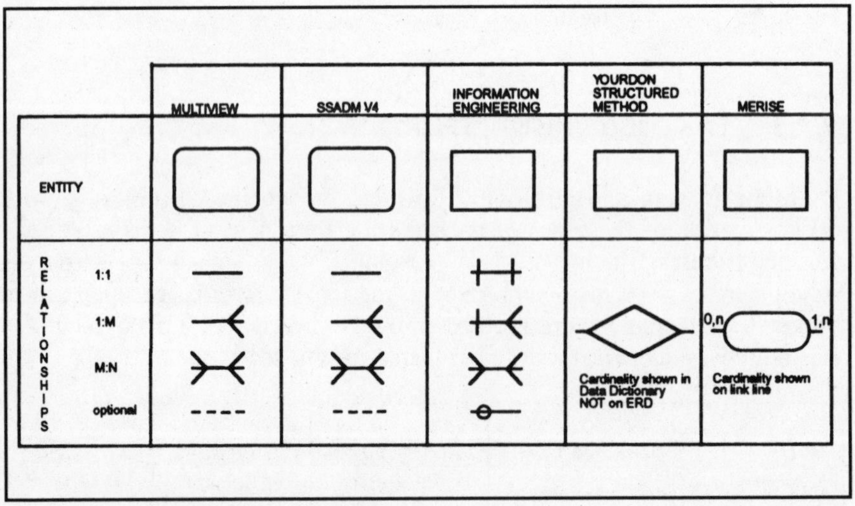

		MULTIVIEW	SSADM V4	INFORMATION ENGINEERING	YOURDON STRUCTURED METHOD	MERISE
ENTITY						
RELATIONSHIPS	1:1					0,n 1,n
	1:M					
	M:N				Cardinality shown in Data Dictionary NOT on ERD	Cardinality shown on link line
	optional					

Figure 9.12 Comparative Entity Modelling

9.7.5 Entity Life Cycles/Histories

The methods adopt different diagramming techniques to represent the states of an entity from the point when it enters the system to the point when it leaves. Multiview shows the valid states diagrammatically as does SSADM but in SSADM optional changes are represented using Jackson notation. IE and YSM use State Transition Diagrams to represent the entity states but as with Multiview optional state changes are not represented. MERISE records state changes on the predicate petri nets.

Soft Systems /Multiview	SSADM (V4)	Information Engineering	Yourdon Structured Method	MERISE
Y	Y	N	N	N

Figure 9.13 Entity Life Cycles/Histories Used

9.7.6 Event/Process/Entity Matrices

The methods use a number of different matrices showing the correlation between events, processes and entities:

- **Entity/Event** matrix is used by Multiview to check the Entity Life Cycle and by SSADM to generate the Entity Life History and Effect Correspondence Diagram;
- **Event/Function** matrix is used by Multiview to confirm the Data Flow Diagrams;
- **Entity/Function** matrix is used by Multiview to check the Entity Life Cycle. IE uses such a matrix in ISP at a high level and at a lower level as an Elementary Process/Entity matrix in preparation for BSD. Cluster Analysis is applied to this matrix to form groupings of elementary processes into Business Systems.

Soft Systems /Multiview	SSADM (V4)	Information Engineering	Yourdon Structured Method	MERISE
Y	Y	Y	N	N

Figure 9.14 Event/Process/Entity Matrix Used

9.7.7 Function Definition

Function Definition has to be treated with care, as the term function has subtly different meanings depending on the method in question. SSADM specifies the function as a user-defined unit of processing. It employs Function Definition as a technique to define functions textually in terms of their described actions, volumes and service levels and diagrammatically in terms of their I/O structures. By contrast both IE and Multiview treat a function just as a high-level process and adopt a diagrammatic approach to their definition. Both use Functional Decomposition to produce hierarchically structured Function Diagrams (IE terms them Process Hierarchy Diagrams and uses soft boxes to represent functions where Multiview calls them Function Charts and uses hard boxes). IE also diagrams Function Dependencies using Process Dependency Diagrams. Multiview essentially creates the same diagram but only for the lowest level functions identifying the diagram as a simplified Data Flow Diagram. This accounts for the diagramming discrepancy between Multiview's simplified and sophisticated DFDs. MERISE and Yourdon do not specifically address processing in terms of functions, although Yourdon refers to Data Flow Diagrams as a Function Model. In MERISE, the units of processing are most frequently referred to as operations and are documented within Petri Nets.

Soft Systems /Multiview	SSADM (V4)	Information Engineering	Yourdon Structured Method	MERISE
Y	Y	Y	N	N

Figure 9.15 Function Definition Used

9.7.8 Normalisation

Normalisation is used by all five of the methods. However, in SSADM it is termed Relational Data Analysis. SSADM employs normalisation to produce a model which can be used to check the Entity Model produced by Logical Data Structuring. IE, in addition to checking the Entity Model can check the Bubble Charts via normalisation. Elsewhere this technique is termed Third Normal Form analysis.

Soft Systems /Multiview	SSADM (V4)	Information Engineering	Yourdon Structured Method	MERISE
Y	Y	Y	Y	Y

Figure 9.16 Normalisation (RDA, TNF) Used

9.7.9 Predicate Petri Nets

Predicate Petri Nets are used in MERISE to model processing and allowable State Changes. In purpose and effect these predicate petri nets combine the function of DFDs and ELHs.

Soft Systems /Multiview	SSADM (V4)	Information Engineering	Yourdon Structured Method	MERISE
N	N	N	N	Y

Figure 9.17 Predicate Petri Nets Used

9.7.10 Process/Procedure Logic Analysis

Process/Procedure Logic Analysis using Process/Procedure Action Diagrams is a feature of IE. It is a structured language, with formal syntax, which is mirrored in Multiview by Structured English for program specification in Technical Design. SSADM accomplishes the same task by using Enquiry Access Paths and Effect Correspondence Diagrams which lead to Process Models with operations documented on them. YSM and MERISE also use Structured English (although in the latter case it is probably more correctly 'Structured French'!)

Soft Systems /Multiview	SSADM (V4)	Information Engineering	Yourdon Structured Method	MERISE
N	N	Y	N	N

Figure 9.18 Process Logic Analysis Used

9.7.11 Prototyping

Prototyping is allowable alongside all five methods, although only formally included within the core framework of SSADM and IE. By Yourdon, Multiview and MERISE, it is acknowledged as a useful technique for giving a 'feel' for certain aspects of the system and for checking that the elements necessary for the required dialogue are present, although it is not necessarily allocated a particular slot in the framework. Thus, we have accepted that each method includes it, although they give differing degrees of support to the concept.

Soft Systems /Multiview	SSADM (V4)	Information Engineering	Yourdon Structured Method	MERISE
Y	Y	Y	Y	Y

Figure 9.19 Prototyping Used

9.7.12 Rich Pictures

Rich pictures feature only in Multiview, this being by virtue of the Soft Systems influence. They equate to Context Diagrams in Data-flow Modelling but exceed Context Diagrams by being loaded with 'soft' information from the problem area.

Soft Systems /Multiview	SSADM (V4)	Information Engineering	Yourdon Structured Method	MERISE
Y	N	N	N	N

Figure 9.20 Rich Pictures Used

9.7.13 Root Definition

A Root Definition features in Multiview but not in any of the other methods. However IE does develop agreed 'Mission and Purpose' statements which as with Root Definitions embody the essence of the area being studied. YSM gives a concise definition of the purpose of the system in 'The Statement of Purpose' which forms part of the Environmental Model. SSADM has the Project Initiation Document (PID) which should hold the same information.

Soft Systems /Multiview	SSADM (V4)	Information Engineering	Yourdon Structured Method	MERISE
Y	N	N	N	N

Figure 9.21 Root Definition Used

9.7.14 Socio-technical Analysis and Design

Socio-technical Analysis and Design is represented within Multiview since the method is influenced by ETHICS (Mumford). MERISE does not have this technique although it includes the concerns and responsibilities of the human 'actors' in the system. IE, SSADM and YSM do not formalise such considerations.

Soft Systems /Multiview	SSADM (V4)	Information Engineering	Yourdon Structured Method	MERISE
Y	N	N	N	N

Figure 9.22 Socio-technical Analysis and Design Used

9.7.15 State Transition Diagrams

State Transition Diagrams are used in Yourdon Structured Method and IE. The diagram shows the 'legal' changes of state and the events which trigger the state change sequence. (Note: See Entity Life Cycles for reference to IEs use of State Transition Diagrams.)

Soft Systems /Multiview	SSADM (V4)	Information Engineering	Yourdon Structured Method	MERISE
N	N	Y	Y	N

Figure 9.23 State Transition Diagrams Used

9.7.16 Structure Charts/Diagrams

Yourdon generates Yourdon specific structure charts with defined characteristics from DFDs to develop the program modules, these charts including selection, sequence and iteration constructs as well as bearing flows of application and control data. SSADM uses Jackson Structure Diagrams for a variety of techniques.

Soft Systems /Multiview	SSADM (V4)	Information Engineering	Yourdon Structured Method	MERISE
N	Y	N	Y	N

Figure 9.24 Structure Charts/Diagrams Used

This concludes our consideration of the major techniques of each method. Other techniques may be present in a method which we have not included here. The reader is advised that the absence of a particular technique from a method may not be a deficiency since the purpose of that technique may be achieved by the method in another way.

9.8 CASE Tools and Methods

CASE Tools can be defined as:

> "Any computer software which aids a software engineer in the specification, design, development, testing or maintenance of computer software, or any aspect of the management of this process."
>
> *Institute of Software Engineering*

Crozier *et al.* (1989) in their analysis of CASE Tools point out that Tools may be classified as either upper or lower CASE where the former addresses the early stages of the system life cycle (Project Initiation, Analysis, Design) and the latter is a tool to aid programming (e.g. Application Generator). It is the upper CASE Tool (such as Analyst Workbenches) for System Analysis and Design which automates and supports the techniques of structured methods. This distinction is, however, being eroded as CASE Tools embody both upper and lower CASE functionality as well as project management, control and testing. This all embracing CASE Tool is known as an Integrated CASE Tool (iCASE).

Theoretically, CASE Tools should enable low defect systems to be developed faster, achieving quality and productivity. CASE Tools are rapidly maturing and the support offered to a particular structured method by CASE Tools is a significant feature of comparison between methods. CASE Tool cost is naturally important and the cost of iCASE Tools may currently be prohibitively expensive for all but the larger organisations. The CASE arena is a rapidly changing one and therefore our assessment can only be made upon the situation at the time of writing. Because of the fluidity of the CASE market, where a feature offered by one CASE supplier today will be offered by all tomorrow, and in fairness to the suppliers, since we cannot mention them all, we have resisted the temptation to make a named comparison of specific tools. We have preferred instead to distil the types of features from the best offerings available for each method. The organisation undertaking its own methods comparison would, in any event, need to make a detailed evaluation of the CASE tools on the market at the time of evaluation.

9.8.1 CASE Tools and Soft Systems/Multiview

Soft Systems/Multiview currently has no specific tool to support its activities although certain tools of general applicability such as Diagramming and Prototyping tools may be employed to improve quality and productivity. General drawing packages may be used in the Soft Systems area to prepare the rich picture. Diagrams such as DFDs, ERDs, Structure Charts and State Transition Diagrams can be generated with the general diagramming tools which are available. Tools at this level require diagram objects to be entered into the underlying dictionary of objects, automation of this function being limited. Prototyping is supported through screen building and the ability to allow the user to enter data to the screen observing the results via the screen design 'inspect' option. Certain tools would allow these screens to be used to generate screen definitions in a target language, e.g. COBOL, BASIC, C, or PL/1

9.8.2 CASE Tools and SSADM

SSADM has evolved through incorporation of tried and tested techniques for system analysis and design. The highly structured nature of, and the increasing complexity of, the method both render it possible and indeed desirable, from the point of view of practicality, to work toward automation. Having started from a state of competitive disadvantage with respect to Tool support compared to IE, it is apparent that the CASE Tools currently available are rapidly closing the gap.

Low cost CASE Tools based in the Upper CASE arena and offering coverage of all the major diagrams except for process models are available. They provide facilities for production of diagrams which are checked automatically for consistency via the central repository. The level of automation is often limited to consistency and completeness checks. However some tools now generate products which can be picked up by 4GLs and database programs and exported into other CASE tools, prototyping tools and Visual Basic. There are iCASE tools which also generate code.

9.8.3 CASE Tools and Information Engineering

Information Engineering originated with the firm commitment to automation via CASE Tools. For this reason it has pioneered the way for others to emulate.

The highly integrated nature of the tools means that changes to diagrams, which impact on other diagrams, are automatically reflected in those diagrams. Notably absent from many of the IE CASE Tools is support for Data-flow Diagramming, which is acknowledged as a secondary technique in IE, useful purely for confirming the existing system with the users.

The Construction stage can be accommodated by automatic generation of source code, data manipulation language statements, screen definition statements, and data-definition statements which after compilation and link creation form the application system.

9.8.4 CASE Tools and Yourdon Structured Method

Whilst there are CASE Tools which specifically support YSM, there are many designed to support other methods which may be used. Since these methods were influenced by YSM the diagram sets in the tools which support them are reasonably consistent with YSM's diagrams.

9.8.5 CASE Tools and MERISE

MERISE specific CASE Tools are available with the level of maturity comparable to those which support SSADM.

9.8.6 CASE Tools Compared

IE emerges, in this aspect of comparison, as the clear leader, with CASE Tools extending over the entire area of the Business Life Cycle. This should not, however, be regarded as any guarantee of production of superior systems since it is still dependent upon the information put in. It could produce more quickly a system which technically works but misses the user requirement! Equally it could generate a technically correct system which meets the functional requirements but does so too slowly. With the release of SSADM (V4) and the accompanying increase in volume of documentation embodied in this version it is inconceivable that CASE tool support will not become an essential requirement for any organisation wishing to adopt SSADM. The availability of iCASE tools which support SSADM addresses the overhead which the method imposes. Multiview does not have any specific Tool support relying on general CASE Tools for this function. YSM has both specific tools as well as being able to be modelled with a general tool.

Soft Systems /Multiview	SSADM (V4)	Information Engineering	Yourdon Structured Method	MERISE
L	M	H	M	M

Figure 9.25 Level of CASE Tool Support

9.9 Feature Analysis Table

The following feature analysis tables do not attempt to quantify the features considered in any significant way. As noted before, such quantification will depend upon the level of importance which an organisation places upon the individual features. Where some level of quantification is possible this has been represented with a H(igh), M(edium) and L(ow) valuation. Other entries ought to be self-explanatory in relation to the information given earlier in the chapter. The techniques featured in the later tables are marked in the tables as Y(es) to indicate their use within a method and N(o) when not used.

METHOD FEATURE	Soft Systems /Multiview	SSADM (V4)	Information Engineering	Yourdon Structured Method	MERISE
Life-Cycle Coverage	M	M	H	M	H
Underlying Philosophy	System/ (Science)	Science	Science	Science	Science/ (System)
Structured-ness	L	H	H	M	H
User Involvement	H/M	M	M	L/M	M
Size of System	Small	Large	Large	Large, Medium & Small	Large
CASE TOOLS	L	M	H	M	M

Figure 9.26 Feature Analysis Table Excluding Techniques

In practice, a good way for this analysis to be set up is to use a spreadsheet program with numerical scores and weightings built into the feature analysis model. The scores for the features for the structured method are entered into the model and the analysis of features made graphically. It is not advisable to be tempted to derive a total score from these tables, since to do so will have the tendency to conceal weaknesses in a particular method.

METHOD **TECH-NIQUE**	Soft Systems /Multiview	SSADM (V4)	Information Engineering	Yourdon Structured Method	MERISE
Canonical Synthesis	N	N	Y	N	N
DFDs	Y	Y	y (minimal)	Y	Y
Dialogue Design	Y	Y	Y	Y	Y
Entity Modelling	Y	Y	Y	Y	Y
Entity Life-Cycles	Y	Y	N	N	N
Event/Process/ Entity Matrix	Y	Y	Y	N	N
Function Definition	Y	Y	Y	N	N
Normalisation	Y	Y	Y	Y	Y
Predict Petri Nets	N	N	N	N	N
Process Logic Analysis	N	N	Y	N	N
Prototyping	Y	Y	Y	Y	Y
Rich Pictures	Y	N	N	N	N
Root Definit.	Y	N	N	N	N
Socio-Tech-nical A & D	Y	N	N	N	N
State Transition Diagrams	N	N	Y	Y	N
Structure Charts	N	Y	N	Y	N

Figure 9.27 Feature Analysis of Techniques Used

9.10 Summary

In this Chapter we have discussed the problems associated with comparing structured methods. Nevertheless structured methods may usefully be compared to identify and take account of any strengths and weaknesses that they may have. They may also be the subject of comparative analysis to be matched against the skills already available in an organisation thereby reducing the learning curve associated with a chosen method.

For the purposes of the comparison presented here, we laid down a limited number of features which could form the basis of a feature analysis, in line with our purpose of highlighting the differences in techniques, approach and framework, between the five methods, in order to give the reader a better understanding of the methods. Readers would be advised to consider extending this list in line with the features of a structured method which are important to them or their organisation, and relevant to the purpose of their evaluation. Chapter 2 contains a list of the types of additional areas for consideration. Having identified our chosen features for comparison, we then presented each feature and examined each method in relation to it, using a simple scale of evaluation or a Y/N in the techniques featured where they were or were not used. In practice more complexity can be introduced by using such a first pass Y/N evaluation to eliminate structured methods on the basis of failing to include techniques or features which the organisation considers to be a mandatory feature. The remaining features would then be quantified in the light of the importance placed upon each feature by the organisation. Since some features will be of greater importance than others to the organisation, some form of weighting factor must be included in the scores assigned to each structured method's feature evaluation. In the absence of this, each feature would be treated as being of equal importance which is an unlikely scenario.

Finally, we presented the results of our comparisons as completed tables, which we hope will provide a useful checklist for the analyst and a quick cross-reference to the techniques and other features of the methods.

10

Focus on Methods and Tools – The Future

10.1 Overview

In this final chapter, we have undertaken to examine the future of structured methods. It is always risky to attempt to predict the future. Take, for example, these two heroic failures:

> "My department is in full knowledge of the details of the invention [the telephone], and the possible use of the telephone is limited."
> *Engineer-in-Chief, British Post Office, 1877.*

> "One day, every town in America will have a telephone."
> *Mayor of small American town , circa 1880.*

However, the trees of the future are the saplings of the present, and if we look at the things which are starting to impact on the structured methods of today, we should gain an insight into the future without danger of barking up too many wrong trees! In this final chapter, we take a look at some of the 'hot topics' around at the moment, and consider their likely effect on structured methods in the future. We look at the problems which appear to remain with systems development in spite of the existence of structured methods. Finally, we consider the approach needed in an organisation to get the best out of its structured method.

10.2 What Other Considerations Are Just Arriving or Are Already Around

The key influences which we have identified are listed below. This is not an exhaustive list, but rather our own personal list, based on consultancy experience of the topics people are asking about in relation to structured methods. The topics we shall consider are:

- Object Orientation;
- CASE Tools and Expert Systems;
- Euromethod;
- Rapid Application Development;

- Business Process Redesign;
- Reverse Engineering.

More emphasis has been given to Object Orientation, CASE Tools and Euromethod, this being in proportion to their likely greater effect on the future of structured methods as we now know them.

10.3 Object Orientation (OO)

Object-oriented Analysis and Design is concerned with modelling objects from the real world and then using the model to build a language-independent design around those objects. In this brief overview we shall look at:

- the terminology and concepts of object orientation;
- the three levels of modelling and the major techniques within these;
- the interaction between the three views;
- the benefits claimed for OO;
- the future implications of OO.

10.3.1 OO Terminology and Concepts

There is some dispute over exactly what characteristics are required to define an object-oriented system, but the 'magic words' in Object Modelling Technique (OMT), the method described by Rumbaugh, Blaha, Premerlani, Eddy and Lorensen (1991), are:

- identity;
- classification;
- polymorphism;
- inheritance.

These concepts are the properties of the mainstream object-oriented languages. They will be explained below, but in order to understand them it is necessary first to introduce some other important terminology. This involves discussion of:

- the object, object instance and object class;
- the operation;
- encapsulation.

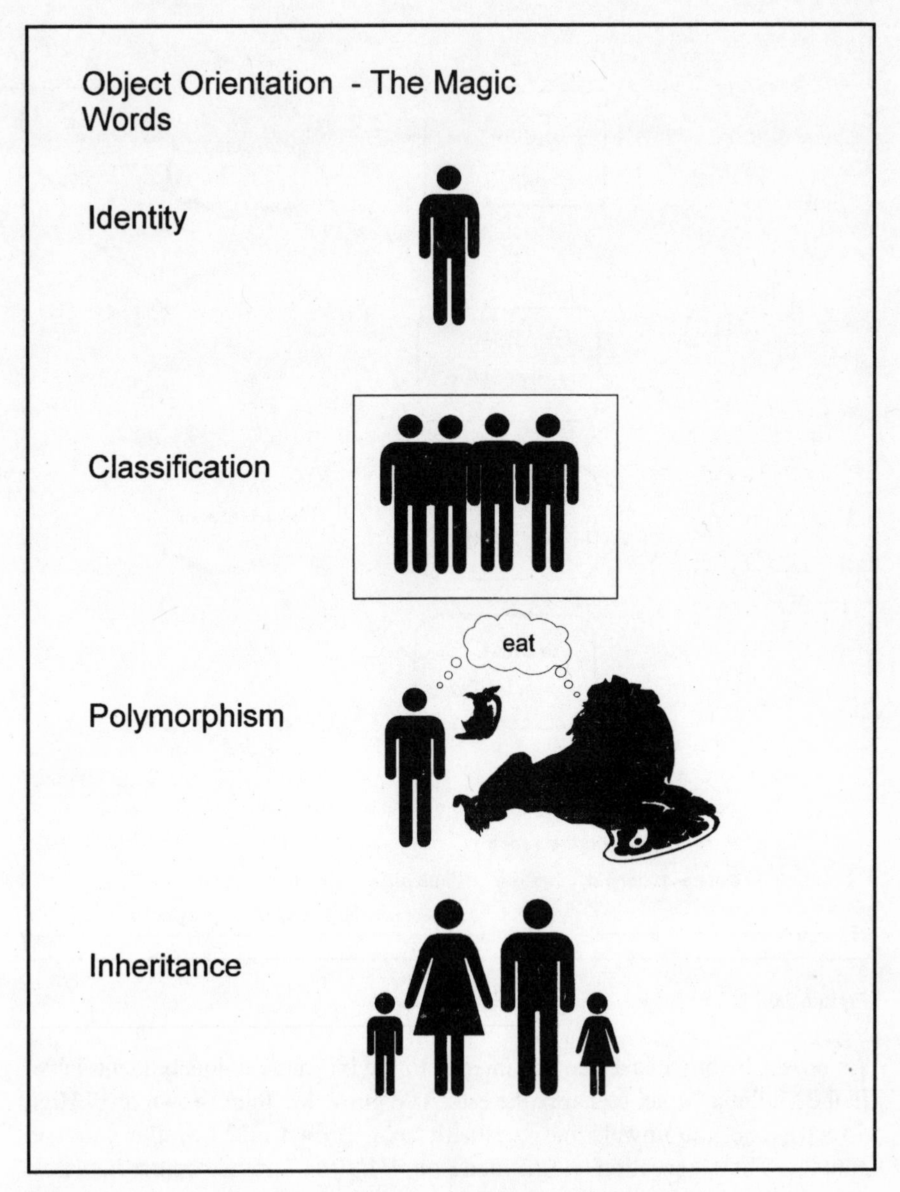

Figure 10.1 The Key Concepts of Object Oriented Systems.

10.3.1.1 The Object, Object Instance and Object Class

The fundamental construct of OO is the **object**, which combines both data structure and behaviour. The traditional structured-methods aficionado will instantly recognise the similarity with the entity. This is not wrong. The entity is an object, but the object goes a little further than the entity, as will be seen. In discussing the object, we must make the distinction between **object instance** and **object class**.

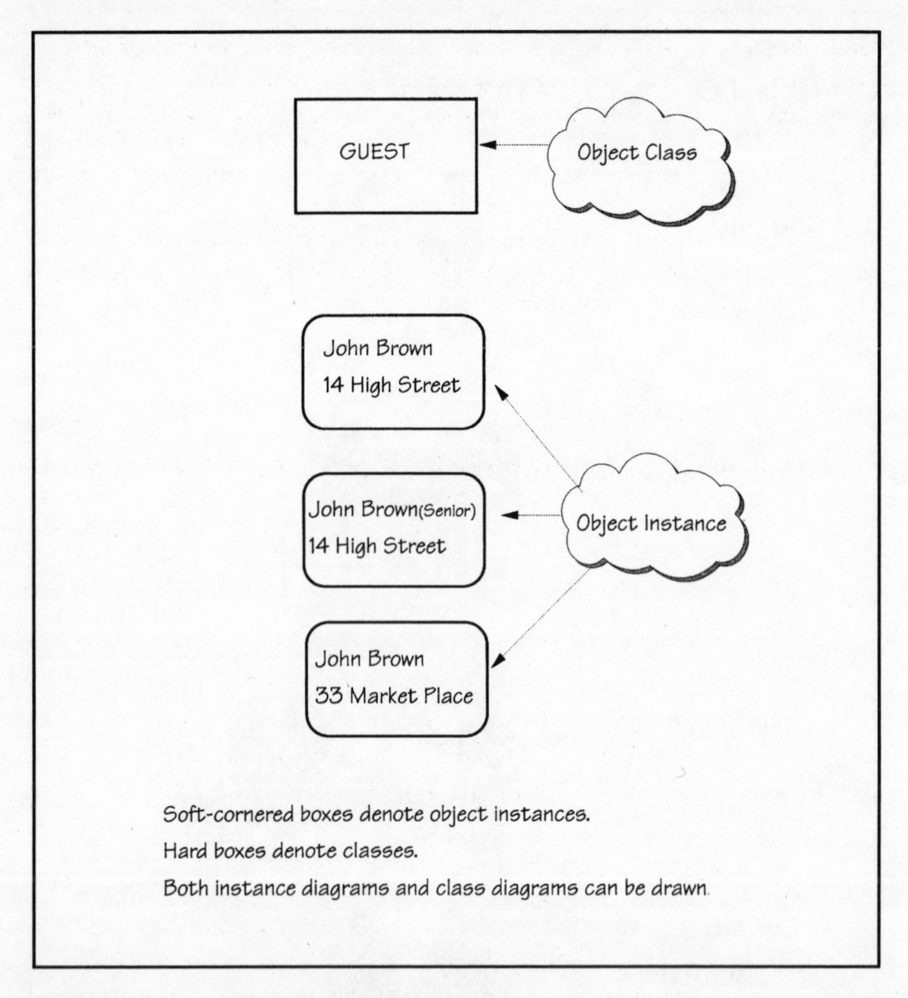

Figure 10.2 Object Class and Object Instance

An **object instance** is a thing of interest to the business, uniquely identifiable. In the Midlinks Motel scenario, the esteemed guest, Mr John Brown of 14 High Street is one, the unwelcome guest, Mr John Brown of 33 Market Place is another. The reservation 11122 made on 4/4/99 is another object instance. (These would all have been identified as entity occurrences by our traditional structured methods. The difference is that, the OO object will eventually also encompass the processing associated with this *thing*.) Here we are talking about individual occurrences of these things. There is a specific individual who is in our mind when we telephone to speak to Mr John Brown of 14 High Street. Even if two Mr John Browns live at the same address, we have in mind a particular one, uniquely identifiable to us. Because object instances are **real** things of interest to the **business**, they are usually easy to elicit from the users. Although the terminology and semantics of OO are foreign to business people, the idea of 'things of interest, with which we work, about which we keep information' are second nature. These things are the objects we seek.

An **object class** is a group of object instances with similar properties (attributes), common behaviour (operations), common relationships with other objects, and common semantics. From investigation of the Midlinks Motel scenario, GUEST, RESERVATION, ROOM TYPE are all object classes. These were all previously discovered as entities (entity types) by our traditional structured methods. The difference here is that we are identifying and grouping together with the object class not only their attributes but also their behaviour (i.e. the processing which can affect them, and the changes which take place as a result of processing).

10.3.1.2 The Operation

An **operation** is an action or transformation which an object performs or is subject to. Object instances are grouped into object classes based on their behaviour, i.e. their affinity for the same operations. The same operation often applies to more than one object class. This gives the concept of inheritance, defined later. Figure 10.3 shows the notation for inclusion of operations with an object.

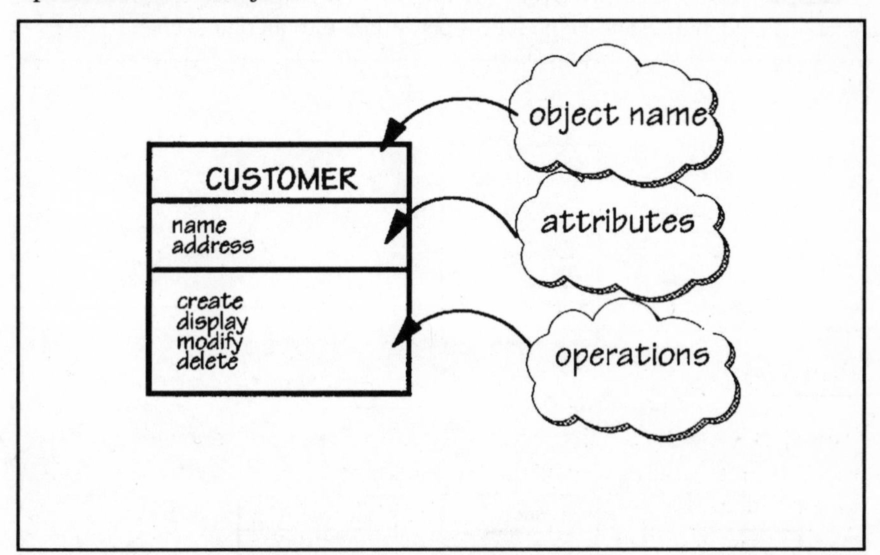

Figure 10.3 The Definition of an Object Class in Terms of Attributes and Operations

Encapsulation describes the linking of operations to their object classes. However, it has a deeper meaning than this. It includes the separation of the 'logical' properties of an object from its physical implementation. The purpose of this is to maintain the independence of processing from data, such that the implementation of an object can be changed without affecting the applications which use it. The concept of abstraction, introduced in Chapter 3, is the essence of this. Encapsulation is akin to surrounding the object class with a *fire wall*, which prevents unauthorised access or tampering.

Encapsulation ensures that the only way an object is allowed to be modified is via its legitimate operation interfaces.

10.3.2 The Magic Words

So we are now in a position to define our four 'magic words' which characterise an object oriented system (see Figure 10.1):

Identity is the expression of data as discrete, distinguishable entities (object instances). These instances of objects are uniquely identifiable, even though all of their attributes may be the same. There is no need in object-oriented analysis or design to give a unique key, unless this is relevant to the real world. The physical implementation will give objects a unique 'handle' anyway. However, the objects of interest in the real world are often the things we have already given keys to so that we can distinguish between them regardless of any computerised system. For example, Reservation Number, Room Number, etc.

Classification is the grouping of object instances with the same data structure (attributes) and behaviour (operations) into an object class.

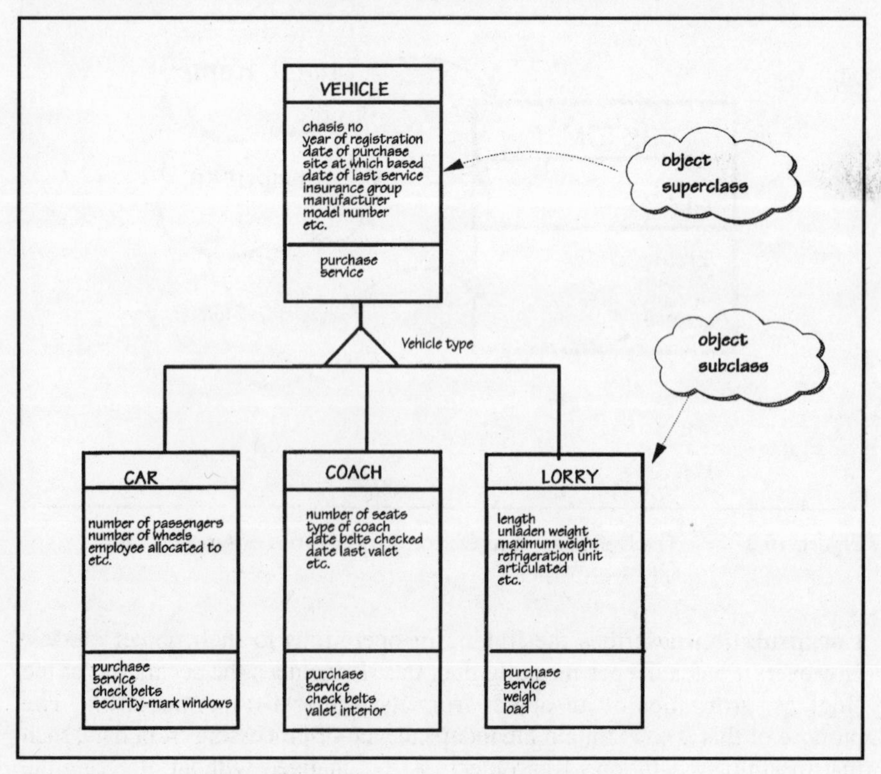

Figure 10.4 Inheritance, Object Subclass and Superclass

Polymorphism is the term used to describe the fact that the same operation may behave differently on different object classes. For example, the operation 'accept sales order' may transform the object order by bringing into being an instance of it, and the object product by altering the attribute 'available quantity'. Each object class knows how to implement its own version of the operation. The version of an operation relevant to an object class is known as a **method**. *(From this point onward, the solo appearance of the word 'method' implies this object-oriented context. We shall use the term 'structured method' where we talk about the structured methods compared in this book.)*

Inheritance is where there is sharing of operations and attributes between object classes. The same attributes can apply to more than one object class, based on a hierarchical relationship. Each **sub-class** inherits all of the properties of its parent **super-class** and adds its own unique properties. Additionally, the same operation can apply to more than one object class. For example, in Figure 10.4, we have defined VEHICLE, CAR, COACH and LORRY. Some attributes are shared by all, others are specific to just one of the object classes. Some operations will apply to them all. An example would be 'service'. Other operations would be specific to just one class. For example, the operation, 'weigh' would probably only apply to the lorry.

10.3.3 The Three Models of OMT

OMT adopts the principle of **abstraction** as a means of focusing on the essential aspects of what an object is, and what behaviour it has, before consideration of how it should be implemented. This is the same principle as the logical (essential, conceptual) abstraction used by the structured methods we have considered earlier in this book. OMT also provides a graphical approach to modelling these different views of the system area.
OMT uses three kinds of model:

- the object model;
- the dynamic model;
- the functional model.

This should be sounding very familiar to the reader who has journeyed with us thus far! These are the models we have grown to know and love from other structured methods, with one major difference. The emphasis is very heavily weighted towards the object model, with the functional model being of least importance. The object model incorporates not only the data structure and relationships, but also the operations (processing) related to each object class. It is the central document to which the other two cross refer. These models are further described below.

10.3.3.1 The Object Model

The Object Model describes the static structure of objects within the system area and their relationships to other objects, their attributes and their operations. The object model is represented graphically as shown in Figure 10.5. This also shows some of the notation for types of associations. In the object diagram illustrated in Figure 10.6, object classes are arranged into hierarchies sharing common behaviour and structure. The associations (relationships) between object classes shown on these diagrams are the familiar combination of one to many , one to one, many to many, optional, exclusive and recursive (see Chapter 3 for the structured methods interpretation of these). The concepts of sub-classes and super-classes, very similar to entity sub-types and super-types, are well-developed in object modelling.

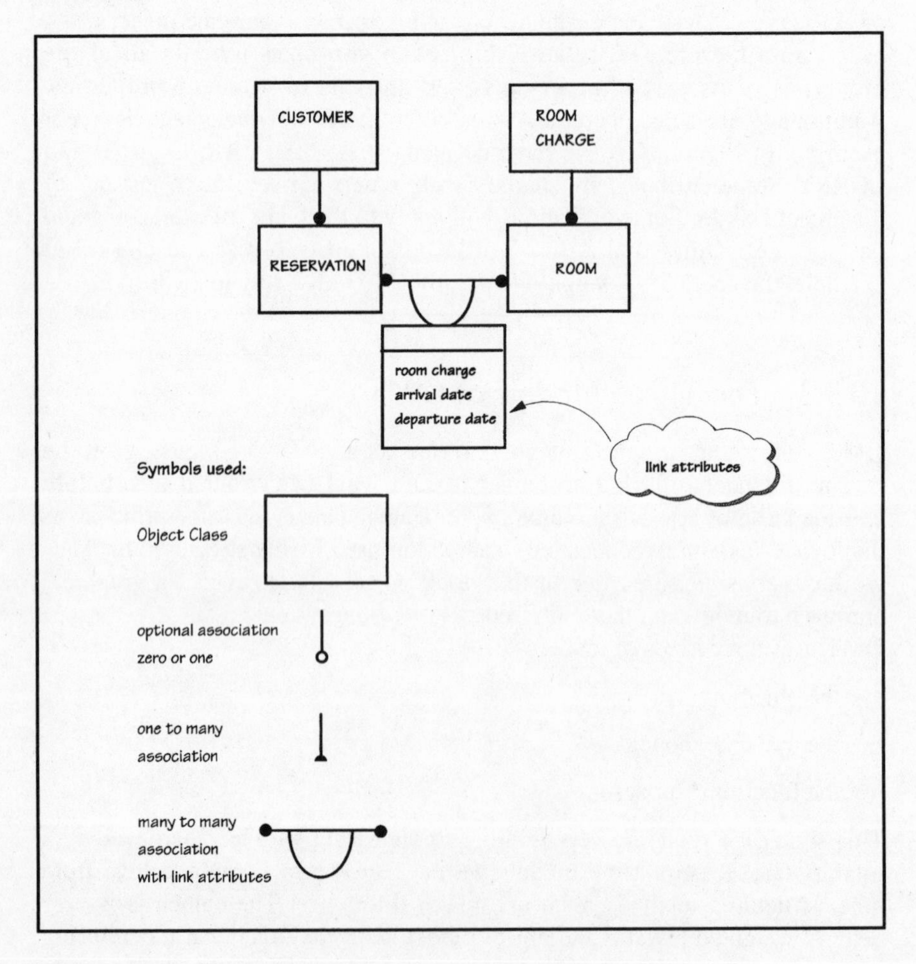

Figure 10.5 The Object Model – Symbols Used and Association

The important additional concepts for the object model are:

- link;
- association;

- aggregation;
- generalisation.

A **link** is a connection between object instances. For example, guest John Brown has reservation number 11223. There is a link between John Brown and his reservation.

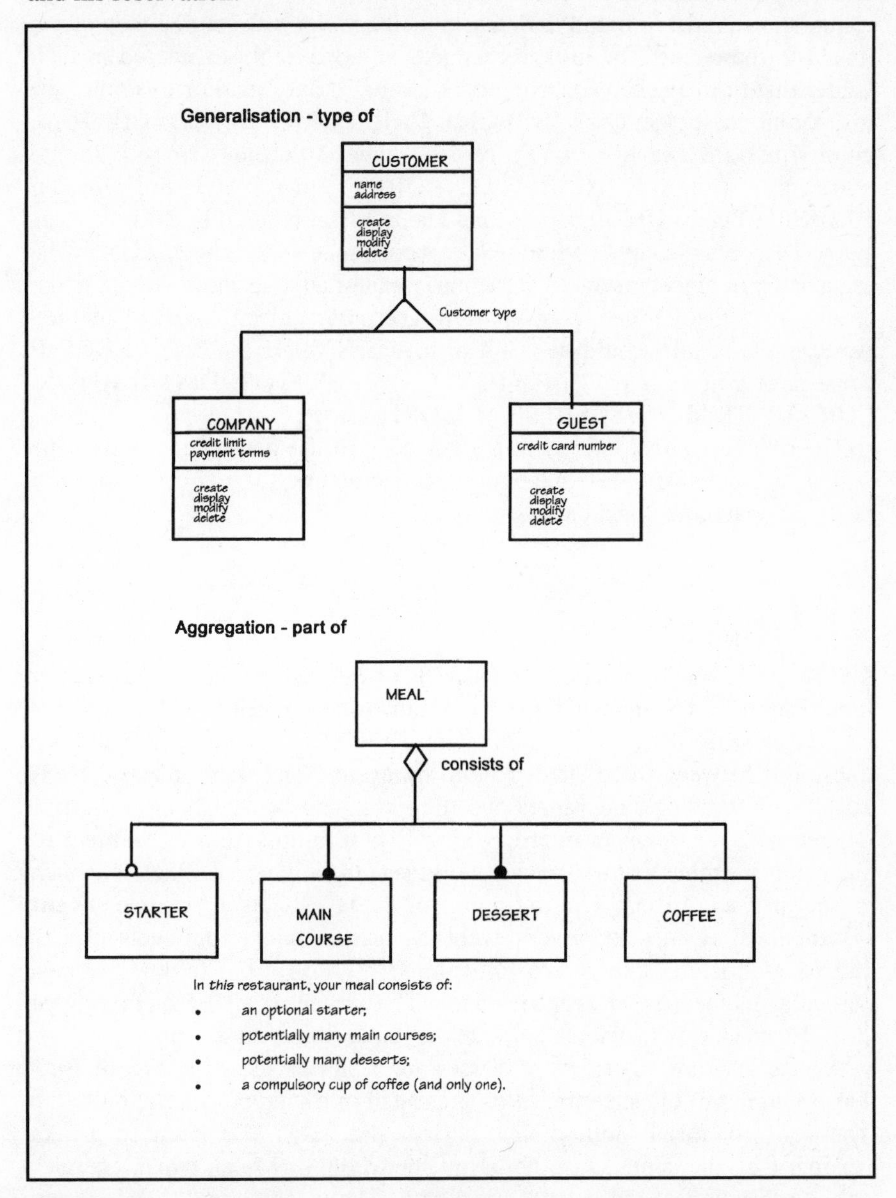

Figure 10.6 Generalisation and Aggregation

An **association** describes a group of links. For example, a customer makes a reservation (or a set of reservations). The notations for one to many, one to one and many to many associations are shown in Figure 10.5. This is referred to as the *multiplicity* of the relationship (cardinality is the structured method term for it) and detailed volumetrics can be shown on the diagram. Many-to-many associations can be defined (resolved) in terms of **link attributes** which can be related to a new object class. Again, the similarity with link entities (intersection entities) found in structured methods is apparent. The recursive relationship found in entity modelling within structured methods is linked to the concept of **inheritance** and is more richly expressed in OO. **Generalisation** is where one object is a *type of* another. For example, we may define the object class VEHICLE, CAR, LORRY and COACH. These were illustrated earlier in Figure 10.4. The superclass (parent in the hierarchy) is the VEHICLE. CAR, LORRY and COACH are *types of* VEHICLE. Figure 10.6 illustrates these concepts in relation to the hotel case study. There are also other additional concepts, such as the ability to show the sequencing of object instances at the many end of an association, to refine the semantics of the diagram. **Aggregation** where one object is *part of* another. An example of this would be a bill of materials where the object ENGINE consists of objects VALVE, PISTON, CARBURETTOR, DISTRIBUTOR, COIL, SPARK PLUG, STARTER MOTOR, etc.)

The object classes defined relate not only to the data stores within the system area but also to external entities, called **actors** (the sources or sinks of the functional model) and data flows.

10.3.3.2 The Dynamic Model

The Dynamic Model describes the aspects of the system which change over time. The dynamic model is used to control the integrity of the data and comprises **state diagrams** showing the valid states of an object and the transitions between these states caused by events. The state of an object is visible from its attribute values and the links held by it. Changes of state happen as a result of an **event**, which is a stimulus from one object to another. Remember that the outside world is defined in terms of actors, which are defined as objects. The definition of event is seen in terms of **event instance** and **event class**. Every event instance is unique, but event classes can be identified. The arrival of John Brown on a particular day at a particular moment is an event instance. The arrival of GUEST is an event class. Event classes may have hierarchies, just as object classes do.

A state diagram relates event classes and states as shown in Figure 10.7. This is similar to the state transition presented in Chapter 7 in relation to the Yourdon Structured Method. It describes the states of a licensed drinks vending machine, situated in the bar of the Midlinks Motel, which operates only on insertion of the guest's room card. The machine will dispense any number of drinks until the 'end' button is pressed, at which point it will return the card with a message 'Have a nice day'. The reader may wish to attempt a diagram describing the states of guest after using the drinks machine!

Both states and actions in the notation presented are represented by lozenges (soft boxes) and movement between states and actions are shown by arrows, labelled where appropriate with events. More complexity can be built into these diagrams, if required. For details, the reader is referred to Rumbaugh *et al.* (1991).

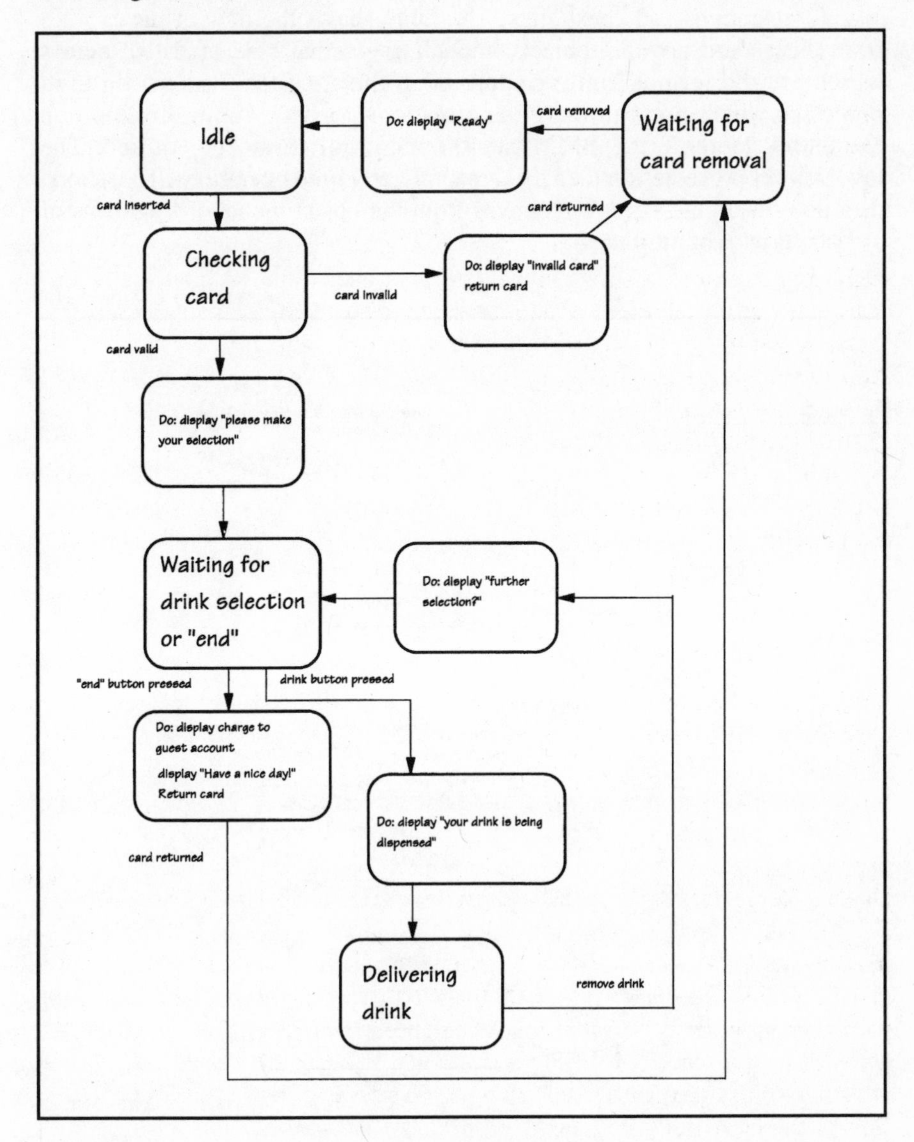

Figure 10.7 The Dynamic Model – a State Diagram

10.3.3.3 The Functional Model

The Functional Model describes the transformations caused to the data by the operations. It models the functions using data flow diagrams (DFDs). These show what the system must achieve, without regard to how, when, where or by whom the processing is done. They are akin to the logical data flow diagrams described in Chapter 3.

The functional model consists of a hierarchy of data flow diagrams which specify operations and constraints. The DFD shows the flow of data values from their sources within objects (including external objects, called **actors** which are the terminators, or sources or sinks, of the system) through processes which perform the transformations on data. As in the Yourdon Structured Method, the DFDs can show control flows, if required. The lowest-level processes on the DFDs map directly onto operations. The actors, data flows and data stores are objects from the object model or fragments of objects (groups of attributes).

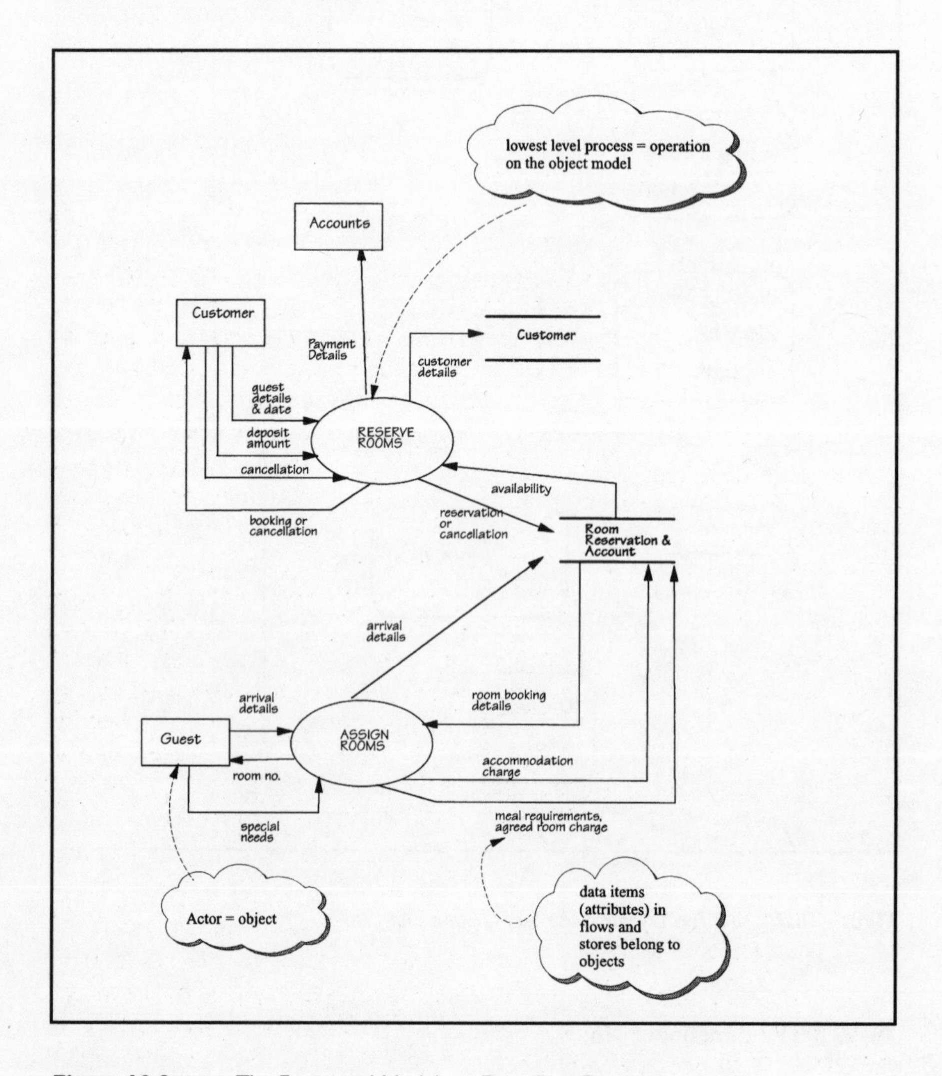

Figure 10.8 The Functional Model – a Data-flow Diagram

10.3.4 The Interrelationship Between the Three Models

The object model shows the structure of the actors, data stores, data flows contained in the functional model. It also shows the attachment to object classes of the operations shown in the functional model. The object model shows the detail of the objects for which the dynamic model shows state changes.

The dynamic model shows the operations performed in response to events and state changes, the sequence in which the operations are performed, and the states of each object class.

The functional model shows the detail of the operations associated with each object class, including the arguments of those operations. It also shows the detailed rules of the operations and constraints under which they operate (e.g. synchronisation of events and states).

10.3.5 The Benefits claimed for OO

OO Analysis and Design covers the software development life-cycle from problem formulation, through requirements analysis and design, to implementation, and can be used with rapid prototyping, incremental and evolutionary development approaches. Its emphasis on data before function should give the development process a more stable base, since the data structure is less likely to undergo fundamental change than the functional requirements of the business. This argument would have been a valid point scored for OO over the early function-oriented structured methods. However, the importance of data is now recognised by most modern versions of the structured methods, and certainly by the structured methods covered in this book. The object model, with operations and attributes incorporated, is an appropriate product to convert directly into a design and implementation. The process is 'seamless, and there are no discontinuities in which the notation of one phase is replaced by a different notation at another phase'. (Rumbaugh *et al.* (1991)). The development process involves the continual refinement of the same model and as such its development is iterative rather than linear, which reduces the chance of introducing inconsistencies and errors once the requirements have been established. The major cost-justification factor for the OO approach, however, is the lure of 'new systems for old' provided by prospect of re-use of code. The realisation of this dream, however, by the effective management of code libraries, is by no means simple;

10.3.6 OO Conclusions

There is little in object orientation which is new to the structured methodologist. Indeed, the hardened SSADMer has the content of the object model and dynamic model within the LDS and ELHs. However, certain factors will undoubtedly motivate increased migration towards object oriented techniques:

- the perceived opportunity for re-use of code provided by the OO approach makes systems development look a faster and cheaper undertaking;
- the concepts of structure and hierarchy of objects is better-defined in the object-oriented approach than in the traditional structured methods;
- the diagramming techniques within OMT are clear and relatively intuitive, and have the benefit of being focused on the *things* the business understands;
- the appropriateness of the terminology and diagrams to the implementation languages of the moment.

The reader should particularly 'watch this space' in relation to SSADM as there is an object-oriented extension to its framework, now available, which should considerably assist the transition from the SSADM products to an object-oriented implementation.

Where the object oriented techniques appear to lack substance is at the 'softer' end of the development spectrum, that of establishing the business requirement. A marriage with Soft Systems principles could considerably strengthen both parties.

There are still problems with the practical implementation of OO concepts onto current database technology. Genuine OO databases, where objects can be stored with their methods tightly linked, are still immature. OO programming languages do not *force* a structured analytical approach to the definition of objects, and thus are capable of suffering from the same data dependency problems which dogged the monolithic file structures of the past. There are experts in the methods field who doubt that OO 'will ever amount to much'! Whether it does or not will depend on the speed with which OO databases develop and the extent to which OO concepts become a part of serious code-constructing CASE tools. We shall look at CASE tools in the next section.

10.4 CASE Tools and Expert Systems

CASE (Computer Aided Software Engineering) was defined in Chapter 9, and has been considered there in relation to the structured methods presented within this book. Therefore, this section will aim to look at future CASE and its likely impact on structured methods. CASE is classified by Yourdon (1992) as the most important of all of the technological advances in the software engineering field in the past decade. Certainly, it offers the solution to some of the major problem with structured methods:

- the amount of documentation;
- the management of complexity;
- the management of the data dictionary/repository;

- the cross-checking of diagrams;
- the eventual production of database schemas and code.

The information overhead imposed by structured methods, together with their diagramming bias, make automated support essential for anything but the smallest of projects. So what will future CASE be capable of in the short term and in the long term? Many CASE Tools currently offer real-time assistance for such aspects as consistency and completeness checking as well as updating all related information when a change to a design object is made. The short-term trend is toward tools which offer greater practical assistance in real-time with the development of diagrammatic models and the underlying data repository. Technological advances in desktop computers and faster communication networks have promoted the early stages of development of a complete knowledgebase of all of the organisation's systems, available to all analysts within the organisation. The tool developers are increasing the number of target environments toward which CASE Tools are directed. Some CASE tools are now producing code, from diagrams and their supporting logic, and many will export the database schema. These code and database generation aspects are still relatively immature, but they will improve. The only difficulty is in predicting how quickly. Various international standards, in existence or currently being developed, will affect the features and data interchange capabilities of CASE tools and will govern such practicalities as the format for interchange of data between CASE tools, and from CASE tools to databases. This will enhance the possibilities of choosing several CASE tools, each aimed at a different aspect of the software development task and 'bolting them together' where necessary through a common interface. This is already possible to a limited extent at the time of writing, and will increase in response to market pressures and the arrival of the standards.

10.4.1 CASE and Expert Systems

In the long term it is likely that Expert CASE Systems, made possible by technological advances, will be developed. Gane (1990) identifies a number of facilities that such Expert Systems should provide to support analysis and design, founded upon the existence of a complete organisational **knowledgebase**. The knowledgebase could offer **generic system descriptions** which would form a discussion start-point to replace the initial vacuum of the early stages of requirements definition. Standard templates for data structures appropriate to a particular organisational function could be part of this knowledgebase. In the area of model representations, diagramming tools should become more intelligent, aligning objects better, routing connections more clearly and making the diagrams more readable by dispersing objects more evenly. Given a central repository of design objects, CASE tools ought to be able to provide intelligent real-time advice when an object is defined, to prevent alias formation for example, and to offer details of existing objects which are very similar to the one being proposed. At design time, the Expert System could offer a suggestion for screen layout which would take into account such factors as the same 'look and feel' and the organisations' standards. It may even call up an existing screen which users have found easy to work with and which could be re-used in the new system. The aim is to use the knowledgebase to reduce the level of new work and eliminate the 're-invention of wheels'.

10.4.2 Implementing a CASE Tool

The implementation of CASE in an organisation is a major project in itself and requires planning, management, training and commitment to make it work. The selection of the appropriate CASE tool will depend upon the structured method already in use by the organisation. Parkinson (1990) identifies the factors which need to be considered to successfully implement the use of CASE within an organisation. An interesting point is his assessment that if an organisation does not **already** actively use a structured method when beginning to implement a CASE tool, the benefits of using the CASE tool (underpinned by that method) will not be realised.

There are special problems to be overcome in the implementation of a networked CASE tool where the same projects and repository are available simultaneously to many analysts. The key issues of:

- security (backup and restore, password protection);
- configuration control;
- identification of logical and physical objects;
- model 'ownership' and control.

must be addressed at the outset of the development project.

As with all major projects, a pilot exercise should be undertaken wherever possible and the results assessed to improve the effectiveness of future use of the tool. The importance of training and expert advice in the use of the tool cannot be overemphasised. Above all, the organisation should progress slowly into the realms of CASE and minimise its risk.

10.5 Euromethod

10.5.1 What Is Euromethod?

Euromethod is a public domain framework for planning and managing services to investigate, develop and amend Information Systems (IS). It is **not** another systems development method of the type we have been considering in detail in this book. Rather, it is complementary to those methods, providing for customer/supplier relationships and management of the IS projects of which structured methods may be a part. It can be used effectively in any IS-related project where the following can be identified:

- customer;
- supplier;
- a contract (formal or informal);
- definable start and end states (which may be defined as part of the contractual negotiation).

Euromethod supports the understanding, planning and management of the contractual relationship between a customer and a supplier *of IS adaptations*. An IS adaptation is any automation, enhancement, modification or improvement to an IS system at any stage of its life cycle. Euromethod supports both the relationship between the contract-authorities of supplier and customer and the subsequent project teams of each organisation which do the agreed work. This is illustrated in Figure 10.9.

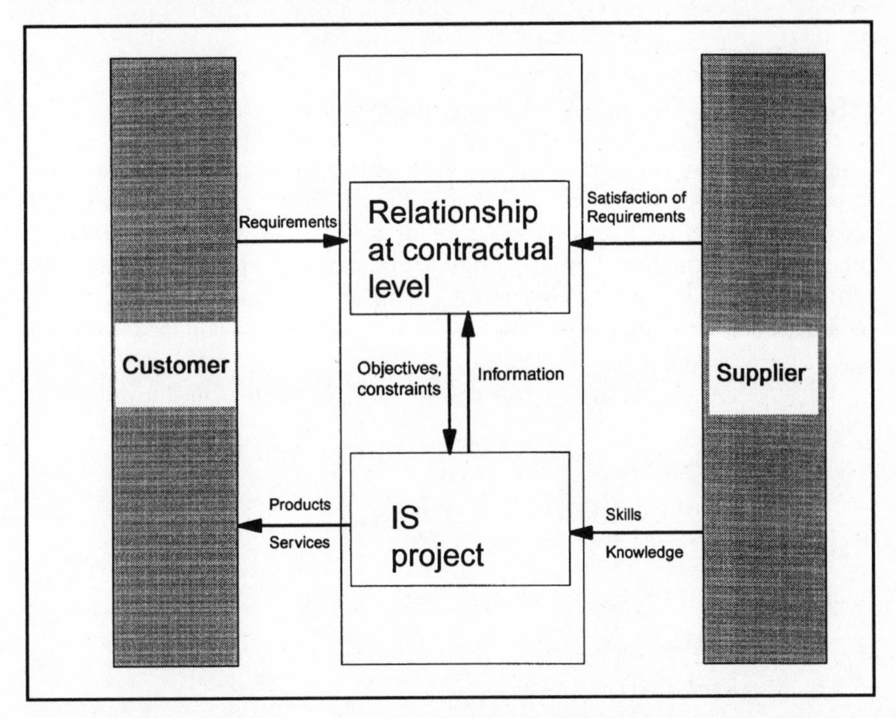

Figure 10.9 The Two Levels of Relationship Between Customer and Supplier

Euromethod will give guidance for both customers and suppliers in an IS relationship, and provide advice on decision points and deliverables for project planning and control. It operates above the level of any individual structured method (for example, it will define general components but not specific techniques).

10.5.2 The History

The Euromethod project, funded by the European Commission, was initiated when, in 1989, EEC member states produced a proposal for Euromethod.

The problem which Euromethod set out to address was the perceived divergence of methods for information systems engineering and the need to be able to compare them, particularly from the procurement point of view. This perceived divergence was confirmed by the Euromethod feasibility study, which concluded that methods currently in use *were* technically divergent, although their purposes and roles were similar.

The work on Euromethod is on-going and is being carried out by a consortium of companies with representatives from eight European countries. The project has been planned, and is being progressed, through the following phases:

- Phase 1 (1988–89): establishment of the requirement by member states;
- Phase 2 (1989–91): feasibility study undertaken and plans for phase 3;
- Phase 3 (1991–ongoing): phased development of Euromethod.

10.5.3 The Purpose of Euromethod

The purpose of Euromethod is to help those involved in IS planning, procurement and development to choose the most cost-effective approach to meeting their IS problems. It will provide a focus for an open competitive market in Information Systems Engineering across Europe and enable the different methods used throughout the European Community to be compared and harmonised, exploiting current skills and providing a common target for future development of these methods.

Euromethod has been developed with specific input from the following methods:

- SSADM (UK);
- MERISE (FRANCE);
- DAFNE (ITALY);
- IE (US);
- MEIN (SPAIN);
- SDM (HOLLAND);
- VORGEHENSMODELL (GERMANY).

10.5.4 The Aims of Euromethod

Euromethod aims to provide a framework which will improve the effectiveness of IS development and maintenance approaches by promoting:

- understanding: Euromethod aims to improve the customer–supplier relationship in information systems procurement, development and maintenance, through the presentation of guidelines, models and the use of a common terminology;
- harmonisation: Euromethod provides framework and concept manuals to assist in the conceptual harmonisation of development methods from procurement and management points of view;
- flexibility: Euromethod will assist in the assessment of the appropriate methods for particular problem situations and in the planning and control of the use of methods.

Thus, across Europe, software development proposals should be able to be made in the preferred method of the proposer. Software procurers should be able to evaluate fairly suppliers' proposals based on a variety of methods. The subsequent development projects should benefit from improved communication, with the definition of terms and concepts, and from a well-planned and controlled customer–supplier relationship. Euromethod should also encourage the introduction of new methodological concepts to areas poorly covered by current methods, by providing a reference framework which highlights such deficiencies.

10.5.5 Euromethod Basic Principles

The design of Euromethod is based on the following principles:

- **IS adaptation:** Any project on which Euromethod operates is termed an *IS adaptation*;
- **Variety of contracts:** Euromethod applies to a wide variety of IS adaptations, where there is a customer-supplier relationship, a contract, and clearly–definable start and end states. Euromethod can be used on the same information system many times, at different stages of its life cycle;
- **Customer–supplier relationship:** Euromethod defines customer and supplier as *roles* and has role guides to address the viewpoints of each role. Euromethod supports the understanding, planning, and management of contractual relationships between customer and supplier, defining the transactions between them and incorporating these into the project plans;
- **Situation driven:** Euromethod gives support in the development of a strategy and plans for the particular IS adaptation. These plans are tailored to the characteristics of the individual problem situation;
- **Focus on key decisions:** Euromethod focuses on the transactions between customer and supplier and the key decisions to be taken by each party. It assists with the positioning of these key decisions within the project plans.

- **Focus on deliverables:** Euromethod focuses on the deliverables exchanged between customer and supplier, and their position in the plans for the IS adaptation;
- **Method bridging:** Guidelines are available within Euromethod to allow for the provision of a bridge between the concepts and the terminology of Euromethod and that of development methods.

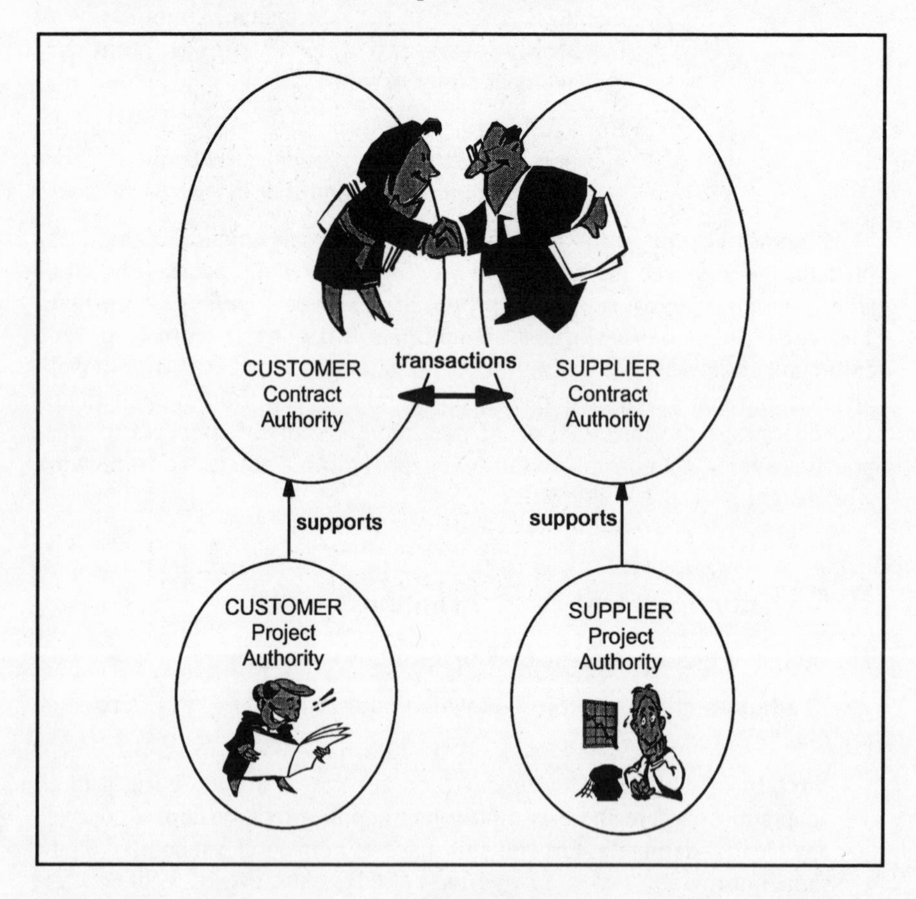

Figure 10.10 The Customer – Supplier Relationship

10.5.6 What Guidance Euromethod Will Provide

Euromethod will provide guidance in the form of ten major documents, illustrated in Figure 10.11. These fall into four categories:

- Euromethod Overview;
- Euromethod Guides;
- Euromethod Concept Manuals;
- Euromethod Concept Dictionary.

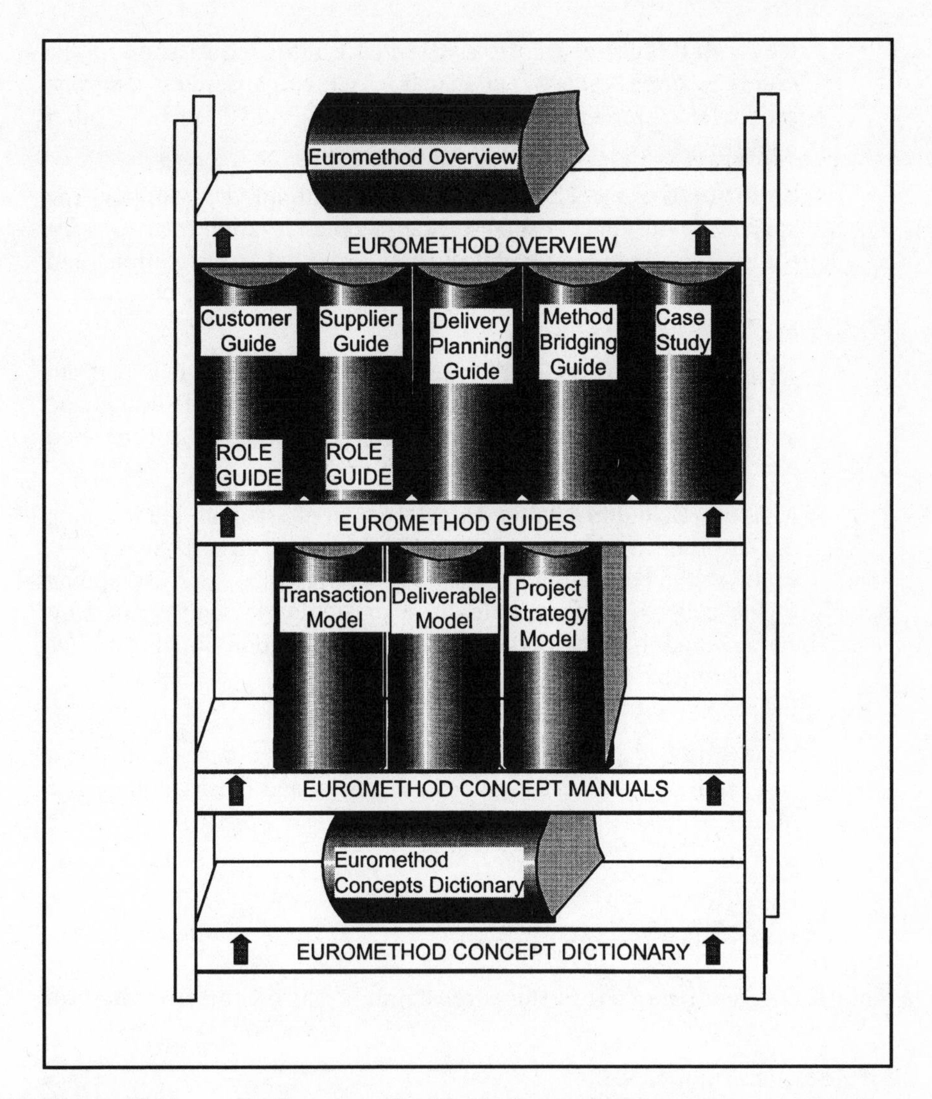

Figure 10.11 The Documents Provided by Euromethod

10.5.6.1 Euromethod Overview

This document is a general introduction to the principles of Euromethod. It serves as a guide to the other documents and should be the first document to be read.

10.5.6.2 Euromethod Guides

This set of four guides and a case study are provided to facilitate and explain the management of the customer – supplier relationship at contractual and project level. The set consists of:

- **Customer Guide:** This is a role guide intended to support the customer in customer – supplier transactions. It defines concepts, products, customer – supplier transactions, and the use of other guidelines to assess and express a plan;
- **Supplier Guide:** This is a role guide intended to support the supplier in customer – supplier transactions. It is intended to assist the supplier in the interpretation of technical specifications and delivery plans. This also defines concepts, products, customer – supplier transactions;
- **Delivery Planning Guide:** This document is intended for both customer and supplier. Its aim is to give guidance on the design and assessment of project plans and deliverables using Euromethod concepts for projects of various types;
- **Methods Bridging Guide:** This document assists with the mapping of the products of a method onto Euromethod. It will also prove useful in the population of plans with deliverables and the mapping of the proposed deliverables of a particular IS adaptation onto Euromethod. It helps in the assessment of suitability of IS methods to a particular problem situation and to determine the suitability of the plans;
- **Case Study:** This comprises a series of project examples related to a case study, to support the guides and to test and illustrate the usage of Euromethod concepts.

10.5.6.3 Euromethod Concepts Manuals

This set of manuals covers the three models within the Euromethod Architecture:

- The Transaction Model;
- The Deliverable Model;
- The Project Strategy Model.

The Transaction Model
The aim of the Transaction Model is to provide a suitable model of the customer – supplier relationship within which to manage interactions between customer and supplier. It identifies the transactions between customer and supplier and the position of these transactions within the project plans. It also assists in the identification of deliverables and inclusion of these in the plans. Euromethod identifies three main processes:

- the tendering process;
- the IS production process;
- the completion process.

These are defined in terms of a sequence of decision points at which customer – supplier transactions take place and decisions are taken.

The Deliverable Model
The Deliverable Model provides a means of describing the products exchanged between customer and supplier. This will be the major mechanism for harmonisation between methods. It defines products as related to three domains:

- the *target* domain, which contains the products from systems development methods such as we have considered in this book;
- the *project* domain, which contains products related to project management and control activities;
- the *delivery plan* domain, which contains the sequence of delivery points for the products to be delivered by the project.

The Project Strategy Model
The Project Strategy Model aims to support the identification and management of project risks and the definition of a project strategy. It gives guidance in the setting up and planning of IS projects.

10.5.6.4 Euromethod Dictionary

The Euromethod Dictionary is a summary of terms defined in Euromethod, classified in alphabetical order. It is intended that each method's vendor will supply a bridging guide for the method, mapping the method's terminology to the standard definitions in the Euromethod dictionary.

10.6 Impact of Euromethod on Structured Methods

Euromethod is in the public domain and has no obvious competitors. The methods which have contributed to it are also committed to intercept Euromethod. Euromethod will define general method components, independent of specific techniques. These components will include current method concepts and be open to future improvements. They should allow, and in fact encourage, the migration of current methods towards Euromethod.

10.6.1 Rapid Application Development (RAD)

Rapid Application Development has been used to describe everything from a well-structured 'fastpath' to small systems development to the 'quick and dirty' system thrown together during a lunch hour in response to a sudden commercial pressure. It should be the former.

RAD is aimed at the speedy delivery of business systems. This approach is usually only applicable when the project is:

- relatively small;
- of limited scope and impact upon the organisation (low risk);
- does not require new or special hardware or software platform;
- able to use fourth generation (or later) tools for the rapid development of the software;
- able to be developed by practitioners highly skilled both in the use of RAD and in the development and implementation of systems in the target environment.

The key features of good RADs are:

- the use of structured techniques and prototyping;
- extensive use of CASE tools for project support, system definition and generation;
- firm project management, including some form of 'time-boxing', i.e. the project is divided into precise tasks, constrained to be completed within a pre-set maximum (and short!) time-slot;
- a sound project plan with tasks derived from the framework of a structured method;
- a determination to re-use existing products (e.g. existing analysis documents as well as code and data structures) wherever feasible;
- clearly defined roles, responsibilities and levels of authority for the project personnel;
- the involvement of users through Joint Application Development (JAD) workshops, organised within a well-defined framework.

JAD workshops are used in RAD because, if used properly, they are a quicker way of discovering facts and requirements, obtaining decisions and resolving conflict. A JAD workshop is a meeting of people involved with a project:

- in specific roles;
- with specific responsibilities;
- with particular knowledge and expertise.

The workshop must have:

- a clearly defined purpose;
- clearly defined deliverables.

The JAD roles include a chairperson who is the facilitator for the JAD. That person must ensure that the meeting knows and achieves its objectives and that the objectives are not over ambitious. If the objectives are not met, action must be agreed to rapidly rectify this. The results of JADs must be documented clearly.

The risks of RAD are many, but from the structured methods viewpoint include the possibility that products of those methods vital to the success of the project will be, what one organisation of our acquaintance once described as, 'tuned out'. ('We tune the method to fit the pragmatic requirements of each individual project. The Entity Model and Function Definitions have been tuned out.'!)

Another risk results from prototyping, where the number of iterations to arrive at a satisfactory result can turn into a seemingly endless loop. Tight control of the number of prototyping iterations of any particular aspect of the system must be controlled. The key to this lies in the clear definition, **beforehand**, of the aspects to be prototyped, and the objectives of each prototyping session. The results of each session should be well documented.

The RAD approach is not an alternative to structured methods. Indeed some structured methods contain details of specific fastpath frameworks for the purpose. The increased use of prototyping is covered by some, but not all, structured methods. RAD techniques are a short-term expedient, pending the arrival of CASE tools which can handle a far greater proportion of the systems development life cycle for us. The future impact is likely to be that the CASE tool support will automate much of the physical design and construction and that JAD workshops will be used to elicit and define requirements, which will be able to be almost immediately prototyped. This will be possible, however, only when CASE tools are sufficiently slick in operation and friendly of face not to interfere with, and dominate, the requirements discovery session.

10.7 Business Process Redesign (BPR)

Business Process Redesign (Business Process Re-engineering) became a 'hot topic' a few years ago when an article by Michael Hammer appeared in the *Harvard Business Review* entitled, 'Re-engineering Work: Don't Automate, Obliterate'. The central message of the article was businesses should be using the power of modern computer technology to radically redesign business processes in order to achieve dramatic improvements in performance. Many national and multinational companies have embraced BPR, with claimed quantum leaps in their business performance. But how does this affect structured methods? Even with structured methods, IT Departments have always had a tendency to automate existing practices and procedures. This is, of course, the users' fault (!!). Users, from their departmental perspectives, are often not in a position to see the potential for complete change to the processes they perform, and not empowered to initiate complete restructuring of the organisation (if indeed it would be in their individual interests to do so). Structured methods prompt the analyst to ask, 'why do we do it this way' but seldom raise the question, 'why are we doing it at all?' (although Soft Systems provides a fairly comprehensive approach to BPR which examines processes logically required by the organisation's mission before considering current processes).

Business processes are logically-related tasks performed to achieve a defined business result. For example, *Handle sales orders, Control stock.* Each process has customers, and may have suppliers, inside or outside the organisation under study. A set of business processes forms a business system. The aims of BPR is to completely rethink and redesign the business processes by:

- questioning perceived business rules;
- questioning current functional/departmental divisions;
- identifying current problems and considering innovative, often radical solutions;
- removing duplication and redundancy of effort;
- integrating and streamlining processes.

How does BPR affect structured methods? The logical (conceptual, essential level of abstraction, gives an ideal point at which to reconsider business processes. It brings together logically grouped tasks and removes references to organisational boundaries. What must be encouraged at the transition from the existing logical model to the required logical model is the use of techniques such as **critical success factors** (see Chapter 6) to streamline and rationalise the logical processes which are built into the required logical models. The increased emphasis of structured techniques on data rather than processes is also appropriate here for although the business processes may be redesigned, the underlying data structure remains fundamentally unchanged.

10.8 Reverse Engineering and Re-engineering

Reverse Engineering is the attempt to 'unearth' the original design and specification of a system, by examination of the content and structure of its code. Re-engineering is the subsequent rebuilding of the system, in a more structured way, from the products of Reverse Engineering. The process is rather like an archaeological dig (some of the code can be very old!). We only find broken artefacts and have to try to reconstruct something meaningful from them.

Considerable effort has been expended in recent years on the development of Reverse Engineering tools. Such products which are currently on the market are able to produce low-level process specifications or Jackson Structure Diagrams from source code. Some will produce a data dictionary from the declared data items in the source code. A few tools are able to produce a data structure (entity model) from the structure and usage of files/database. The results of these need careful re-analysis, however, as so much of the required structure will not be apparent from the source, or will have been masked by changes made to improve performance.

Figure 10.12 Reverse Engineering: Making Sense of the Broken Artefacts

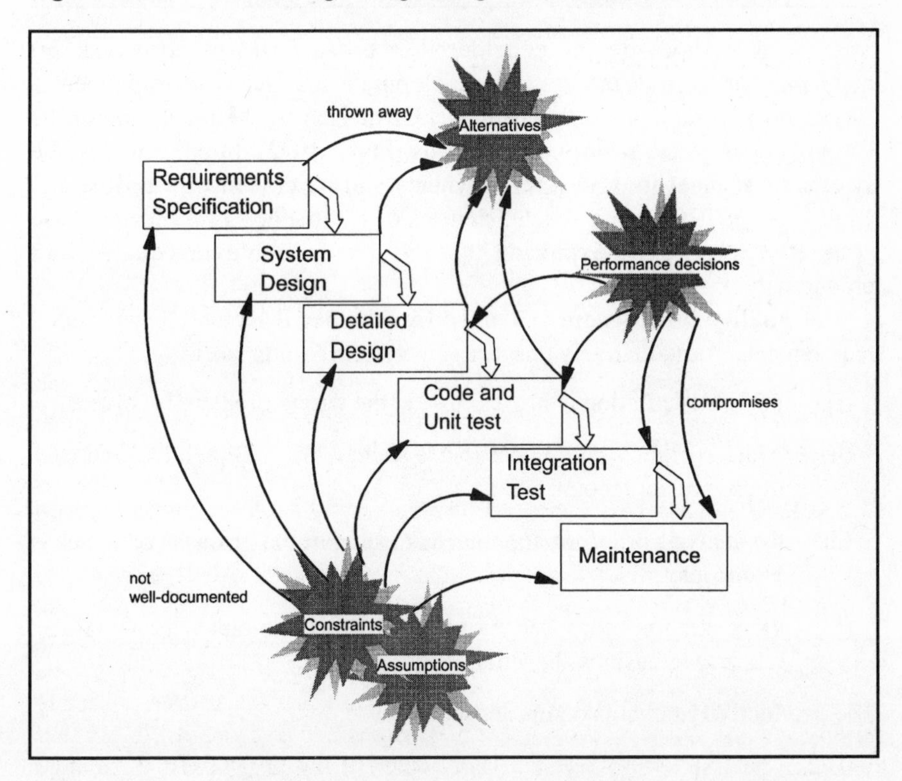

Figure 10.13 The Factors Which Cannot Easily Be Rediscovered by Reverse Engineering

Sabanis and Darlison (1992) in their discussion of the REDO project and Reverse Engineering Toolset (COBOL source) state that Reverse Engineering is not fully automatable because many programs have inadequate documentation of the *semantics* of the information used and the *relationships* between various data objects. However such tools, with human intervention, will be a useful way forward in spite of the problems. Commercial demand for such tools with their potential cost-saving will fuel vendor development effort.

In relation to structured methods, such tools will eventually be able to reconstruct an albeit incomplete specification of processes, data and events, which must be validated and verified before it can be re-engineered into a more structured, more easily maintainable system. Structured methods will be the means of verifying and validating the specification and highlighting the conflicts and omissions. Eventually, with CASE tool advances, much of this process should be automatable, with the tool merely asking for human intervention to supply missing information and to resolve discrepancies.

10.9 What Problems Remain in Spite of Structured Methods?

Structured methods can be considered in broad terms of *hard* and *soft* methods. The aim of any system development method is to implement a successful system, this success being acknowledged by the realisation of the system's **quality** and **productivity** targets (Flynn 1992). In other words, the system must meet its stated requirements and be developed on time and within budget. To identify the deficiencies of any method in achieving these aims, it is worthwhile exploring the obstacles to achieving quality and productivity.

The **quality** of a system is judged by whether it meets clearly stated requirements. A qualitatively unsuccessful system results when:

(i) problem definition is incorrect – i.e. the wrong problem is solved;

(ii) wider social/psychological issues such as the acceptability and ease of use are not recognised;

(iii) the analysis of information needs to support the business activities is erroneous;

(iv) the system is motivated by political or technological issues alone, rather than business benefits.

The **productivity** target is compromised because:

(i) user requirements at the later stages of the project are at variance with their stated requirements at the project outset or at variance with other users' requirements;

(ii) the requirements are a moving target affected by external changes such as statutory factors, the competitive market environment of the organisation or political circumstances;

(iii) the requirements may embody an unperceived implementation problem which is discovered late in the development process;

(iv) there is poor project control with gross underestimates of resource requirement and check point/critical path analysis.

Structured methods must therefore seek to overcome these obstacles. How far have current structured methods been able to address the identified problems? In spite of the considerable effort directed toward hard approaches to development, a significant number of failed systems still occur. Peat Marwick McLintock (1990) reported that 30% of all major projects, within a group of 252 companies, failed to meet quality and/or productivity targets. Clearly the *hard* approaches have not yet reached maturity and/or they do not provide a complete methodological solution for information systems development.

By considering the above quality and productivity issues in relation to the hard and soft structured methods, some light may be thrown upon the future of such methods for system development. We consider each of the quality and productivity issues listed above in turn.

10.9.1 Solving the wrong problem [Quality Problem (i)]

If the problem, identified by the analyst, is not the problem which the users perceive or one which will help achieve the organisation's mission then although the resulting system may function it will not *usefully* function. Hard approaches assume that the problem is clear and well defined and as such do not address the difficulties of problem definition. Soft approaches confront 'fuzzy' problem areas by using a variety of techniques for problem definition.

10.9.2 Consideration of wider social/psychological issues [Quality Problem (ii)]

Hard approaches do not formally consider the wider sphere of the social and psychological impact of system development. The penalty for lack of such considerations is exemplified in Harrington (1991) wherein job performance after computerisation was worse than with the old manual system. This arose because the team spirit borne out of personal interactions was destroyed by the personal isolation created by the computer system. Soft approaches, through attention to the whole organisational environment and the context in which the system will function, directly address these wider issues. The system should be *usable* and *acceptable*, forming a tool with which the users can comfortably operate to more easily achieve their work-related goals, thereby improving job satisfaction. Hard approaches are acknowledging the importance of these factors to some extent through the use of prototyping (prototyping has been formally included in SSADM Version 4 and the Soft Systems approach is being discussed by the CCTA in relation to the front-end of SSADM).

10.9.3 The Analysis of Information [Quality Problem (iii)]

Hard approaches to system development are designed to eliminate errors in development by formalising the techniques to be used, by improving communication and by preventing *ad hoc* intuitive problem solutions. One of the criticisms levelled at soft approaches is that they are directed toward problem formulation and not solution design.

10.9.4 Political and Technological Motivation [Quality Problem (iv)]

Hard approaches do not consider wider socio-political issues which could lead to the development of systems which deliver little or no organisational benefit. However, where the structured method encompasses an Information Strategy Plan based on the organisation's underlying Business Strategy (e.g. IE) the impact of individuals using a system development project to satisfy their political or technological desires will be diluted. This dilution is proportional to the position held by the individual. Soft approaches, whilst focusing on the wider organisational issues, do seek to counter individual goals by consensus and participation.

10.9.5 Changing Requirements [Productivity Problems (i) and (ii)]

The definition of requirements for a new system is pivotal to successful system development and user acceptance. If the definition of requirements is flawed then so must be the developed system. Often the requirements change after the development work has been initiated for a number of reasons:

- the users do not know their requirements;
- users disagree over their requirements;
- communication problems between the analyst and the users foster misunderstanding of what the proposed system will deliver in relation to the requirements;
- the organisational environment has changed in the time during which the system has been under development.

The hard approach to the variable nature of requirements is epitomised by Boehm in his observation that 'developing software from requirements is like walking on water – it's easier if it's frozen'. The approach is to freeze the requirements at the outset of the project via some form of contractual document whereby the requirements stated by the analyst are 'signed off' by the user manager. However, this is a defensive approach on the part of the IS department, which only corroborates the analysts' stance that the system does what it was specified to do, which still may not be what the user really needs it to do. The incorporation of prototyping is a move which will allow some intermediate feedback into the development process to accommodate shifts in requirements.

Another possibly fruitful area (especially related to large systems) in which variable requirements can be accommodated is that of incremental development and incremental delivery (Gilb 1988). This involves the division of the system to be developed/delivered into self contained functional and useful units which can be delivered in a relatively short time period. Since the amount of deviation of the requirements from their initial state is, in part, time-related, a shortened delivery time should be accompanied by fewer requirement changes. These changes can be more easily incorporated in the developed unit and the impact of these leads to a modified requirement base on which to develop the subsequent functional unit. Moreover the psychological effect is to improve development team morale through positive user feedback and increased user confidence in the developers as they receive something useful (signifying progress) in short time frames. The difficulties with incremental development are concerned with how to partition the system and determine the priorities and scheduling of the functional units.

The soft approach to requirement variability is to define consensus requirements for the new system via participation, discussion and merging of user views with the resolution of differences between users. Land (1982) describes a technique called Future Analysis which provides a means for analysts and users to formally explore the future possible environments in which the system may have to function. The areas of the system which are susceptible to these changes can be identified and contingency plans can be laid down to deal with them.

10.9.6 Unperceived Implementation Problems [Productivity Problem (iii)]

Neither the hard nor soft Approaches specifically handle this type of problem but it can be seen that Incremental Development may allow adjustment of the system on a smaller scale before the rest of the system is built.

10.9.7 Poor Project Control [Productivity Problem (iv)]

Hard approaches which formally define the phases of the system development also define the products of each phase. These products can thus be used as checks upon progress in combination with a project management method. For example, the Structural Model of SSADM was amended in Version 4 to allow it to be integrated with the Project Management Method, PRINCE.

10.10 The Future Direction for Structured Methods?

It is apparent from the above discussion of quality and productivity parameters that the hard and soft approaches counter different obstacles to successful Information System development. Although they do not provide a solution to all problems likely to be encountered, in the areas on which they do have an impact, they complement each other. It is logical, therefore, to seek ways to combine the two approaches to take advantage of the benefits of both. It is also possible that the combined benefits may be greater than the sum of the individual benefits of the two approaches! We now need to ask whether such a combination is possible and, if so, how it will be achieved.

Burrell and Morgan (1979) argue that it is theoretically and pragmatically impossible to move between the differing underlying paradigms of hard and soft approaches for information system development. However, this tunnel vision ignores the potential benefits from methodological co-operation. The challenge is to define and refine the points of interface between hard and soft approaches. Miles (1988) argues that grafting the SSM onto the front of a hard method loses the value of soft method in later phases of information systems development. Miles therefore promotes the view that the soft systems method should guide the project, with elements of the hard method embedded in it. This approach would, however, have the effect of breaking up the highly structured and coherent nature of hard methods, which lend themselves to the benefits of automation by CASE tools. A less disruptive scenario can be envisaged by using soft techniques to impact upon the established products of hard methods. At the front-end of the project, soft methods would facilitate better problem definition and requirements specification whilst, further into the project, they would incorporate wider issues into the analysis which would influence the design. The structured method of the future could encompass both **soft** and **hard** approaches with defined interfaces between them to maintain the possibility of CASE tool automation of the construction of the system.

The impact of **Euromethod**, in the provision of a common language for discussion of and procurement of software development solutions, will allow structured methods to be more easily compared. This may, as discussed earlier in this chapter, bring about harmonisation of the terminology of structured methods possibly of the techniques within the structured methods. The European Modelling Language is an extension to Euromethod which attempts to provide a common underlying model of application.

CASE tools may have a similar harmonising influence, since there may come a point where the better tools offerings will start to drive the methods. This may be offset by the possibility that CASE tools will become user-customisable (there is at least one such 'build your own CASE tool' product currently on the market).

The **object-oriented approach** is also likely to have a major impact. More rigorous definition of objects has begun to enter the doctrines of more than one of the traditional structured approaches, and 'bridges' between the traditional structured methods and the object-oriented approaches have already started to appear. OO, with its comfortable relationship with the modern graphical user interfaces (GUIs), may absorb the traditional approaches. Perhaps the analyst of the not too distant future will sit down with a user at a personal workstation and design objects, in response to pertinent questions from the computerised whizz kit which the CASE tool will have become. The CASE tool will supply increasingly more automation to the mechanics of the structured method. The objects entered by the analyst will be instantly turned into code and the system will be prototyped before their very eyes. As the well-known scientist and author, Arthur C. Clarke, observes, 'Any sufficiently advanced technology is indistinguishable from magic.'

10.11 Making Methods Work

If structured methods are to work **effectively** within an organisation, it requires more than just the decision about which one to adopt. They are all suitable in the appropriate environment. They can also **all** fail to deliver the required benefits if circumstances work against them. Therefore, to achieve success, you need a cunning plan! The following features are the key factors which need to be in place. (The acronym METHODSPLAN has proved useful):

- **M** Management commitment from the very top, borne of management understanding of the benefits which can be expected to accrue and the investments in time and money which must be made to make this so;
- **E** Early in the project, more time must be allowed for analysis. This comes as a shock to many managers, to whom no work is being done until someone is churning out lines of code at a terminal;
- **T** Training is essential at the start of the project, to overcome the learning curve, but there must be an on-going commitment to training throughout the project, and for subsequent projects;
- **H** 'Holes' identification is vital, to assess the areas of the project life cycle which your chosen structured method does not cover, or where the coverage is incomplete. Other methods and tools can then be identified to support these areas, if appropriate;

- **O** Objectives of building a system to the quality and productivity standards discussed above must be kept in mind. In order to measure performance against these, and compare productivity with other projects both past and future, information must be captured. You should plan to keep selected information for this purpose, but keep this in proportion. Too great a weight of metrics will sink the project boat!
- **D** Determination and commitment to the success of the structured methods approach plays a large part in the acceptance and use of the methods. The introduction of any new way of working changes the culture of the organisation, with the attendant effects on the people involved. These effects can encompass fear of change, insecurity, even hostility. People must be encouraged and cajoled to try the methods in order to personally experience the benefits;
- **S** Support from a good CASE tool or tools. There are several aspects of the project which will benefit from tool support, the drawing of structured methods models being only one. Others include project management, risk analysis and configuration control. The stability of the tools should be ensured at the start of the project. The reader should be aware of the potential mental overload of taking on a new method and new CASE tools simultaneously;
- **P** Planning of the project must be clearly based on the products of the structured method. However, it must also allow for the learning curve, metrics collection, user involvement, quality reviews and iteration in the production of products;
- **L** Liaison with users is fundamental to the success of the project, and their understanding of the modelling process, in outline, will better equip them to answer the analysts questions in an informed and useful way. They do not need to be structured methods experts but should be given an overview of the development process. The analyst should take time to really involve the users and ensure they understand what is going on. It is often too easy to obtain acquiescence to Structured methods products because user understanding of them is limited;
- **A** Availability has several aspects. Availability of help in using an unfamiliar method or tool, availability of time in which to gain experience in it and consolidate the training and develop the models sufficiently, and even availability of a PC on which to use the CASE tools whenever appropriate have proved to be significant problems on 'live' projects. These have resulted in unfinished and unreliable models, and CASE tools used at the end of the project as a documentation tool, rather than as an active, cross-checking tool during the development of models. The reader must actively argue for availability in these areas;
- **N** New developments which will make the software development process better and faster are already over the horizon and steaming at full speed towards us. The structured methodologist/project

manager should keep a keen eye on these, especially on methods advances and future case tools. In no way is it safe to select your method and tool and then relax.

The above list is considered to be the minimum set of considerations for the success of the structured development process.

10.12 Conclusion

The study presented in this book has shown that whilst the popular structured methods of the day do exhibit some fundamental differences of approach there are many similarities. As in most specialist areas an array of technical terms is generated, in many cases representing the same fundamental object within different methods. These aliases have been identified to allow the structured methods to be compared. We have presented the reasons for adopting structured methods and the problems they set out to address. We saw what areas of similarity and difference the chosen methods exhibited and briefly looked at a methodology for comparing and evaluating methods. The mainstream common techniques were explained. Then each chosen structured method was presented in turn. We compared the methods against our list of features. However, we hope that there is sufficient information in the book for a comparison based on the reader's own features list. In this final chapter, we have tried to deal with the main topics which may affect structured methods to a greater or lesser extent over the next few years.

The reader should now be in a better position to appreciate the similarities and differences between structured methods and, if appropriate, to assess them for potential adoption by his/her organisation. Armed with this information, the METHODSPLAN above, and the chosen structured method and tool, we hope that the reader now feels equipped to go out and tame the software development process.

Figure 10.14 Taming the Systems Development Monster

Appendix A

Midlinks Motel Scenario
Case Study Background Information

1. Company Profile

Midlinks Motel is a large modern hotel situated in the heart of the Midlands, close to the Motorway links. It has grown over the past twenty years from an original 40 bedrooms to its current 150 bedrooms. No further expansion is planned. The hotel derives most of its custom from short-stay business travellers, although it hopes to encourage more families during the holiday periods to obtain better utilisation of its predominantly double rooms. There is a Conference Centre attached to the hotel, but this is run as a separate business, the hotel merely providing rooms to the Conference Manager as a customer.

2. The Project

The hotel's 'back-office' accounting systems were computerised two years ago, largely successfully. Management now feel that the time is right to consider improving the efficiency of the front desk functions such as reservations, check-ins and check-outs. They also feel the need for additional features, such as detailed guest histories in order that direct mailing can be accurately targeted. However, they would only consider computerisation at the front desk if it allowed them to retain flexibility in their approach to their customers.

3. Current Computer Configuration

The current accounting systems are run as a single-user system, using integrated software packages for sales ledger, purchase ledger and general ledger. These were written, and are maintained, by a local software company who would be willing to handle the integration needed between any new front-desk system and their accounting system provided that the interface is clearly specified by the hotel.

4. Staff Organisation Within the Hotel

The Staff organisation within the hotel is shown in Figure A.1.0.

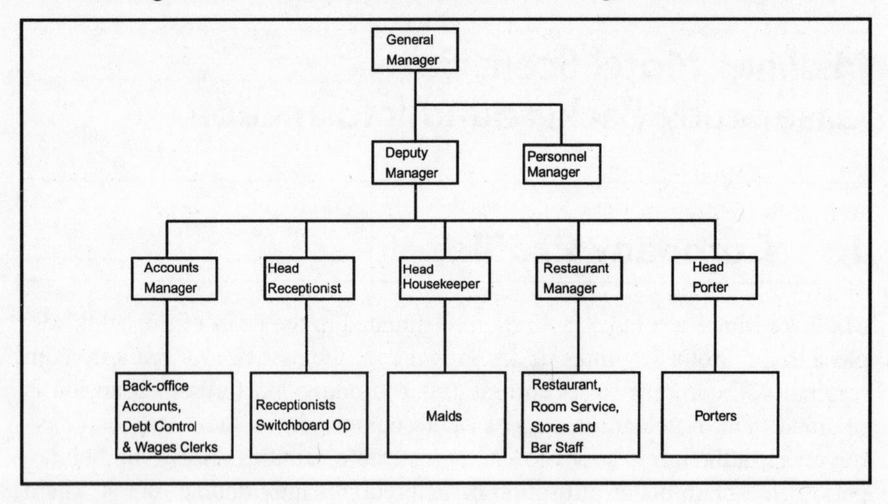

Figure A.1.0 Midlinks Motel Staff Organisation Chart

5. Staff Responsibilities

The individual responsibilities of the key members of the hotel staff are described below. Also given below are the notes taken at interview detailing the current system and the problems of the current system as identified by the interviewee.

5.1 General Manager

Assisted by the deputy manager, in charge of the day-to-day smooth running of the hotel, its finances and stores. Also responsible for future planning.

5.2 Accounts Manager

Responsible for all back-office accounting functions including acceptance of account-customers, credit-checking, debt-chasing and the maintenance of the hotel's financial ledgers. Also responsible for payment of staff wages.

5.3 Head Receptionist

In charge of day-to-day running of reception: acceptance of bookings, greeting and checking-in of guests, maintenance of an up-to-date room chart, preparation of accounts for guests, notification to the housekeeper of rooms for service. Also responsible for the hotel switchboard.

6. Interview Notes

The documents which follow detail the main points which were obtained by interview.

Project: Front-desk System	**Sheet** 1 **of** 2
Person Interviewed: David Wise	**Document Ref:** FDS1/AL/1
Position Held: General Manager	**Date:** 7/11/99
Subject: Existing systems: overview, problems and objectives	**Time:** 10.00
Interviewer: Anna List	**Location:** Midlinks Motel

Notes:

Current Systems Overview

Guests may book rooms in advance or turn up without prior booking. Bookings made by telephone must be confirmed in writing. A deposit is usually paid for block bookings.

On checking in, the guest is asked to complete an index-card listing personal details and car registration number before being given a room-key. The room-charge and whether full-board, half-board, or bed-and-breakfast, are recorded on the guest's account (one account per room). Any special dietary needs are notified to the restaurant immediately.

On checking out, bills are calculated to include restaurant receipts, bar receipts, and telephone calls as recorded on the Switchboard Operator's Log. A guest who has a customer-account with the hotel may pay none, part or all of the bill, the balance being sent to Accounts to be put to the sales ledger for normal invoicing and statements.

Reception prepare daily lists for the Housekeeper to advise which rooms require service, and for the restaurant to notify room-numbers and number of guests expected for breakfast, lunch and dinner.

Problem Areas

The accounting system was brought in mainly to address the problem of bad debts. This has been successful and the staff are now becoming comfortable with the use of the system, although there were some initial teething problems due to inexperience with computers. This system must not be changed by the current project.

Cross-reference:

Project: Front-desk System	**Sheet** 2 **of** 2
Person Interviewed: David Wise	**Document Ref:** FDS1/AL/1
Position Held: General Manager	**Date:** 7/11/99
Subject: Existing systems: overview, problems and objectives	**Time:** 10.00
Interviewer: Anna List	**Location:** Midlinks Motel

Notes:	**Cross-reference:**
The front-desk systems are too slow: guests frequently experience unacceptably large delays on checking-out. The reconciliation of cash for passing through to accounts is often delayed and inaccurate which results in angry words between the Accountant and the Head Receptionist. Some problems have been experienced with the reservations procedure, which has resulted in double booking. Additionally, information about guest histories, required for planning and marketing purposes, is not readily available. No information is available to assess unsatisfied demand for rooms. Other information is time-consuming to extract.	

Project: Front-desk System	**Sheet** 1 **of** 3
Person Interviewed: Francis Davison	**Document Ref:** FDS1/AL/2
Position Held: Head Receptionist	**Date:** 7/11/99
Subject: Existing systems, problems and objectives	**Time:** 14.00
Interviewer: Anna List	**Location:** Midlinks Motel

Notes:	**Cross-reference:**
<u>Current System Overview</u> Guests usually book their accommodation, and may do so anything up to two years in advance, either by telephone or letter. Some guests, or their companies, are customer-account holders with the hotel. Guests must always confirm in writing if they are not already customer-account holders with the hotel. For customer-account holders, and for written bookings, reception send a written confirmation of the booking. Customer-account holders may be asked to pay a deposit, particularly in the case of large block-bookings. Bookings may at first be provisional and be confirmed later. Bookings (provisional or confirmed) and are all recorded on the room-booking chart. The original documents related to the booking are annotated with a reservation number (next sequential number) and filed in arrival-date order in an office behind reception. A reservation card is completed and filed in reservation number order. On checking in, guests are asked to complete an index card (if a booking has previously been made, this will be the reservation card) with personal details. This is filed in guest-name sequence in reception and discarded once the bill has been paid. The guest's room-account is annotated with the room charge and meals required.	

Project: Front-desk System	**Sheet** 2 **of** 3
Person Interviewed: Francis Davison	**Document Ref:** FDS1/AL/2
Position Held: Head Receptionist	**Date:** 7/11/99
Subject: Existing systems, problems and objectives	**Time:** 14.00
Interviewer: Anna List	**Location:** Midlinks Motel

Notes:	**Cross-reference:**
On checking out the bill is finalised with newspaper charges and last-day telephone, bar, and restaurant charges. The guest may pay the whole bill or have all or part of it allocated to his company's customer-account. All cash received is recorded in the daily cash book. At the start of each day restaurant, bar, room service and telephone charges are allocated to the appropriate room accounts. Each day, lists are prepared for the housekeeper to identify rooms for service and rooms changing occupancy the following day, for the night-porter to notify newspaper requirements, and for the restaurant to give room-numbers of guests expected for each meal. These are prepared from the Room Booking Chart and the guest's account. At the end of the day, all payments received are reconciled against the daily cash book and sent, together with the paid-up bills and bills deferred for company payment, to accounts. <u>Problems</u> Some problems occur with prior bookings as the Room Bookings chart becomes illegible after it has been manually changed several times. The Room Bookings chart is in the form of 5 books (1 for each floor of the hotel) each with details of 30 rooms. Although the policy is to fill the hotel from the lower floors upwards, block bookings often disturb this pattern, and determining room availability can be a slow process.	

Project: Front-desk System	**Sheet** 3 **of** 3
Person Interviewed: Francis Davison	**Document Ref:** FDS1/AL/2
Position Held: Head Receptionist	**Date:** 7/11/99
Subject: Existing systems, problems and objectives	**Time:** 14.00
Interviewer: Anna List	**Location:** Midlinks Motel

Notes:	**Cross-reference:**
Checking out can be time-consuming, as many bills are queried and errors are often found, for example, bar receipts allocated to wrong accounts, or simple addition errors (bills are calculated manually, with the help of a calculator). Occasionally, room charges, held on the Room Booking chart, are found to be in error. Customer-account numbers given by guests are not checked until the final bill arrives in the accounts office, after the guest has left the premises. The reconciliation of cash at the end of the day means that substantial sums of money are often held at Reception, which is recognised as a security risk. More frequent release of cash to Accounts would mean that reception was without the cash-book for periods of time, which is unacceptable. The use of separate loose sheets for cash recording has been tried but failed as these often got lost.	

Project: Front-desk System	**Sheet** 1 **of** 2
Person Interviewed: Rebecca Johnson	**Document Ref:** FDS1/AL/3
Position Held: Receptionist in charge of reservations	**Date:** 8/11/99
Subject: Bookings	**Time:** 11.00
Interviewer: Anna List	**Location:** Midlinks Motel

Notes:

Current System

Guests may make enquiries and bookings either in writing or by telephone. Telephoned bookings are treated as provisional bookings unless a credit-card number or one of the hotel's customer-account numbers is quoted. A written communication may result in either a firm booking or a provisional booking.

Guests making provisional bookings are required to confirm the booking in writing within 2 weeks of the reservation date, (although we always ask them to confirm within 7 days). If the booking is made too close to the arrival-date for this to be possible, the accommodation will be released for re-letting at 6pm on the date of arrival. Provisional bookings are identified in the Room Booking Chart with a 'P' against the date of reservation and reservation number. A reservation card is completed and marked 'P'. The next reservation number is kept on a card at the front of the reservations tray.

A written booking may be taken as a firm booking if the guest's accommodation requirements can be matched exactly. For a firm booking, a written confirmation is always sent by the hotel. Such bookings may include a deposit, which is recorded as part-payment against the guest's account. The accommodation is booked as 'reserved' in the room Booking Chart and the reservation number, date and 'F' entered against the dates required. A reservation card is completed and filed in the reservations tray. The original confirmation or booking letter is filed in date-of-accommodation sequence.

Cross-reference:

Project: Front-desk System	**Sheet** 2 **of** 2
Person Interviewed: Rebecca Johnson	**Document Ref:** FDS1/AL/3
Position Held: Receptionist in charge of reservations	**Date:** 8/11/99
Subject: Bookings	**Time:** 11.00
Interviewer: Anna List	**Location:** Midlinks Motel

Notes:	**Cross-reference:**
Either a provisional or firm booking may be cancelled. Provided that this is 14 days or more before the date accommodation is required, no charge will be made. Within 14 days, if a deposit has been paid or a customer-account is involved, the guest's account is referred to Accounts for decision on whether a refund, or a charge, is appropriate. All appropriate records are amended to cancel the booking.	

7. Volumes and Trends

The statistical data was collected as accurately as possible given the limitations of the current system and is detailed below.

7.1 Volumes:

Number of rooms:	150 Bedrooms
Average stay:	2 nights
Average room occupants:	1.1
Average bookings/month:	1800 (for 2000 guests approx.)
Prior reservations/month:	1400
Average % full:	80% (i.e. 120 rooms /night)
Number of customer accounts:	300

7.2 Trends

Most residents are on business. Therefore, the majority of trade is weekdays, with weekends and holidays being slack (no figures available to exactly quantify trends).

Bibliography

Ashworth, C. and Goodland, M. (1990) *SSADM: A Practical Approach,* McGraw-Hill, London.

Avison, D. E., Fitzgerald G. and Wood-Harper, A. T. (1988) 'Information Systems Development: A toolkit is not enough', *The Computer Journal,* Vol. 31, No. 4.

Avison, D. E. and Fitzgerald, G. (1990) *Information Systems Development – Methodologies, Techniques and Tools,* Blackwell, Oxford.

Avison, D. E. and Wood-Harper, A. T. (1990) *Multiview: An Exploration in Information Systems Development,* Blackwell Scientific Publications, Oxford.

Avison, D. E. and Wood-Harper, A. T. (1991) 'Information systems development research: An exploration of ideas in practice', *The Computer Journal,* Vol. 34, No. 2.

Bachman, C. W. (1969) 'Data structure diagrams', *Data Base,* Vol. 1, No. 2.

Benyon, D. and Skidmore, S. (1987) 'Towards a tool kit for the systems analyst', *The Computer Journal,* Vol. 30, No. l.

Boehm,B. W. (1981) *Software Engineering Economics,* Prentice-Hall, N. J.

Burrell, G. and Morgan, G. 1979 *Sociological Paradigms and Organisational Analysis,* Heinemann, London

CCTA (1989) *COMPACT Manual,* V1.1, Norwich.

CCTA (1993) *Euromethod Phase 3a Information Pack,* Norwich.

CCTA (1994) *Euromethod Overview Version 0,* June, Norwich.

Checkland, P. (1981) *Systems Thinking, Systems Practice,* John Wiley, Chichester.

Checkland, P. 1988 'Information systems and systems thinking: Time to unite?' *International Journal of Information Management,* 8, 239–48.

Checkland, P. and Scholes, J. (1990) *Soft Systems: Methodology in Action,* John Wiley, Chichester.

Chen, P. (1976) *The Entity Relationship Model – Towards a Unified View of Data ACM Transactions on Database Systems,* Vol. 1, No. 1.

Codd, E. F. (1974) *Recent investigations into relational data base systems,* Proc.ifip congress.

Coad, P. and Yourdon, E. (1990) *Object Oriented Analysis,* Yourdon Press/Prentice-Hall, N. J.

Connor, D. (1985) *Information System Specification & Design Road Map,* Prentice-Hall, N.J.

Cutts, G. (1988) *Structured Systems Analysis and Design Methodology,* Paradigm, London.

Date, C. J. (1985) *An Introduction to Database Systems,* Vol. l, Addison-Wesley, Wokingham.

Date, C. J. (1986) *An Introduction to Database Systems,* Addison-Wesley, Wokingham.

Date, C. J. (1990) *An Introduction to Database Systems,* 5th edn, Addison-Wesley, Wokingham.

Davids, A. (1992) *Practical Information Engineering – The Management Challenge,* Pitman, London.

De Marco, T. (1979) *Structured Analysis and System Specification,* Prentice-Hall, N.J.

DESMET Handbooks 1994, National Computing Centre, Oxford.

Downs, E., Clare, P. and Coe, I. (1988) *Structured Systems Analysis and Design –* Method, Application and Context, Prentice-Hall, London.

Finkelstein, C. (1990) *An Introduction to Information Engineering,* Addison-Wesley, Wokingham.

Fitzgerald, G., Stokes, N. and Wood, J. R. G. (1985) 'Feature analysis of contemporary information systems methodologies', *The Computer Journal,* Vol. 28, No. 3.

Flavin, J. M. (1981) *Fundamental Concepts of Information Modeling,* Yourdon Press, N.Y.

Flynn, D. J, (1992) *Information System Requirements: Determination and Analysis,* McGraw-Hill, London.

Gane, C. (1990) *Computer-Aided Software Engineering,* Prentice-Hall, N.J.

Gane, C. and Sarson, T. (1979) *Structured Systems Analysis: Tools and Techniques,* Prentice-Hall, N. J.

Gilb, T. (1988) *Principles of Software Engineering Management,* Addison-Wesley, Wokingham.

Graham, D. R. (1989) 'Incremental development: Review of nonmonolithic life-cycle development models', *Information and Software Technology,* Vol. 31, 7–20.

Graham, D. R. (1990) 'Incremental development and delivery for large software systems', in CSR: *Sixth Annual Increase Conference on Large Software Systems,* Elsevier, Oxford.

Harrington, J. (1991) *Organisational Structure and Information Technology,* Prentice-Hall, N.J.

Henderson-Sellers, B. (1991) *A Book of Object-oriented Knowledge,* Prentice-Hall, N.J.

Holloway, S. (1989) *Methodology Handbook for Information Managers,* Gower Technical, Aldershot.

Jackson, M. A. (1975) *Principles of Program Design,* Academic Press, London.

Kim, W. and Lochovsky, F. H. (eds) (1989) *Object-oriented Concepts, Databases and Applications,* ACM Press, N.Y.

Land, F. (1982) *Adapting to Changing User Requirements in Information Analysis: Selected Readings,* Galliers, R. (ed.), Addison-Wesley, Wokingham.

Land, F. (1985) 'Is an information theory enough?' *The Computer Journal,* Vol. 28, No. 3.

Land, F. and Hirschheim, R. (1983) 'Participative systems design: Rationale, tools and techniques', *Journal of Applied Systems Analysis,* Vol. 10.

Law, D. (1988) *Methods for Comparing Methods,* NCC Blackwell, Oxford.

Martin, J. (1982) *Computer Database Organisation,* Prentice-Hall, N.J.

Martin, J. (1988) *Information Strategy Planning,* James Martin Associates, Dublin.

Martin, J. (1988) *Business Systems Design,* James Martin Associates, Dublin.

Martin, J. (1989) *Information Engineering,* Vols. 1–4, Savant, Carnforth.

Martin, J. (1991) *Rapid Application Development,* Macmillan, N.Y.

Martin, J. and McClure, C. (1985) *Structured Techniques For Computing,* Prentice-Hall, N.J.

Miller, G. A. (1956) 'The magical number seven, plus or minus two: some limits on our capacity for processing information', *Psychological Review,* Vol. 63, pp. 81–97.

Mumford, E. (1983a) *Designing Human Systems,* Manchester Business School.

Mumford, E. (1983b) *Designing Participatively,* Manchester Business School.

Mumford, E. (1985) 'Defining system requirements to meet business needs: A case study example', *The Computer Journal,* Vol. 28, No. 2, 97–104.

NCC Training (1991) *SSADM Version 4,* Vols 1, 2 & 3, Hobbs, Southampton.

Olle, T. W. et al. (1989) *Information Systems Methodologies: A Framework for Understanding,* Addison-Wesley, Wokingham.

Parkinson, J. (1990) 'Making CASE work in CASE on trial', Spurr, K. and Layzell, P. (eds), John Wiley, Chichester.

Peat, Marwick, McLintock (1990) *Runaway Computer Systems – A Business Issue for the 199Os,* KPMG Peat, Marwick, McLintock, London.

Peterson, J. L. (1977) 'Petri Nets ACM', *Computing Surveys,* Vol. 9 (3), 223–52.

Prevost, P. (1976) '"Soft" systems methodology, functionalism and the social sciences', *Journal of Applied Systems Analysis,* Vol. 5, No. l.

Rochfeld, A. and Tardieu, H. (1983) 'Merise: An information system design and development methodology', *Information Management,* Vol. 6(3) 143–59.

Rochfeld, A. *et al.* (1985) *La Méthode merise – Démarche et Pratiques,* Les Editions d'organisation, Paris.

Rock-Evans, R. (1981) *Data Analysis,* IPC Business Press, Surrey.

Rudman, B. (1992) *The Secret of Successful Feasibility Studies in SSADM,* SSADM User Group Newsletter, 10–12.

Rumbaugh, J. *et al.* (1991) *Object-oriented Modelling and Design,* Prentice-Hall, N.J.

Sabanis, N. and Darlison, A. (1992) *In CASE: Current Practice, Future Prospects,* Spurr, K. and Layzell, P. (eds), John Wiley, Chichester.

Smyth, D. S. and Checkland, P. B. (1976) 'Using a systems approach: The structure of root definitions', *Journal of Applied Systems Analysis,* Vol. 5, No. 1.

Sommerville, I. (1992) *Software Engineering,* Addison-Wesley, Wokingham.

SSADM Version 4 Manual, (1990) NCC, Oxford, Crown Copyright.

Tardieu, H. *et al.* (1980) *A Method, a Formalism and Tools for Database Design: Three Years of Experimental Practice in Entity Relationship Approach to System Analysis,* North Holland, Oxford.

Thompson,K. (1993) *What's in Store: Euromethod – Esperanto Or Eldorado?,* Software Development Monitor, Elsevier.

Tsichritzis, D. and Klug, A. (1978) 'The ANSI Spark report of the study group on database management systems', *Information Systems,* Vol. 3.

Turner, P. (1994) *An Overview of Euromethod,* International SSADM User Group Newsletter, No. 22, June.

Ward, P. and Mellor, S. (1986) *Structured Development for Real Time Systems,* Yourdon Press, N.J.

Wood-Harper, A. T. and Fitzgerald, G. (1982) 'A taxonomy of current approaches to systems analysis', *Computer Journal,* Vol. 25, No. 1.

Wood-Harper, A. T., Antill, L. and Avison, D. E. (1985) *Information Systems Definition: The Multiview Approach,* Blackwell Scientific Publications, Oxford.

Wood-Harper, T. and Episcopou, D. M. (1984) 'Towards a framework to choose appropriate IS approaches', *The Computer Journal,* Vol. 29, No. 3.

Yourdon, E. (ed.) (1982) *Writings of the Revolution – Selected Readings on Software Engineering,* Yourdon Press, N.Y.

Yourdon, E. and Constantine, L. (1989) *Structured Design: Fundamentals of a Discipline of Computer Program and Systems Design,* Englewood Cliffs, Prentice-Hall, N.J.

Yourdon, E. (1989) *Modern Structured Analysis,* Prentice-Hall, N.J.

Yourdon, E. (1992) *The Decline and Fall of the American Programmer,* Yourdon Press, N.J.

Index